Springer Texts in Statistics

For other titles published in this series, go to
http://www.springer.com/series/417

Peter D. Hoff

A First Course in Bayesian Statistical Methods

Springer

Peter D. Hoff
Department of Statistics
University of Washington
Seattle WA 98195-4322
USA
hoff@stat.washington.edu

ISSN 1431-875X
ISBN 978-0-387-92299-7 e-ISBN 978-0-387-92407-6
DOI 10.1007/978-0-387-92407-6
Springer Dordrecht Heidelberg London New York

Library of Congress Control Number: 2009929120

Printed on acid-free paper

(Corrected at 2nd printing 2010)

Springer is part of Springer Science+Business Media (www.springer.com)

Preface

This book originated from a set of lecture notes for a one-quarter graduate-level course taught at the University of Washington. The purpose of the course is to familiarize the students with the basic concepts of Bayesian theory and to quickly get them performing their own data analyses using Bayesian computational tools. The audience for this course includes non-statistics graduate students who did well in their department's graduate-level introductory statistics courses and who also have an interest in statistics. Additionally, first- and second-year statistics graduate students have found this course to be a useful introduction to statistical modeling. Like the course, this book is intended to be a self-contained and compact introduction to the main concepts of Bayesian theory and practice. By the end of the text, readers should have the ability to understand and implement the basic tools of Bayesian statistical methods for their own data analysis purposes. The text is not intended as a comprehensive handbook for advanced statistical researchers, although it is hoped that this latter category of readers could use this book as a quick introduction to Bayesian methods and as a preparation for more comprehensive and detailed studies.

Computing

Monte Carlo summaries of posterior distributions play an important role in the way data analyses are presented in this text. My experience has been that once a student understands the basic idea of posterior sampling, their data analyses quickly become more creative and meaningful, using relevant posterior predictive distributions and interesting functions of parameters. The open-source R statistical computing environment provides sufficient functionality to make Monte Carlo estimation very easy for a large number of statistical models, and example R-code is provided throughout the text. Much of the example code can be run "as is" in R, and essentially all of it can be run after downloading the relevant datasets from the companion website for this book.

Acknowledgments

The presentation of material in this book, and my teaching style in general, have been heavily influenced by the diverse set of students taking CSSS-STAT 564 at the University of Washington. My thanks to them for improving my teaching. I also thank Chris Hoffman, Vladimir Minin, Xiaoyue Niu and Marc Suchard for their extensive comments, suggestions and corrections for this book, and to Adrian Raftery for bibliographic suggestions. Finally, I thank my wife Jen for her patience and support.

Seattle, WA

Peter Hoff
March 2009

Contents

1

Introduction and examples

1.1 Introduction

We often use probabilities informally to express our information and beliefs about unknown quantities. However, the use of probabilities to express information can be made formal: In a precise mathematical sense, it can be shown that probabilities can numerically represent a set of rational beliefs, that there is a relationship between probability and information, and that Bayes' rule provides a rational method for updating beliefs in light of new information. The process of inductive learning via Bayes' rule is referred to as *Bayesian inference*.

More generally, *Bayesian methods* are data analysis tools that are derived from the principles of Bayesian inference. In addition to their formal interpretation as a means of induction, Bayesian methods provide:

- parameter estimates with good statistical properties;
- parsimonious descriptions of observed data;
- predictions for missing data and forecasts of future data;
- a computational framework for model estimation, selection and validation.

Thus the uses of Bayesian methods go beyond the formal task of induction for which the methods are derived. Throughout this book we will explore the broad uses of Bayesian methods for a variety of inferential and statistical tasks. We begin in this chapter with an introduction to the basic ingredients of Bayesian learning, followed by some examples of the different ways in which Bayesian methods are used in practice.

Bayesian learning

Statistical induction is the process of learning about the general characteristics of a population from a subset of members of that population. Numerical values of population characteristics are typically expressed in terms of a parameter θ, and numerical descriptions of the subset make up a dataset y. Before a dataset

P.D. Hoff, *A First Course in Bayesian Statistical Methods*,
Springer Texts in Statistics, DOI 10.1007/978-0-387-92407-6_1,

is obtained, the numerical values of both the population characteristics and the dataset are uncertain. After a dataset y is obtained, the information it contains can be used to decrease our uncertainty about the population characteristics. Quantifying this change in uncertainty is the purpose of Bayesian inference.

The *sample space* $\mathcal{Y}$ is the set of all possible datasets, from which a single dataset y will result. The *parameter space* Θ is the set of possible parameter values, from which we hope to identify the value that best represents the true population characteristics. The idealized form of Bayesian learning begins with a numerical formulation of joint beliefs about y and θ, expressed in terms of probability distributions over $\mathcal{Y}$ and Θ.

1. For each numerical value $\theta \in \Theta$, our *prior distribution* $p(\theta)$ describes our belief that θ represents the true population characteristics.
2. For each $\theta \in \Theta$ and $y \in \mathcal{Y}$, our *sampling model* $p(y|\theta)$ describes our belief that y would be the outcome of our study if we knew θ to be true.

Once we obtain the data y, the last step is to update our beliefs about θ:

3. For each numerical value of $\theta \in \Theta$, our *posterior distribution* $p(\theta|y)$ describes our belief that θ is the true value, having observed dataset y.

The posterior distribution is obtained from the prior distribution and sampling model via *Bayes' rule*:

$$p(\theta|y) = \frac{p(y|\theta)p(\theta)}{\int_\Theta p(y|\tilde{\theta})p(\tilde{\theta})\, d\tilde{\theta}}.$$

It is important to note that Bayes' rule does not tell us what our beliefs should be, it tells us how they should change after seeing new information.

1.2 Why Bayes?

Mathematical results of Cox (1946, 1961) and Savage (1954, 1972) prove that if $p(\theta)$ and $p(y|\theta)$ represent a rational person's beliefs, then Bayes' rule is an optimal method of updating this person's beliefs about θ given new information y. These results give a strong theoretical justification for the use of Bayes' rule as a method of quantitative learning. However, in practical data analysis situations it can be hard to precisely mathematically formulate what our prior beliefs are, and so $p(\theta)$ is often chosen in a somewhat ad hoc manner or for reasons of computational convenience. What then is the justification of Bayesian data analysis?

A famous quote about sampling models is that "all models are wrong, but some are useful" (Box and Draper, 1987, pg. 424). Similarly, $p(\theta)$ might be viewed as "wrong" if it does not accurately represent our prior beliefs. However, this does not mean that $p(\theta|y)$ is not useful. If $p(\theta)$ approximates our beliefs, then the fact that $p(\theta|y)$ is optimal under $p(\theta)$ means that it will also generally serve as a good approximation to what our posterior beliefs should be. In other situations it may not be *our* beliefs that are of interest. Rather, we may want to use Bayes' rule to explore how the data would update the beliefs of a variety of people with differing prior opinions. Of particular interest might be the posterior beliefs of someone with weak prior information. This has motivated the use of "diffuse" prior distributions, which assign probability more or less evenly over large regions of the parameter space.

Finally, in many complicated statistical problems there are no obvious non-Bayesian methods of estimation or inference. In these situations, Bayes' rule can be used to generate estimation procedures, and the performance of these procedures can be evaluated using non-Bayesian criteria. In many cases it has been shown that Bayesian or approximately Bayesian procedures work very well, even for non-Bayesian purposes.

The next two examples are intended to show how Bayesian inference, using prior distributions that may only roughly represent our or someone else's prior beliefs, can be broadly useful for statistical inference. Most of the mathematical details of the calculations are left for later chapters.

1.2.1 Estimating the probability of a rare event

Suppose we are interested in the prevalence of an infectious disease in a small city. The higher the prevalence, the more public health precautions we would recommend be put into place. A small random sample of 20 individuals from the city will be checked for infection.

Parameter and sample spaces

Interest is in θ, the fraction of infected individuals in the city. Roughly speaking, the parameter space includes all numbers between zero and one. The data y records the total number of people in the sample who are infected. The parameter and sample spaces are then as follows:

$$\Theta = [0,1] \qquad \mathcal{Y} = \{0, 1, \ldots, 20\}.$$

Sampling model

Before the sample is obtained the number of infected individuals in the sample is unknown. We let the variable Y denote this to-be-determined value. If the value of θ were known, a reasonable sampling model for Y would be a binomial$(20, \theta)$ probability distribution:

$$Y|\theta \sim \text{binomial}(20, \theta).$$

The first panel of Figure 1.1 plots the binomial$(20, \theta)$ distribution for θ equal to 0.05, 0.10 and 0.20. If, for example, the true infection rate is 0.05, then the probability that there will be zero infected individuals in the sample ($Y = 0$) is 36%. If the true rate is 0.10 or 0.20, then the probabilities that $Y = 0$ are 12% and 1%, respectively.

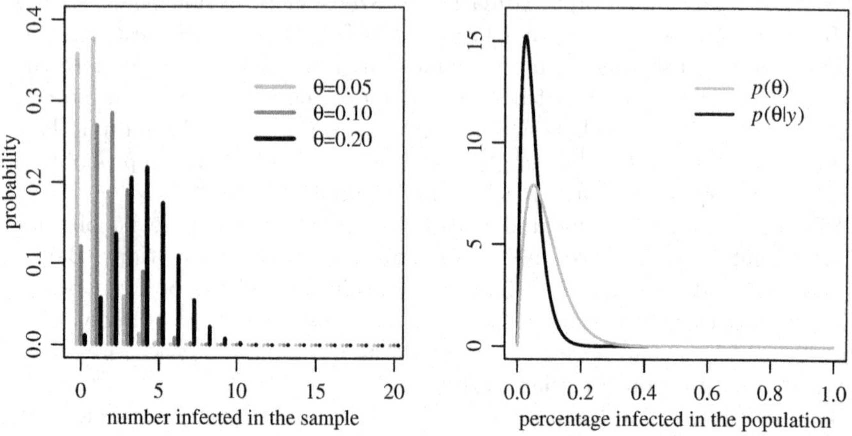

Fig. 1.1. Sampling model, prior and posterior distributions for the infection rate example. The plot on the left-hand side gives binomial$(20, \theta)$ distributions for three values of θ. The right-hand side gives prior (gray) and posterior (black) densities of θ.

Prior distribution

Other studies from various parts of the country indicate that the infection rate in comparable cities ranges from about 0.05 to 0.20, with an average prevalence of 0.10. This prior information suggests that we use a prior distribution $p(\theta)$ that assigns a substantial amount of probability to the interval (0.05, 0.20), and that the expected value of θ under $p(\theta)$ is close to 0.10. However, there are infinitely many probability distributions that satisfy these conditions, and it is not clear that we can discriminate among them with our limited amount of prior information. We will therefore use a prior distribution $p(\theta)$ that has the characteristics described above, but whose particular mathematical form is chosen for reasons of computational convenience. Specifically, we will encode the prior information using a member of the family of beta distributions. A beta distribution has two parameters which we denote as a and b. If θ has a beta(a, b) distribution, then the expectation of θ is $a/(a+b)$ and the most probable value of θ is $(a-1)/(a-1+b-1)$. For our problem where θ is the infection rate, we will represent our prior information about θ with a beta(2,20) probability distribution. Symbolically, we write

$$\theta \sim \text{beta}(2, 20).$$

This distribution is shown in the gray line in the second panel of Figure 1.1. The expected value of θ for this prior distribution is 0.09. The curve of the prior distribution is highest at $\theta = 0.05$ and about two-thirds of the area under the curve occurs between 0.05 and 0.20. The prior probability that the infection rate is below 0.10 is 64%.

$$
\begin{aligned}
\mathrm{E}[\theta] &= 0.09 \\
\mathrm{mode}[\theta] &= 0.05 \\
\Pr(\theta < 0.10) &= 0.64 \\
\Pr(0.05 < \theta < 0.20) &= 0.66\,.
\end{aligned}
$$

Posterior distribution

As we will see in Chapter 3, if $Y|\theta \sim \text{binomial}(n,\theta)$ and $\theta \sim \text{beta}(a,b)$, then if we observe a numeric value y of Y, the posterior distribution is a $\text{beta}(a+y, b+n-y)$ distribution. Suppose that for our study a value of $Y=0$ is observed, i.e. none of the sample individuals are infected. The posterior distribution of θ is then a beta(2, 40) distribution.

$$\theta|\{Y=0\} \sim \text{beta}(2,40)$$

The density of this distribution is given by the black line in the second panel of Figure 1.1. This density is further to the left than the prior distribution, and more peaked as well. It is to the left of $p(\theta)$ because the observation that $Y=0$ provides evidence of a low value of θ. It is more peaked than $p(\theta)$ because it combines information from the data and the prior distribution, and thus contains more information than in $p(\theta)$ alone. The peak of this curve is at 0.025 and the posterior expectation of θ is 0.048. The posterior probability that $\theta < 0.10$ is 93%.

$$
\begin{aligned}
\mathrm{E}[\theta|Y=0] &= 0.048 \\
\mathrm{mode}[\theta|Y=0] &= 0.025 \\
\Pr(\theta < 0.10|Y=0) &= 0.93.
\end{aligned}
$$

The posterior distribution $p(\theta|Y=0)$ provides us with a model for learning about the city-wide infection rate θ. From a theoretical perspective, a rational individual whose prior beliefs about θ were represented by a beta(2,20) distribution now has beliefs that are represented by a beta(2,40) distribution. As a practical matter, if we accept the beta(2,20) distribution as a reasonable measure of prior information, then we accept the beta(2,40) distribution as a reasonable measure of posterior information.

Sensitivity analysis

Suppose we are to discuss the results of the survey with a group of city health officials. A discussion of the implications of our study among a diverse group of people might benefit from a description of the posterior beliefs corresponding to a variety of prior distributions. Suppose we were to consider beliefs represented by $\text{beta}(a,b)$ distributions for values of (a,b) other than (2,20). As mentioned above, if $\theta \sim \text{beta}(a,b)$, then given $Y=y$ the posterior distribution of θ is $\text{beta}(a+y, b+n-y)$. The posterior expectation is

$$\begin{aligned}
\mathrm{E}[\theta|Y=y] &= \frac{a+y}{a+b+n} \\
&= \frac{n}{a+b+n}\frac{y}{n} + \frac{a+b}{a+b+n}\frac{a}{a+b} \\
&= \frac{n}{w+n}\bar{y} + \frac{w}{w+n}\theta_0,
\end{aligned}$$

where $\theta_0 = a/(a+b)$ is the prior expectation of θ and $w = a+b$. From this formula we see that the posterior expectation is a weighted average of the sample mean $\bar{y}$ and the prior expectation θ_0. In terms of estimating θ, θ_0 represents our prior guess at the true value of θ and w represents our confidence in this guess, expressed on the same scale as the sample size. If

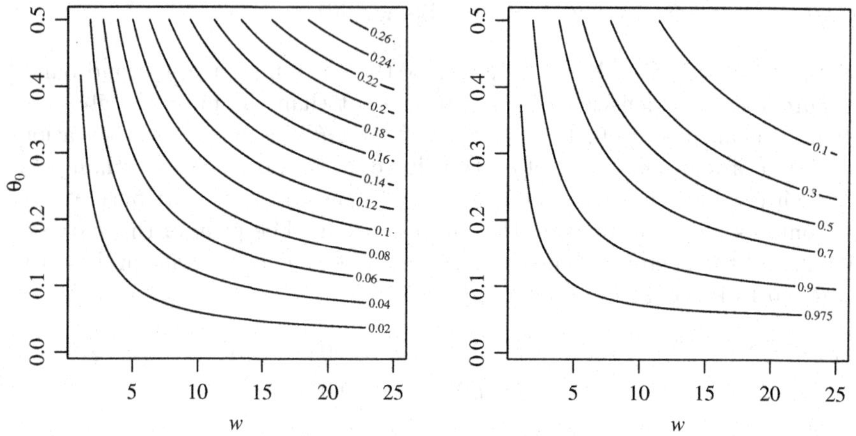

Fig. 1.2. Posterior quantities under different beta prior distributions. The left- and right-hand panels give contours of $\mathrm{E}[\theta|Y=0]$ and $\Pr(\theta < 0.10|Y=0)$, respectively, for a range of prior expectations and levels of confidence.

someone provides us with a prior guess θ_0 and a degree of confidence w, then we can approximate their prior beliefs about θ with a beta distribution having parameters $a = w\theta_0$ and $b = w(1-\theta_0)$. Their approximate posterior beliefs are then represented with a beta$(w\theta_0 + y, w(1-\theta_0) + n - y)$ distribution. We can compute such a posterior distribution for a wide range of θ_0 and w values to perform a *sensitivity analysis*, an exploration of how posterior information is affected by differences in prior opinion. Figure 1.2 explores the effects of θ_0 and w on the posterior distribution via contour plots of two posterior quantities. The first plot gives contours of the posterior expectation $\mathrm{E}[\theta|Y=0]$, and the second gives the posterior probabilities $\Pr(\theta < 0.10|Y=0)$. This latter plot may be of use if, for instance, the city officials would like to recommend a vaccine to the general public unless they were reasonably sure that the current infection rate was less than 0.10. The plot indicates, for example, that people with weak prior beliefs (low values of w) or low prior expectations are generally 90% or more certain that the infection rate is below 0.10. However, a high degree of certainty (say 97.5%) is only achieved by people who already thought the infection rate was lower than the average of the other cities.

Comparison to non-Bayesian methods

A standard estimate of a population proportion θ is the sample mean $\bar{y} = y/n$, the fraction of infected people in the sample. For our sample in which $y = 0$ this of course gives an estimate of zero, and so by using $\bar{y}$ we would be estimating that zero people in the city are infected. If we were to report this estimate to a group of doctors or health officials we would probably want to include the caveat that this estimate is subject to sampling uncertainty. One way to describe the sampling uncertainty of an estimate is with a confidence interval. A popular 95% confidence interval for a population proportion θ is the *Wald interval*, given by

$$\bar{y} \pm 1.96\sqrt{\bar{y}(1-\bar{y})/n}.$$

This interval has *correct asymptotic frequentist coverage*, meaning that if n is large, then with probability approximately equal to 95%, Y will take on a value y such that the above interval contains θ. Unfortunately this does not hold for small n: For an n of around 20 the probability that the interval contains θ is only about 80% (Agresti and Coull, 1998). Regardless, for our sample in which $\bar{y} = 0$ the Wald confidence interval comes out to be just a single point: zero. In fact, the 99.99% Wald interval also comes out to be zero. Certainly we would not want to conclude from the survey that we are 99.99% certain that no one in the city is infected.

People have suggested a variety of alternatives to the Wald interval in hopes of avoiding this type of behavior. One type of confidence interval that performs well by non-Bayesian criteria is the "adjusted" Wald interval suggested by Agresti and Coull (1998), which is given by

$$\hat{\theta} \pm 1.96\sqrt{\hat{\theta}(1-\hat{\theta})/n}\ , \text{ where}$$
$$\hat{\theta} = \frac{n}{n+4}\bar{y} + \frac{4}{n+4}\frac{1}{2}.$$

While not originally motivated as such, this interval is clearly related to Bayesian inference: The value of $\hat{\theta}$ here is equivalent to the posterior mean for θ under a beta(2,2) prior distribution, which represents weak prior information centered around $\theta = 1/2$.

General estimation of a population mean

Given a random sample of n observations from a population, a standard estimate of the population mean θ is the sample mean $\bar{y}$. While $\bar{y}$ is generally a reliable estimate for large sample sizes, as we saw in the example it can be statistically unreliable for small n, in which case it serves more as a summary of the sample data than as a precise estimate of θ.

If our interest lies more in obtaining an estimate of θ than in summarizing our sample data, we may want to consider estimators of the form

$$\hat{\theta} = \frac{n}{n+w}\bar{y} + \frac{w}{n+w}\theta_0,$$

where θ_0 represents a "best guess" at the true value of θ and w represents a degree of confidence in the guess. If the sample size is large, then $\bar{y}$ is a reliable estimate of θ. The estimator $\hat{\theta}$ takes advantage of this by having its weights on $\bar{y}$ and θ_0 go to one and zero, respectively, as n increases. As a result, the statistical properties of $\bar{y}$ and $\hat{\theta}$ are essentially the same for large n. However, for small n the variability of $\bar{y}$ might be more than our uncertainty about θ_0. In this case, using $\hat{\theta}$ allows us to combine the data with prior information to stabilize our estimation of θ.

These properties of $\hat{\theta}$ for both large and small n suggest that it is a useful estimate of θ for a broad range of n. In Section 5.4 we will confirm this by showing that, under some conditions, $\hat{\theta}$ outperforms $\bar{y}$ as an estimator of θ for all values of n. As we saw in the infection rate example and will see again in later chapters, $\hat{\theta}$ can be interpreted as a Bayesian estimator using a certain class of prior distributions. Even if a particular prior distribution $p(\theta)$ does not exactly reflect our prior information, the corresponding posterior distribution $p(\theta|y)$ can still be a useful means of providing stable inference and estimation for situations in which the sample size is low.

1.2.2 Building a predictive model

In Chapter 9 we will discuss an example in which our task is to build a predictive model of diabetes progression as a function of 64 baseline explanatory variables such as age, sex and body mass index. Here we give a brief synopsis of that example. We will first estimate the parameters in a regression model using a "training" dataset consisting of measurements from 342 patients. We will then evaluate the predictive performance of the estimated regression model using a separate "test" dataset of 100 patients.

Sampling model and parameter space

Letting Y_i be the diabetes progression of subject i and $\boldsymbol{x}_i = (x_{i,1}, \ldots, x_{i,64})$ be the explanatory variables, we will consider linear regression models of the form

$$Y_i = \beta_1 x_{i,1} + \beta_2 x_{i,2} + \cdots + \beta_{64} x_{i,64} + \sigma\epsilon_i.$$

The sixty-five unknown parameters in this model are the vector of regression coefficients $\boldsymbol{\beta} = (\beta_1, \ldots, \beta_{64})$ as well as σ, the standard deviation of the error term. The parameter space is 64-dimensional Euclidean space for $\boldsymbol{\beta}$ and the positive real line for σ.

Prior distribution

In most situations, defining a joint prior probability distribution for 65 parameters that accurately represents prior beliefs is a near-impossible task. As an alternative, we will use a prior distribution that only represents some aspects of our prior beliefs. The main belief that we would like to represent is that most of the 64 explanatory variables have little to no effect on diabetes progression, i.e. most of the regression coefficients are zero. In Chapter 9 we will discuss a prior distribution on $\boldsymbol{\beta}$ that roughly represents this belief, in that each regression coefficient has a 50% prior probability of being equal to zero.

Posterior distribution

Given data $\boldsymbol{y} = (y_1, \ldots, y_{342})$ and $\mathbf{X} = (\boldsymbol{x}_1, \ldots, \boldsymbol{x}_{342})$, the posterior distribution $p(\boldsymbol{\beta}|\boldsymbol{y}, \mathbf{X})$ can be computed and used to obtain $\Pr(\beta_j \neq 0|\boldsymbol{y}, \mathbf{X})$ for each regression coefficient j. These probabilities are plotted in the first panel of Figure 1.3. Even though each of the sixty-four coefficients started out with a 50-50 chance of being non-zero in the prior distribution, there are only six β_j's for which $\Pr(\beta_j \neq 0|\boldsymbol{y}, \mathbf{X}) \geq 0.5$. The vast majority of the remaining coefficients have high posterior probabilities of being zero. This dramatic increase in the expected number of zero coefficients is a result of the information in the data, although it is the prior distribution that allows for such zero coefficients in the first place.

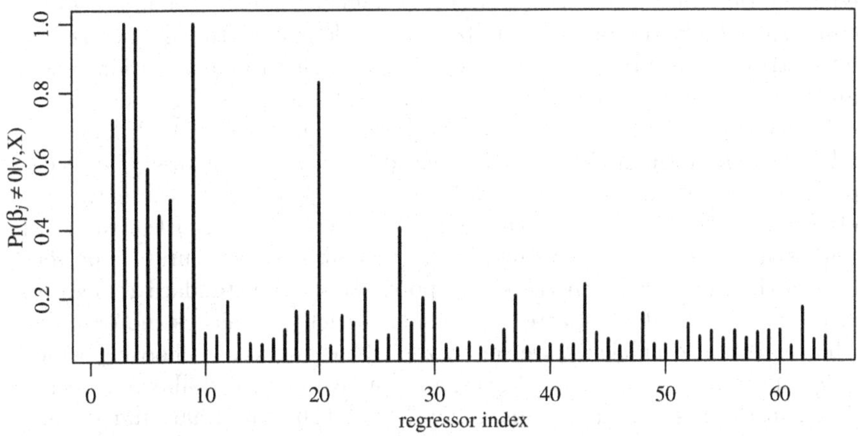

Fig. 1.3. Posterior probabilities that each coefficient is non-zero.

Predictive performance and comparison to non-Bayesian methods

We can evaluate how well this model performs by using it to predict the test data: Let $\hat{\boldsymbol{\beta}}_{\text{Bayes}} = \text{E}[\boldsymbol{\beta}|\boldsymbol{y}, \mathbf{X}]$ be the posterior expectation of $\boldsymbol{\beta}$, and let $\mathbf{X}_{\text{test}}$ be the 100×64 matrix giving the data for the 100 patients in the test dataset. We can compute a predicted value for each of the 100 observations in the test set using the equation $\hat{\boldsymbol{y}}_{\text{test}} = \mathbf{X}\hat{\boldsymbol{\beta}}_{\text{Bayes}}$. These predicted values can then be compared to the actual observations $\boldsymbol{y}_{\text{test}}$. A plot of $\boldsymbol{y}_{\text{test}}$ versus $\hat{\boldsymbol{y}}_{\text{test}}$ appears in the first panel of Figure 1.4, and indicates how well $\hat{\boldsymbol{\beta}}_{\text{Bayes}}$ is able to predict diabetes progression from the baseline variables.

How does this Bayesian estimate of $\boldsymbol{\beta}$ compare to a non-Bayesian approach? The most commonly used estimate of a vector of regression coefficients is the ordinary least squares (OLS) estimate, provided in most if not all statistical software packages. The OLS regression estimate is the value $\hat{\boldsymbol{\beta}}_{\text{ols}}$ of $\boldsymbol{\beta}$ that minimizes the sum of squares of the residuals (SSR) for the observed data,

$$\text{SSR}(\boldsymbol{\beta}) = \sum_{i=1}^{n} (y_i - \boldsymbol{\beta}^T \boldsymbol{x}_i)^2,$$

and is given by the formula $\hat{\boldsymbol{\beta}}_{\text{ols}} = (\mathbf{X}^T\mathbf{X})^{-1}\mathbf{X}^T\boldsymbol{y}$. Predictions for the test data based on this estimate are given by $\mathbf{X}\hat{\boldsymbol{\beta}}_{\text{ols}}$ and are plotted against the observed values in the second panel of Figure 1.4. Notice that using $\hat{\boldsymbol{\beta}}_{\text{ols}}$ gives a weaker relationship between observed and predicted values than using $\hat{\boldsymbol{\beta}}_{\text{Bayes}}$. This can be quantified numerically by computing the average squared prediction error, $\sum(y_{\text{test},i} - \hat{y}_{\text{test},i})^2/100$, for both sets of predictions. The prediction error for OLS is 0.67, about 50% higher than the value of 0.45 we obtain using the Bayesian estimate. In this problem, even though our ad hoc prior distribution for $\boldsymbol{\beta}$ only captures the basic structure of our prior beliefs (namely, that many of the coefficients are likely to be zero), this is enough to provide a large improvement in predictive performance over the OLS estimate.

The poor performance of the OLS method is due to its inability to recognize when the sample size is too small to accurately estimate the regression coefficients. In such situations, the linear relationship between the values of $\boldsymbol{y}$ and $\mathbf{X}$ in the dataset, quantified by $\hat{\boldsymbol{\beta}}_{\text{ols}}$, is often an inaccurate representation of the relationship in the entire population. The standard remedy to this problem is to fit a "sparse" regression model, in which some or many of the regression coefficients are set to zero. One method of choosing which coefficients to set to zero is the Bayesian approach described above. Another popular method is the "lasso," introduced by Tibshirani (1996) and studied extensively by many others. The lasso estimate is the value $\hat{\boldsymbol{\beta}}_{\text{lasso}}$ of $\boldsymbol{\beta}$ that minimizes $\text{SSR}(\boldsymbol{\beta} : \lambda)$, a modified version of the sum of squared residuals:

$$\text{SSR}(\boldsymbol{\beta} : \lambda) = \sum_{i=1}^{n} (y_i - \boldsymbol{x}_i^T \boldsymbol{\beta})^2 + \lambda \sum_{j=1}^{p} |\beta_j| \,.$$

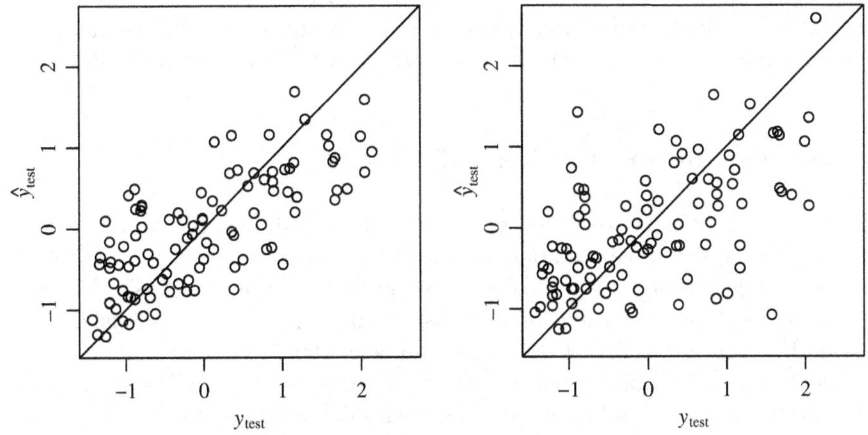

Fig. 1.4. Observed versus predicted diabetes progression values using the Bayes estimate (left panel) and the OLS estimate (right panel).

In other words, the lasso procedure penalizes large values of $|\beta_j|$. Depending on the size of λ, this penalty can make some elements of $\hat{\boldsymbol{\beta}}_{\text{lasso}}$ equal to zero. Although the lasso procedure has been motivated by and studied in a non-Bayesian context, in fact it corresponds to a Bayesian estimate using a particular prior distribution: The lasso estimate is equal to the posterior mode of $\boldsymbol{\beta}$ in which the prior distribution for each β_j is a double-exponential distribution, a probability distribution that has a sharp peak at $\beta_j = 0$.

1.3 Where we are going

As the above examples indicate, the uses of Bayesian methods are quite broad. We have seen how the Bayesian approach provides

- models for rational, quantitative learning;
- estimators that work for small and large sample sizes;
- methods for generating statistical procedures in complicated problems.

An understanding of the benefits and limits of Bayesian methods comes with experience. In the chapters that follow, we will become familiar with these methods by applying them to a large number of statistical models and data analysis examples. After a review of probability in Chapter 2, we will learn the basics of Bayesian data analysis and computation in the context of some simple one-parameter statistical models in Chapters 3 and 4. Chapters 5, 6 and 7 discuss Bayesian inference with the normal and multivariate normal models. While important in their own right, normal models also provide the building blocks of more complicated modern statistical methods, such as hierarchical modeling, regression, variable selection and mixed effects models. These advanced topics and others are covered in Chapters 8 through 12.

1.4 Discussion and further references

The idea of probability as a measure of uncertainty about unknown but deterministic quantities is an old one. Important historical works include Bayes' "An essay towards solving a Problem in the Doctrine of Chances" (Bayes, 1763) and Laplace's "A Philosophical Essay on Probabilities," published in 1814 and currently published by Dover (Laplace, 1995).

The role of prior opinion in statistical inference was debated for much of the 20th century. Most published articles on this debate take up one side or another, and include mischaracterizations of the other side. More informative are discussions among statisticians of different viewpoints: Savage (1962) includes a short introduction by Savage, followed by a discussion among Bartlett, Barnard, Cox, Pearson and Smith, among others. Little (2006) considers the strengths and weaknesses of Bayesian and frequentist statistical criteria. Efron (2005) briefly discusses the role of different statistical philosophies in the last two centuries, and speculates on the interplay between Bayesian and non-Bayesian methods in the future of statistical science.

2

Belief, probability and exchangeability

We first discuss what properties a reasonable belief function should have, and show that probabilities have these properties. Then, we review the basic machinery of discrete and continuous random variables and probability distributions. Finally, we explore the link between independence and exchangeability.

2.1 Belief functions and probabilities

At the beginning of the last chapter we claimed that probabilities are a way to numerically express rational beliefs. We do not prove this claim here (see Chapter 2 of Jaynes (2003) or Chapters 2 and 3 of Savage (1972) for details), but we do show that several properties we would want our numerical beliefs to have are also properties of probabilities.

Belief functions

Let F, G, and H be three possibly overlapping statements about the world. For example:

$F = \{$ a person votes for a left-of-center candidate $\}$
$G = \{$ a person's income is in the lowest 10% of the population $\}$
$H = \{$ a person lives in a large city $\}$

Let Be() be a *belief function*, that is, a function that assigns numbers to statements such that the larger the number, the higher the degree of belief. Some philosophers have tried to make this more concrete by relating beliefs to preferences over bets:

- $\text{Be}(F) > \text{Be}(G)$ means we would prefer to bet F is true than G is true.

We also want Be() to describe our beliefs under certain conditions:

- $\text{Be}(F|H) > \text{Be}(G|H)$ means that if we knew that H were true, then we would prefer to bet that F is also true than bet G is also true.

P.D. Hoff, *A First Course in Bayesian Statistical Methods*,
Springer Texts in Statistics, DOI 10.1007/978-0-387-92407-6_2,

- Be($F|G$) > Be($F|H$) means that if we were forced to bet on F, we would prefer to do it under the condition that G is true rather than H is true.

Axioms of beliefs

It has been argued by many that any function that is to numerically represent our beliefs should have the following properties:

B1 Be(not $H|H$) $\leq$ Be($F|H$) $\leq$ Be($H|H$)
B2 Be(F or $G|H$) $\geq$ max{Be($F|H$), Be($G|H$)}
B3 Be(F and $G|H$) can be derived from Be($G|H$) and Be($F|G$ and H)

How should we interpret these properties? Are they reasonable?

B1 says that the number we assign to Be($F|H$), our conditional belief in F given H, is bounded below and above by the numbers we assign to complete disbelief (Be(not $H|H$)) and complete belief (Be($H|H$)).

B2 says that our belief that the truth lies in a given set of possibilities should not decrease as we add to the set of possibilities.

B3 is a bit trickier. To see why it makes sense, imagine you have to decide whether or not F and G are true, knowing that H is true. You could do this by first deciding whether or not G is true given H, and if so, then deciding whether or not F is true given G and H.

Axioms of probability

Now let's compare **B1, B2** and **B3** to the standard axioms of probability. Recall that $F \cup G$ means "F or G," $F \cap G$ means "F and G" and $\emptyset$ is the empty set.

P1 $0 = \Pr(\text{not } H|H) \leq \Pr(F|H) \leq \Pr(H|H) = 1$
P2 $\Pr(F \cup G|H) = \Pr(F|H) + \Pr(G|H)$ if $F \cap G = \emptyset$
P3 $\Pr(F \cap G|H) = \Pr(G|H)\Pr(F|G \cap H)$

You should convince yourself that a probability function, satisfying **P1, P2** and **P3**, also satisfies **B1, B2** and **B3**. Therefore if we use a probability function to describe our beliefs, we have satisfied the axioms of belief.

2.2 Events, partitions and Bayes' rule

Definition 1 (Partition) *A collection of sets* $\{H_1, \ldots, H_K\}$ *is a partition of another set* $\mathcal{H}$ *if*

1. *the events are disjoint, which we write as* $H_i \cap H_j = \emptyset$ *for* $i \neq j$;
2. *the union of the sets is* $\mathcal{H}$, *which we write as* $\cup_{k=1}^K H_k = \mathcal{H}$.

In the context of identifying which of several statements is true, if $\mathcal{H}$ is the set of all possible truths and $\{H_1, \ldots, H_K\}$ is a partition of $\mathcal{H}$, then exactly one out of $\{H_1, \ldots, H_K\}$ contains the truth.

Examples

- Let $\mathcal{H}$ be someone's religious orientation. Partitions include
  - {Protestant, Catholic, Jewish, other, none};
  - {Christian, non-Christian};
  - {atheist, monotheist, multitheist}.
- Let $\mathcal{H}$ be someone's number of children. Partitions include
  - {0, 1, 2, 3 or more};
  - {0, 1, 2, 3, 4, 5, 6, ...}.
- Let $\mathcal{H}$ be the relationship between smoking and hypertension in a given population. Partitions include
  - {some relationship, no relationship};
  - {negative correlation, zero correlation, positive correlation}.

Partitions and probability

Suppose $\{H_1, \ldots, H_K\}$ is a partition of $\mathcal{H}$, $\Pr(\mathcal{H}) = 1$, and E is some specific event. The axioms of probability imply the following:

Rule of total probability : $$\sum_{k=1}^{K} \Pr(H_k) = 1$$

Rule of marginal probability : $$\Pr(E) = \sum_{k=1}^{K} \Pr(E \cap H_k) = \sum_{k=1}^{K} \Pr(E|H_k)\Pr(H_k)$$

Bayes' rule : $$\Pr(H_j|E) = \frac{\Pr(E|H_j)\Pr(H_j)}{\Pr(E)} = \frac{\Pr(E|H_j)\Pr(H_j)}{\sum_{k=1}^{K} \Pr(E|H_k)\Pr(H_k)}$$

Example

A subset of the 1996 General Social Survey includes data on the education level and income for a sample of males over 30 years of age. Let $\{H_1, H_2, H_3, H_4\}$ be the events that a randomly selected person in this sample is in, respectively, the lower 25th percentile, the second 25th percentile, the third 25th percentile and the upper 25th percentile in terms of income. By definition,

$$\{\Pr(H_1), \Pr(H_2), \Pr(H_3), \Pr(H_4)\} = \{.25, .25, .25, .25\}.$$

Note that $\{H_1, H_2, H_3, H_4\}$ is a partition and so these probabilities sum to 1. Let E be the event that a randomly sampled person from the survey has a college education. From the survey data, we have

$$\{\Pr(E|H_1), \Pr(E|H_2), \Pr(E|H_3), \Pr(E|H_4)\} = \{.11, .19, .31, .53\}.$$

These probabilities do not sum to 1 - they represent the proportions of people with college degrees in the four different income subpopulations H_1, H_2, H_3 and H_4. Now let's consider the income distribution of the college-educated population. Using Bayes' rule we can obtain

$$\{\Pr(H_1|E), \Pr(H_2|E), \Pr(H_3|E), \Pr(H_4|E)\} = \{.09, .17, .27, .47\},$$

and we see that the income distribution for people in the college-educated population differs markedly from $\{.25, .25, .25, .25\}$, the distribution for the general population. Note that these probabilities do sum to 1 - they are the conditional probabilities of the events in the partition, given E.

In Bayesian inference, $\{H_1, \ldots, H_K\}$ often refer to disjoint hypotheses or states of nature and E refers to the outcome of a survey, study or experiment. To compare hypotheses post-experimentally, we often calculate the following ratio:

$$\begin{aligned}
\frac{\Pr(H_i|E)}{\Pr(H_j|E)} &= \frac{\Pr(E|H_i)\Pr(H_i)/\Pr(E)}{\Pr(E|H_j)\Pr(H_j)/\Pr(E)} \\
&= \frac{\Pr(E|H_i)\Pr(H_i)}{\Pr(E|H_j)\Pr(H_j)} \\
&= \frac{\Pr(E|H_i)}{\Pr(E|H_j)} \times \frac{\Pr(H_i)}{\Pr(H_j)} \\
&= \text{“Bayes factor”} \times \text{“prior beliefs”}.
\end{aligned}$$

This calculation reminds us that Bayes' rule does not determine what our beliefs should be after seeing the data, it only tells us how they should change after seeing the data.

Example

Suppose we are interested in the rate of support for a particular candidate for public office. Let

$\mathcal{H} = \{$ all possible rates of support for candidate A $\}$;
$H_1 = \{$ more than half the voters support candidate A $\}$;
$H_2 = \{$ less than or equal to half the voters support candidate A $\}$;
$E = \{$ 54 out of 100 people surveyed said they support candidate A $\}$.

Then $\{H_1, H_2\}$ is a partition of $\mathcal{H}$. Of interest is $\Pr(H_1|E)$, or $\Pr(H_1|E)/\Pr(H_2|E)$. We will learn how to obtain these quantities in the next chapter.

2.3 Independence

Definition 2 (Independence) *Two events F and G are conditionally independent given H if* $\Pr(F \cap G|H) = \Pr(F|H)\Pr(G|H)$.

How do we interpret conditional independence? By Axiom **P3**, the following is always true:

$$\Pr(F \cap G|H) = \Pr(G|H)\Pr(F|H \cap G).$$

If F and G are conditionally independent given H, then we must have

$$\begin{aligned}
\Pr(G|H)\Pr(F|H \cap G) &\overset{\text{always}}{=} \Pr(F \cap G|H) \overset{\text{independence}}{=} \Pr(F|H)\Pr(G|H) \\
\Pr(G|H)\Pr(F|H \cap G) &= \Pr(F|H)\Pr(G|H) \\
\Pr(F|H \cap G) &= \Pr(F|H).
\end{aligned}$$

Conditional independence therefore implies that $\Pr(F|H \cap G) = \Pr(F|H)$. In other words, if we know H is true and F and G are conditionally independent given H, then knowing G does not change our belief about F.

Examples

Let's consider the conditional dependence of F and G when H is assumed to be true in the following two situations:

$F = \{$ a hospital patient is a smoker $\}$
$G = \{$ a hospital patient has lung cancer $\}$
$H = \{$ smoking causes lung cancer$\}$

$F = \{$ you are thinking of the jack of hearts $\}$
$G = \{$ a mind reader claims you are thinking of the jack of hearts $\}$
$H = \{$ the mind reader has extrasensory perception $\}$

In both of these situations, H being true implies a relationship between F and G. What about when H is not true?

2.4 Random variables

In Bayesian inference a random variable is defined as an unknown numerical quantity about which we make probability statements. For example, the quantitative outcome of a survey, experiment or study is a random variable before the study is performed. Additionally, a fixed but unknown population parameter is also a random variable.

2.4.1 Discrete random variables

Let Y be a random variable and let $\mathcal{Y}$ be the set of all possible values of Y. We say that Y is discrete if the set of possible outcomes is *countable*, meaning that $\mathcal{Y}$ can be expressed as $\mathcal{Y} = \{y_1, y_2, \ldots\}$.

Examples

- Y = number of churchgoers in a random sample from a population
- Y = number of children of a randomly sampled person
- Y = number of years of education of a randomly sampled person

Probability distributions and densities

The event that the outcome Y of our survey has the value y is expressed as $\{Y = y\}$. For each $y \in \mathcal{Y}$, our shorthand notation for $\Pr(Y = y)$ will be $p(y)$. This function of y is called the *probability density function* (pdf) of Y, and it has the following properties:

1. $0 \leq p(y) \leq 1$ for all $y \in \mathcal{Y}$;
2. $\sum_{y \in \mathcal{Y}} p(y) = 1$.

General probability statements about Y can be derived from the pdf. For example, $\Pr(Y \in A) = \sum_{y \in A} p(y)$. If A and B are disjoint subsets of $\mathcal{Y}$, then

$$\begin{aligned}\Pr(Y \in A \text{ or } Y \in B) \equiv \Pr(Y \in A \cup B) &= \Pr(Y \in A) + \Pr(Y \in B) \\ &= \sum_{y \in A} p(y) + \sum_{y \in B} p(y).\end{aligned}$$

Example: Binomial distribution

Let $\mathcal{Y} = \{0, 1, 2, \ldots, n\}$ for some positive integer n. The uncertain quantity $Y \in \mathcal{Y}$ has a *binomial distribution with probability* θ if

$$\Pr(Y = y|\theta) = \text{dbinom}(y, n, \theta) = \binom{n}{y} \theta^y (1-\theta)^{n-y}.$$

For example, if $\theta = .25$ and $n = 4$, we have:

$$\Pr(Y = 0|\theta = .25) = \binom{4}{0}(.25)^0(.75)^4 = .316$$

$$\Pr(Y = 1|\theta = .25) = \binom{4}{1}(.25)^1(.75)^3 = .422$$

$$\Pr(Y = 2|\theta = .25) = \binom{4}{2}(.25)^2(.75)^2 = .211$$

$$\Pr(Y = 3|\theta = .25) = \binom{4}{3}(.25)^3(.75)^1 = .047$$

$$\Pr(Y = 4|\theta = .25) = \binom{4}{4}(.25)^4(.75)^0 = .004\,.$$

Example: Poisson distribution

Let $\mathcal{Y} = \{0, 1, 2, \ldots\}$. The uncertain quantity $Y \in \mathcal{Y}$ has a *Poisson distribution with mean* θ if

$$\Pr(Y = y|\theta) = \text{dpois}(y, \theta) = \theta^y e^{-\theta}/y!.$$

For example, if $\theta = 2.1$ (the 2006 U.S. fertility rate),

$$\begin{aligned}
\Pr(Y = 0|\theta = 2.1) &= (2.1)^0 e^{-2.1}/(0!) = .12 \\
\Pr(Y = 1|\theta = 2.1) &= (2.1)^1 e^{-2.1}/(1!) = .26 \\
\Pr(Y = 2|\theta = 2.1) &= (2.1)^2 e^{-2.1}/(2!) = .27 \\
\Pr(Y = 3|\theta = 2.1) &= (2.1)^3 e^{-2.1}/(3!) = .19 \\
&\vdots \qquad\qquad \vdots \qquad\qquad \vdots
\end{aligned}$$

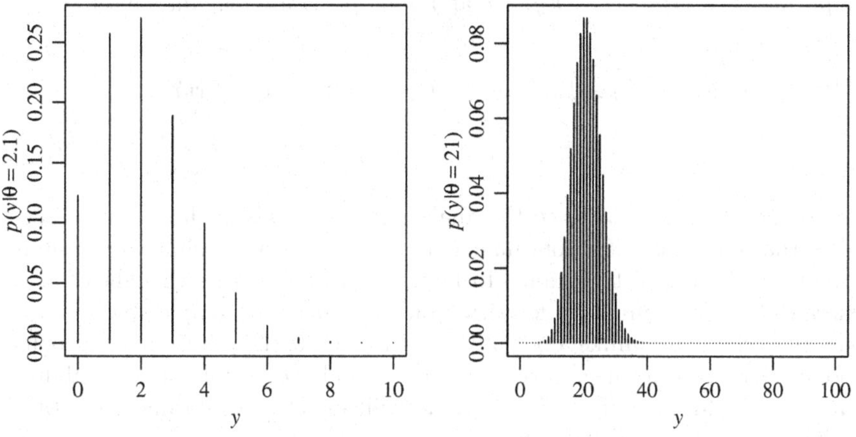

Fig. 2.1. Poisson distributions with means of 2.1 and 21.

2.4.2 Continuous random variables

Suppose that the sample space $\mathcal{Y}$ is roughly equal to $\mathbb{R}$, the set of all real numbers. We cannot define $\Pr(Y \leq 5)$ as equal to $\sum_{y \leq 5} p(y)$ because the sum does not make sense (the set of real numbers less than or equal to 5 is "uncountable"). So instead of defining probabilities of events in terms of a pdf $p(y)$, courses in mathematical statistics often define probability distributions for random variables in terms of something called a *cumulative distribution function*, or cdf:

$$F(y) = \Pr(Y \leq y).$$

Note that $F(\infty) = 1$, $F(-\infty) = 0$, and $F(b) \leq F(a)$ if $b < a$. Probabilities of various events can be derived from the cdf:

- $\Pr(Y > a) = 1 - F(a)$
- $\Pr(a < Y \leq b) = F(b) - F(a)$

If F is continuous (i.e. lacking any "jumps"), we say that Y is a continuous random variable. A theorem from mathematics says that for every continuous and differentiable cdf F there exists a positive function $p(y)$ such that

$$F(a) = \int_{-\infty}^{a} p(y)\, dy.$$

This function is called the probability density function of Y, and its properties are similar to those of a pdf for a discrete random variable:

1. $0 \leq p(y)$ for all $y \in \mathcal{Y}$;
2. $\int_{y \in \mathbb{R}} p(y)\, dy = 1$.

As in the discrete case, probability statements about Y can be derived from the pdf: $\Pr(Y \in A) = \int_{y \in A} p(y)\, dy$, and if A and B are disjoint subsets of $\mathcal{Y}$, then

$$\begin{aligned}\Pr(Y \in A \text{ or } Y \in B) \equiv \Pr(Y \in A \cup B) &= \Pr(Y \in A) + \Pr(Y \in B) \\ &= \int_{y \in A} p(y)\, dy + \int_{y \in B} p(y)\, dy.\end{aligned}$$

Comparing these properties to the analogous properties in the discrete case, we see that integration for continuous distributions behaves similarly to summation for discrete distributions. In fact, integration can be thought of as a generalization of summation for situations in which the sample space is not countable. However, unlike a pdf in the discrete case, the pdf for a continuous random variable is not necessarily less than 1, and $p(y)$ is not "the probability that $Y = y$." However, if $p(y_1) > p(y_2)$ we will sometimes informally say that y_1 "has a higher probability" than y_2.

Example: Normal distribution

Suppose we are sampling from a population on $\mathcal{Y} = (-\infty, \infty)$, and we know that the mean of the population is μ and the variance is σ^2. Among all probability distributions having a mean of μ and a variance of σ^2, the one that is the most "spread out" or "diffuse" (in terms of a measure called entropy), is the normal(μ, σ^2) distribution, having a cdf given by

$$\Pr(Y \leq y|\mu, \sigma^2) = F(y) = \int_{-\infty}^{y} \frac{1}{\sqrt{2\pi}\sigma} \exp\left\{-\frac{1}{2}\left(\frac{y-\mu}{\sigma}\right)^2\right\} dy.$$

Evidently,

$$p(y|\mu, \sigma^2) = \text{dnorm}(y, \mu, \sigma) = \frac{1}{\sqrt{2\pi}\sigma} \exp\left\{-\frac{1}{2}\left(\frac{y-\mu}{\sigma}\right)^2\right\}.$$

Letting $\mu = 10.75$ and $\sigma = .8$ ($\sigma^2 = .64$) gives the cdf and density in Figure 2.2. This mean and standard deviation make the median value of e^Y equal to about 46,630, which is about the median U.S. household income in 2005. Additionally, $\Pr(e^Y > 100000) = \Pr(Y > \log 100000) = 0.17$, which roughly matches the fraction of households in 2005 with incomes exceeding $100,000.

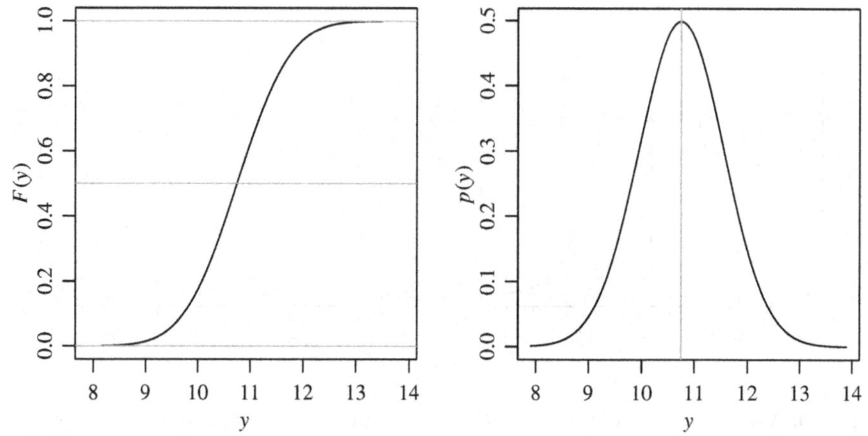

Fig. 2.2. Normal distribution with mean 10.75 and standard deviation 0.8.

2.4.3 Descriptions of distributions

The *mean* or *expectation* of an unknown quantity Y is given by

$\mathrm{E}[Y] = \sum_{y \in \mathcal{Y}} y p(y)$ if Y is discrete;
$\mathrm{E}[Y] = \int_{y \in \mathcal{Y}} y p(y)\, dy$ if Y is continuous.

The mean is the center of mass of the distribution. However, it is not in general equal to either of

the *mode*: "the most probable value of Y," or
the *median*: "the value of Y in the middle of the distribution."

In particular, for skewed distributions (like income distributions) the mean can be far from a "typical" sample value: see, for example, Figure 2.3. Still, the mean is a very popular description of the location of a distribution. Some justifications for reporting and studying the mean include the following:

1. The mean of $\{Y_1, \ldots, Y_n\}$ is a scaled version of the total, and the total is often a quantity of interest.
2. Suppose you are forced to guess what the value of Y is, and you are penalized by an amount $(Y - y_{\text{guess}})^2$. Then guessing $\mathrm{E}[Y]$ minimizes your expected penalty.

3. In some simple models that we shall see shortly, the sample mean contains all of the information about the population that can be obtained from the data.

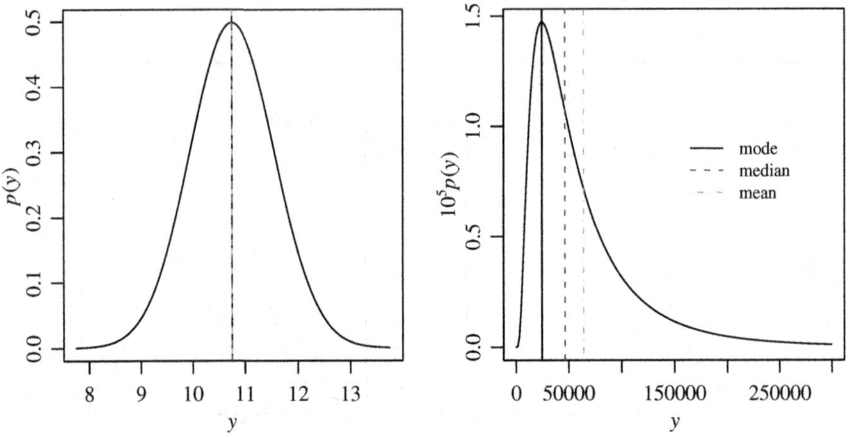

Fig. 2.3. Mode, median and mean of the normal and lognormal distributions, with parameters $\mu = 10.75$ and $\sigma = 0.8$.

In addition to the location of a distribution we are often interested in how spread out it is. The most popular measure of spread is the *variance* of a distribution:

$$\begin{aligned}\text{Var}[Y] &= \text{E}[(Y - \text{E}[Y])^2] \\ &= \text{E}[Y^2 - 2Y\text{E}[Y] + \text{E}[Y]^2] \\ &= \text{E}[Y^2] - 2\text{E}[Y]^2 + \text{E}[Y]^2 \\ &= \text{E}[Y^2] - \text{E}[Y]^2.\end{aligned}$$

The variance is the average squared distance that a sample value Y will be from the population mean $\text{E}[Y]$. The *standard deviation* is the square root of the variance, and has the same units as Y.

Alternative measures of spread are based on *quantiles*. For a continuous, strictly increasing cdf F, the α-quantile is the value y_α such that $F(y_\alpha) \equiv \Pr(Y \leq y_\alpha) = \alpha$. The *interquartile range* of a distribution is the interval $(y_{.25}, y_{.75})$, which contains 50% of the mass of the distribution. Similarly, the interval $(y_{.025}, y_{.975})$ contains 95% of the mass of the distribution.

2.5 Joint distributions

Discrete distributions

Let

- $\mathcal{Y}_1, \mathcal{Y}_2$ be two countable sample spaces;
- Y_1, Y_2 be two random variables, taking values in $\mathcal{Y}_1, \mathcal{Y}_2$ respectively.

Joint beliefs about Y_1 and Y_2 can be represented with probabilities. For example, for subsets $A \subset \mathcal{Y}_1$ and $B \subset \mathcal{Y}_2$, $\Pr(\{Y_1 \in A\} \cap \{Y_2 \in B\})$ represents our belief that Y_1 is in A and that Y_2 is in B. The *joint pdf* or *joint density* of Y_1 and Y_2 is defined as

$$p_{Y_1 Y_2}(y_1, y_2) = \Pr(\{Y_1 = y_1\} \cap \{Y_2 = y_2\}), \text{ for } y_1 \in \mathcal{Y}_1,\ y_2 \in \mathcal{Y}_2.$$

The *marginal density* of Y_1 can be computed from the joint density:

$$\begin{aligned} p_{Y_1}(y_1) &\equiv \Pr(Y_1 = y_1) \\ &= \sum_{y_2 \in \mathcal{Y}_2} \Pr(\{Y_1 = y_1\} \cap \{Y_2 = y_2\}) \\ &\equiv \sum_{y_2 \in \mathcal{Y}_2} p_{Y_1 Y_2}(y_1, y_2) \,. \end{aligned}$$

The *conditional density* of Y_2 given $\{Y_1 = y_1\}$ can be computed from the joint density and the marginal density:

$$\begin{aligned} p_{Y_2|Y_1}(y_2|y_1) &= \frac{\Pr(\{Y_1 = y_1\} \cap \{Y_2 = y_2\})}{\Pr(Y_1 = y_1)} \\ &= \frac{p_{Y_1 Y_2}(y_1, y_2)}{p_{Y_1}(y_1)} \,. \end{aligned}$$

You should convince yourself that

$\{p_{Y_1}, p_{Y_2|Y_1}\}$ can be derived from $p_{Y_1 Y_2}$,
$\{p_{Y_2}, p_{Y_1|Y_2}\}$ can be derived from $p_{Y_1 Y_2}$,
$p_{Y_1 Y_2}$ can be derived from $\{p_{Y_1}, p_{Y_2|Y_1}\}$,
$p_{Y_1 Y_2}$ can be derived from $\{p_{Y_2}, p_{Y_1|Y_2}\}$,

but

$p_{Y_1 Y_2}$ cannot be derived from $\{p_{Y_1}, p_{Y_2}\}$.

The subscripts of density functions are often dropped, in which case the type of density function is determined from the function argument: $p(y_1)$ refers to $p_{Y_1}(y_1)$, $p(y_1, y_2)$ refers to $p_{Y_1 Y_2}(y_1, y_2)$, $p(y_1|y_2)$ refers to $p_{Y_1|Y_2}(y_1|y_2)$, etc.

Example: Social mobility

Logan (1983) reports the following joint distribution of occupational categories of fathers and sons:

	son's occupation				
father's occupation	farm	operatives	craftsmen	sales	professional
farm	0.018	0.035	0.031	0.008	0.018
operatives	0.002	0.112	0.064	0.032	0.069
craftsmen	0.001	0.066	0.094	0.032	0.084
sales	0.001	0.018	0.019	0.010	0.051
professional	0.001	0.029	0.032	0.043	0.130

Suppose we are to sample a father-son pair from this population. Let Y_1 be the father's occupation and Y_2 the son's occupation. Then

$$\begin{aligned}\Pr(Y_2 = \text{professional}|Y_1 = \text{farm}) &= \frac{\Pr(Y_2 = \text{professional} \cap Y_1 = \text{farm})}{\Pr(Y_1 = \text{farm})} \\ &= \frac{.018}{.018 + .035 + .031 + .008 + .018} \\ &= .164\,.\end{aligned}$$

Continuous joint distributions

If Y_1 and Y_2 are continuous we start with a cumulative distribution function. Given a continuous joint cdf $F_{Y_1Y_2}(a,b) \equiv \Pr(\{Y_1 \leq a\} \cap \{Y_2 \leq b\})$, there is a function $p_{Y_1Y_2}$ such that

$$F_{Y_1Y_2}(a,b) = \int_{-\infty}^{a}\int_{-\infty}^{b} p_{Y_1Y_2}(y_1,y_2)\, dy_2 dy_1\,.$$

The function $p_{Y_1Y_2}$ is the joint density of Y_1 and Y_2. As in the discrete case, we have

- $p_{Y_1}(y_1) = \int_{-\infty}^{\infty} p_{Y_1Y_2}(y_1,y_2)\, dy_2$;
- $p_{Y_2|Y_1}(y_2|y_1) = p_{Y_1Y_2}(y_1,y_2)/p_{Y_1}(y_1)$.

You should convince yourself that $p_{Y_2|Y_1}(y_2|y_1)$ is an actual probability density, i.e. for each value of y_1 it is a probability density for Y_2.

Mixed continuous and discrete variables

Let Y_1 be discrete and Y_2 be continuous. For example, Y_1 could be occupational category and Y_2 could be personal income. Suppose we define

- a marginal density p_{Y_1} from our beliefs $\Pr(Y_1 = y_1)$;
- a conditional density $p_{Y_2|Y_1}(y_2|y_1)$ from $\Pr(Y_2 \leq y_2|Y_1 = y_1) \equiv F_{Y_2|Y_1}(y_2|y_1)$ as above.

The joint density of Y_1 and Y_2 is then

$$p_{Y_1 Y_2}(y_1, y_2) = p_{Y_1}(y_1) \times p_{Y_2|Y_1}(y_2|y_1),$$

and has the property that

$$\Pr(Y_1 \in A, Y_2 \in B) = \int_{y_2 \in B} \left\{ \sum_{y_1 \in A} p_{Y_1 Y_2}(y_1, y_2) \right\} dy_2.$$

Bayes' rule and parameter estimation

Let

θ = proportion of people in a large population who have a certain characteristic.
Y = number of people in a small random sample from the population who have the characteristic.

Then we might treat θ as continuous and Y as discrete. Bayesian estimation of θ derives from the calculation of $p(\theta|y)$, where y is the observed value of Y. This calculation first requires that we have a joint density $p(y, \theta)$ representing our beliefs about θ and the survey outcome Y. Often it is natural to construct this joint density from

- $p(\theta)$, beliefs about θ;
- $p(y|\theta)$, beliefs about Y for each value of θ.

Having observed $\{Y = y\}$, we need to compute our updated beliefs about θ:

$$p(\theta|y) = p(\theta, y)/p(y) = p(\theta)p(y|\theta)/p(y).$$

This conditional density is called the *posterior density* of θ. Suppose θ_a and θ_b are two possible numerical values of the true value of θ. The posterior probability (density) of θ_a relative to θ_b, conditional on $Y = y$, is

$$\begin{aligned} \frac{p(\theta_a|y)}{p(\theta_b|y)} &= \frac{p(\theta_a)p(y|\theta_a)/p(y)}{p(\theta_b)p(y|\theta_b)/p(y)} \\ &= \frac{p(\theta_a)p(y|\theta_a)}{p(\theta_b)p(y|\theta_b)}. \end{aligned}$$

This means that to evaluate the relative posterior probabilities of θ_a and θ_b, we do not need to compute $p(y)$. Another way to think about it is that, as a function of θ,

$$p(\theta|y) \propto p(\theta)p(y|\theta).$$

The constant of proportionality is $1/p(y)$, which *could* be computed from

$$p(y) = \int_\Theta p(y, \theta)\, d\theta = \int_\Theta p(y|\theta)p(\theta)\, d\theta$$

giving

$$p(\theta|y) = \frac{p(\theta)p(y|\theta)}{\int_\theta p(\theta)p(y|\theta)\, d\theta}.$$

As we will see in later chapters, the numerator is the critical part.

2.6 Independent random variables

Suppose $Y_1, \ldots, Y_n$ are random variables and that θ is a parameter describing the conditions under which the random variables are generated. We say that $Y_1, \ldots, Y_n$ are conditionally independent given θ if for every collection of n sets $\{A_1, \ldots, A_n\}$ we have

$$\Pr(Y_1 \in A_1, \ldots, Y_n \in A_n|\theta) = \Pr(Y_1 \in A_1|\theta) \times \cdots \times \Pr(Y_n \in A_n|\theta).$$

Notice that this definition of independent random variables is based on our previous definition of independent events, where here each $\{Y_j \in A_j\}$ is an event. From our previous calculations, if independence holds, then

$$\Pr(Y_i \in A_i|\theta, Y_j \in A_j) = \Pr(Y_i \in A_i|\theta),$$

so conditional independence can be interpreted as meaning that Y_j gives no additional information about Y_i beyond that in knowing θ. Furthermore, under independence the joint density is given by

$$p(y_1, \ldots, y_n|\theta) = p_{Y_1}(y_1|\theta) \times \cdots \times p_{Y_n}(y_n|\theta) = \prod_{i=1}^{n} p_{Y_i}(y_i|\theta),$$

the product of the marginal densities.

Suppose $Y_1, \ldots, Y_n$ are generated in similar ways from a common process. For example, they could all be samples from the same population, or runs of an experiment performed under similar conditions. This suggests that the marginal densities are all equal to some common density giving

$$p(y_1, \ldots, y_n|\theta) = \prod_{i=1}^{n} p(y_i|\theta).$$

In this case, we say that $Y_1, \ldots, Y_n$ are *conditionally independent and identically distributed* (i.i.d.). Mathematical shorthand for this is

$$Y_1, \ldots, Y_n|\theta \sim \text{i.i.d. } p(y|\theta).$$

2.7 Exchangeability

Example: Happiness

Participants in the 1998 General Social Survey were asked whether or not they were generally happy. Let Y_i be the random variable associated with this question, so that

$$Y_i = \begin{cases} 1 & \text{if participant } i \text{ reports being generally happy,} \\ 0 & \text{otherwise.} \end{cases}$$

In this section we will consider the structure of our joint beliefs about $Y_1, \ldots, Y_{10}$, the outcomes of the first 10 randomly selected survey participants. As before, let $p(y_1, \ldots, y_{10})$ be our shorthand notation for $\Pr(Y_1 = y_1, \ldots, Y_{10} = y_{10})$, where each y_i is either 0 or 1.

Exchangeability

Suppose we are asked to assign probabilities to three different outcomes:

$$p(1,0,0,1,0,1,1,0,1,1) = ?$$
$$p(1,0,1,0,1,1,0,1,1,0) = ?$$
$$p(1,1,0,0,1,1,0,0,1,1) = ?$$

Is there an argument for assigning them the same numerical value? Notice that each sequence contains six ones and four zeros.

Definition 3 (Exchangeable) *Let $p(y_1, \ldots, y_n)$ be the joint density of Y_1, ..., Y_n. If $p(y_1, \ldots, y_n) = p(y_{\pi_1}, \ldots, y_{\pi_n})$ for all permutations π of $\{1, \ldots, n\}$, then $Y_1, \ldots, Y_n$ are* exchangeable.

Roughly speaking, $Y_1, \ldots, Y_n$ are exchangeable if the subscript labels convey no information about the outcomes.

Independence versus dependence

Consider the following two probability assignments:

$$\Pr(Y_{10} = 1) = a$$
$$\Pr(Y_{10} = 1 | Y_1 = Y_2 = \cdots = Y_8 = Y_9 = 1) = b$$

Should we have $a < b$, $a = b$, or $a > b$? If $a \neq b$ then Y_{10} is NOT independent of $Y_1, \ldots, Y_9$.

Conditional independence

Suppose someone told you the numerical value of θ, the rate of happiness among the 1,272 respondents to the question. Do the following probability assignments seem reasonable?

$$\Pr(Y_{10} = 1|\theta) \overset{?}{\approx} \theta$$
$$\Pr(Y_{10} = 1|Y_1 = y_1, \ldots, Y_9 = y_9, \theta) \overset{?}{\approx} \theta$$
$$\Pr(Y_9 = 1|Y_1 = y_1, \ldots, Y_8 = y_8, Y_{10} = y_{10}, \theta) \overset{?}{\approx} \theta$$

If these assignments are reasonable, then we can consider the Y_i's as conditionally independent and identically distributed given θ, or at least approximately so: The population size of 1,272 is much larger than the sample size of 10, in which case sampling without replacement is approximately the same as i.i.d. sampling with replacement. Assuming conditional independence,

$$\Pr(Y_i = y_i|\theta, Y_j = y_j, j \neq i) = \theta^{y_i}(1-\theta)^{1-y_i}$$
$$\Pr(Y_1 = y_1, \ldots, Y_{10} = y_{10}|\theta) = \prod_{i=1}^{10} \theta^{y_i}(1-\theta)^{1-y_i}$$
$$= \theta^{\sum y_i}(1-\theta)^{10-\sum y_i}.$$

If θ is uncertain to us, we describe our beliefs about it with $p(\theta)$, a prior distribution. The marginal joint distribution of $Y_1, \ldots, Y_{10}$ is then

$$p(y_1, \ldots, y_{10}) = \int_0^1 p(y_1, \ldots, y_{10}|\theta)p(\theta)\, d\theta = \int_0^1 \theta^{\sum y_i}(1-\theta)^{10-\sum y_i} p(\theta)\, d\theta.$$

Now consider our probabilities for the three binary sequences given above:

$$p(1,0,0,1,0,1,1,0,1,1) = \textstyle\int \theta^6(1-\theta)^4 p(\theta)\, d\theta$$
$$p(1,0,1,0,1,1,0,1,1,0) = \textstyle\int \theta^6(1-\theta)^4 p(\theta)\, d\theta$$
$$p(1,1,0,0,1,1,0,0,1,1) = \textstyle\int \theta^6(1-\theta)^4 p(\theta)\, d\theta$$

It looks like $Y_1, \ldots, Y_n$ are exchangeable under this model of beliefs.

Claim:

If $\theta \sim p(\theta)$ and $Y_1, \ldots, Y_n$ are conditionally i.i.d. given θ, then marginally (unconditionally on θ), $Y_1, \ldots, Y_n$ are exchangeable.

Proof:

Suppose $Y_1, \ldots, Y_n$ are conditionally i.i.d. given some unknown parameter θ. Then for any permutation π of $\{1, \ldots, n\}$ and any set of values $(y_1, \ldots, y_n) \in \mathcal{Y}^n$,

$$
\begin{aligned}
p(y_1,\ldots,y_n) &= \int p(y_1,\ldots,y_n|\theta)p(\theta)\;d\theta && \text{(definition of marginal probability)}\\
&= \int\left\{\prod_{i=1}^{n} p(y_i|\theta)\right\}p(\theta)\;d\theta && \text{(Y_i's are conditionally i.i.d.)}\\
&= \int\left\{\prod_{i=1}^{n} p(y_{\pi_i}|\theta)\right\}p(\theta)\;d\theta && \text{(product does not depend on order)}\\
&= p(y_{\pi_1},\ldots y_{\pi_n}) && \text{(definition of marginal probability)}\,.
\end{aligned}
$$

2.8 de Finetti's theorem

We have seen that

$$
\left.\begin{array}{l} Y_1,\ldots,Y_n|\theta \text{ i.i.d} \\ \theta\sim p(\theta)\end{array}\right\} \Rightarrow Y_1,\ldots,Y_n \text{ are exchangeable.}
$$

What about an arrow in the other direction? Let $\{Y_1, Y_2, \ldots\}$ be a potentially infinite sequence of random variables all having a common sample space $\mathcal{Y}$.

Theorem 1 (de Finetti) *Let $Y_i \in \mathcal{Y}$ for all $i \in \{1, 2, \ldots\}$. Suppose that, for any n, our belief model for $Y_1, \ldots, Y_n$ is exchangeable:*

$$
p(y_1,\ldots,y_n) = p(y_{\pi_1},\ldots,y_{\pi_n})
$$

for all permutations π of $\{1,\ldots,n\}$. Then our model can be written as

$$
p(y_1,\ldots,y_n) = \int\left\{\prod_{1}^{n} p(y_i|\theta)\right\}p(\theta)\;d\theta
$$

for some parameter θ, some prior distribution on θ and some sampling model $p(y|\theta)$. The prior and sampling model depend on the form of the belief model $p(y_1,\ldots,y_n)$.

The probability distribution $p(\theta)$ represents our beliefs about the outcomes of $\{Y_1, Y_2, \ldots\}$, induced by our belief model $p(y_1, y_2, \ldots)$. More precisely,

> $p(\theta)$ represents our beliefs about $\lim_{n\to\infty}\sum Y_i/n$ in the binary case;
> $p(\theta)$ represents our beliefs about $\lim_{n\to\infty}\sum(Y_i \le c)/n$ for each c in the general case.

The main ideas of this and the previous section can be summarized as follows:

$$
\left.\begin{array}{l} Y_1,\ldots,Y_n|\theta \text{ are i.i.d.} \\ \theta\sim p(\theta)\end{array}\right\} \Leftrightarrow Y_1,\ldots,Y_n \text{ are exchangeable for all } n\,.
$$

When is the condition "$Y_1, \ldots, Y_n$ are exchangeable for all n" reasonable? For this condition to hold, we must have exchangeability and repeatability. Exchangeability will hold if the labels convey no information. Situations in which repeatability is reasonable include the following:

$Y_1, \ldots, Y_n$ are outcomes of a repeatable experiment;
$Y_1, \ldots, Y_n$ are sampled from a finite population with replacement;
$Y_1, \ldots, Y_n$ are sampled from an infinite population without replacement.

If $Y_1, \ldots, Y_n$ are exchangeable and sampled from a finite population of size $N >> n$ without replacement, then they can be modeled as approximately being conditionally i.i.d. (Diaconis and Freedman, 1980).

2.9 Discussion and further references

The notion of subjective probability in terms of a coherent gambling strategy was developed by de Finetti, who is of course also responsible for de Finetti's theorem (de Finetti, 1931, 1937). Both of these topics were studied further by many others, including Savage (Savage, 1954; Hewitt and Savage, 1955).

The concept of exchangeability goes beyond just the concept of an infinitely exchangeable sequence considered in de Finetti's theorem. Diaconis and Freedman (1980) consider exchangeability for finite populations or sequences, and Diaconis (1988) surveys some other versions of exchangeability. Chapter 4 of Bernardo and Smith (1994) provides a guide to building statistical models based on various types of exchangeability. A very comprehensive and mathematical review of exchangeability is given in Aldous (1985), which in particular provides an excellent survey of exchangeability as applied to random matrices.

3

One-parameter models

A one-parameter model is a class of sampling distributions that is indexed by a single unknown parameter. In this chapter we discuss Bayesian inference for two one-parameter models: the binomial model and the Poisson model. In addition to being useful statistical tools, these models also provide a simple environment within which we can learn the basics of Bayesian data analysis, including conjugate prior distributions, predictive distributions and confidence regions.

3.1 The binomial model

Happiness data

Each female of age 65 or over in the 1998 General Social Survey was asked whether or not they were generally happy. Let $Y_i = 1$ if respondent i reported being generally happy, and let $Y_i = 0$ otherwise. If we lack information distinguishing these $n = 129$ individuals we may treat their responses as being exchangeable. Since 129 is much smaller than the total size N of the female senior citizen population, the results of the last chapter indicate that our joint beliefs about $Y_1, \ldots, Y_{129}$ are well approximated by

- our beliefs about $\theta = \sum_{i=1}^{N} Y_i/N$;
- the model that, conditional on θ, the Y_i's are i.i.d. binary random variables with expectation θ.

The last item says that the probability for any potential outcome $\{y_1, \ldots, y_{129}\}$, conditional on θ, is given by

$$p(y_1, \ldots, y_{129}|\theta) = \theta^{\sum_{i=1}^{129} y_i}(1-\theta)^{129-\sum_{i=1}^{129} y_i}.$$

What remains to be specified is our prior distribution.

P.D. Hoff, *A First Course in Bayesian Statistical Methods*,
Springer Texts in Statistics, DOI 10.1007/978-0-387-92407-6_3,

A uniform prior distribution

The parameter θ is some unknown number between 0 and 1. Suppose our prior information is such that all subintervals of $[0, 1]$ having the same length also have the same probability. Symbolically,

$$\Pr(a \leq \theta \leq b) = \Pr(a + c \leq \theta \leq b + c) \text{ for } 0 \leq a < b < b + c \leq 1.$$

This condition implies that our density for θ must be the uniform density:

$$p(\theta) = 1 \text{ for all } \theta \in [0, 1].$$

For this prior distribution and the above sampling model, Bayes' rule gives

$$\begin{aligned} p(\theta|y_1, \ldots, y_{129}) &= \frac{p(y_1, \ldots, y_{129}|\theta)p(\theta)}{p(y_1, \ldots, y_{129})} \\ &= p(y_1, \ldots, y_{129}|\theta) \times \frac{1}{p(y_1, \ldots, y_{129})} \\ &\propto p(y_1, \ldots, y_{129}|\theta) . \end{aligned}$$

The last line says that in this particular case $p(\theta|y_1, \ldots, y_{129})$ and $p(y_1, \ldots, y_{129}|\theta)$ are proportional to each other as functions of θ. This is because the posterior distribution is equal to $p(y_1, \ldots, y_{129}|\theta)$ divided by something that does not depend on θ. This means that these two functions of θ have the same shape, but not necessarily the same scale.

Data and posterior distribution

- 129 individuals surveyed;
- 118 individuals report being generally happy (91%);
- 11 individuals do not report being generally happy (9%).

The probability of these data for a given value of θ is

$$p(y_1, \ldots, y_{129}|\theta) = \theta^{118}(1 - \theta)^{11}.$$

A plot of this probability as a function of θ is shown in the first plot of Figure 3.1. Our result above about proportionality says that the posterior distribution $p(\theta|y_1, \ldots, y_{129})$ will have the same shape as this function, and so we know that the true value of θ is very likely to be near 0.91, and almost certainly above 0.80. However, we will often want to be more precise than this, and we will need to know the scale of $p(\theta|y_1, \ldots, y_n)$ as well as the shape. From Bayes' rule, we have

$$\begin{aligned} p(\theta|y_1, \ldots, y_{129}) &= \theta^{118}(1 - \theta)^{11} \times p(\theta)/p(y_1, \ldots, y_{129}) \\ &= \theta^{118}(1 - \theta)^{11} \times 1/p(y_1, \ldots, y_{129}). \end{aligned}$$

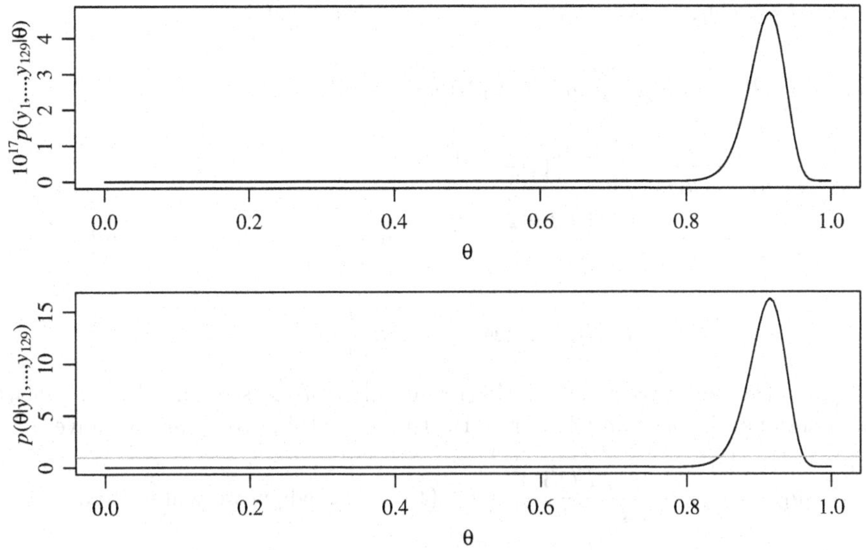

Fig. 3.1. Sampling probability of the data as a function of θ, along with the posterior distribution. Note that a uniform prior distribution (plotted in gray in the second panel) gives a posterior distribution that is proportional to the sampling probability.

It turns out that we can calculate the scale or "normalizing constant" $1/p(y_1, \ldots, y_{129})$ using the following result from calculus:

$$\int_0^1 \theta^{a-1}(1-\theta)^{b-1}\, d\theta = \frac{\Gamma(a)\Gamma(b)}{\Gamma(a+b)}.$$

(the value of the gamma function $\Gamma(x)$ for any number $x > 0$ can be looked up in a table, or with R using the `gamma()` function). How does the calculus result help us compute $p(\theta|y_1, \ldots, y_{129})$? Let's recall what we know about $p(\theta|y_1, \ldots, y_{129})$:

(a) $\int_0^1 p(\theta|y_1, \ldots, y_{129})\, d\theta = 1$, since all probability distributions integrate or sum to 1;
(b) $p(\theta|y_1, \ldots, y_{129}) = \theta^{118}(1-\theta)^{11}/p(y_1, \ldots, y_{129})$, from Bayes' rule.

Therefore,

$$1 = \int_0^1 p(\theta|y_1,\ldots,y_{129})\,d\theta \qquad \text{using (a)}$$
$$1 = \int_0^1 \theta^{118}(1-\theta)^{11}/p(y_1,\ldots,y_{129})\,d\theta \qquad \text{using (b)}$$
$$1 = \frac{1}{p(y_1,\ldots,y_{129})}\int_0^1 \theta^{118}(1-\theta)^{11}\,d\theta$$
$$1 = \frac{1}{p(y_1,\ldots,y_{129})}\frac{\Gamma(119)\Gamma(12)}{\Gamma(131)} \qquad \text{using the calculus result, and so}$$

$$p(y_1,\ldots,y_{129}) = \frac{\Gamma(119)\Gamma(12)}{\Gamma(131)}.$$

You should convince yourself that this result holds for any sequence $\{y_1,\ldots,y_{129}\}$ that contains 118 ones and 11 zeros. Putting everything together, we have

$$\begin{aligned} p(\theta|y_1,\ldots,y_{129}) &= \frac{\Gamma(131)}{\Gamma(119)\Gamma(12)}\theta^{118}(1-\theta)^{11} \text{ , which we will write as} \\ &= \frac{\Gamma(131)}{\Gamma(119)\Gamma(12)}\theta^{119-1}(1-\theta)^{12-1}. \end{aligned}$$

This density for θ is called a *beta distribution* with parameters $a = 119$ and $b = 12$, which can be calculated, plotted and sampled from in R using the functions dbeta() and rbeta() .

The beta distribution

An uncertain quantity θ, known to be between 0 and 1, has a beta(a, b) distribution if

$$p(\theta) = \text{dbeta}(\theta, a, b) = \frac{\Gamma(a+b)}{\Gamma(a)\Gamma(b)}\theta^{a-1}(1-\theta)^{b-1} \quad \text{for } 0 \le \theta \le 1.$$

For such a random variable,

$\text{mode}[\theta] = (a-1)/[(a-1)+(b-1)]$ if $a > 1$ and $b > 1$;
$\text{E}[\theta] = a/(a+b)$;
$\text{Var}[\theta] = ab/[(a+b+1)(a+b)^2] = \text{E}[\theta] \times \text{E}[1-\theta]/(a+b+1)$.

For our data on happiness in which we observed $(Y_1,\ldots,Y_{129}) = (y_1,\ldots,y_{129})$ with $\sum_{i=1}^{129} y_i = 118$,

$\text{mode}[\theta|y_1,\ldots,y_{129}] = 0.915$;
$\text{E}[\theta|y_1,\ldots,y_{129}] = 0.908$;
$\text{sd}[\theta|y_1,\ldots,y_{129}] = 0.025$.

3.1.1 Inference for exchangeable binary data

Posterior inference under a uniform prior

If $Y_1, \ldots, Y_n|\theta$ are i.i.d. binary(θ), we showed that

$$p(\theta|y_1, \ldots, y_n) = \theta^{\sum y_i}(1-\theta)^{n-\sum y_i} \times p(\theta)/p(y_1, \ldots, y_n).$$

If we compare the relative probabilities of any two θ-values, say θ_a and θ_b, we see that

$$\frac{p(\theta_a|y_1, \ldots, y_n)}{p(\theta_b|y_1, \ldots, y_n)} = \frac{\theta_a^{\sum y_i}(1-\theta_a)^{n-\sum y_i} \times p(\theta_a)/p(y_1, \ldots, y_n)}{\theta_b^{\sum y_i}(1-\theta_b)^{n-\sum y_i} \times p(\theta_b)/p(y_1, \ldots, y_n)}$$
$$= \left(\frac{\theta_a}{\theta_b}\right)^{\sum y_i} \left(\frac{1-\theta_a}{1-\theta_b}\right)^{n-\sum y_i} \frac{p(\theta_a)}{p(\theta_b)}.$$

This shows that the probability density at θ_a relative to that at θ_b depends on $y_1, \ldots, y_n$ only through $\sum_{i=1}^n y_i$. From this, you can show that

$$\Pr(\theta \in A|Y_1 = y_1, \ldots, Y_n = y_n) = \Pr\left(\theta \in A|\sum_{i=1}^n Y_i = \sum_{i=1}^n y_i\right).$$

We interpret this as meaning that $\sum_{i=1}^n Y_i$ contains all the information about θ available from the data, and we say that $\sum_{i=1}^n Y_i$ is a *sufficient statistic* for θ and $p(y_1, \ldots, y_n|\theta)$. The word "sufficient" is used because it is "sufficient" to know $\sum Y_i$ in order to make inference about θ. In this case where $Y_1, \ldots, Y_n|\theta$ are i.i.d. binary(θ) random variables, the sufficient statistic $Y = \sum_{i=1}^n Y_i$ has a *binomial distribution* with parameters (n, θ).

The binomial distribution

A random variable $Y \in \{0, 1, \ldots, n\}$ has a binomial(n, θ) distribution if

$$\Pr(Y = y|\theta) = \text{dbinom}(y, n, \theta) = \binom{n}{y}\theta^y(1-\theta)^{n-y}, \quad y \in \{0, 1, \ldots, n\}.$$

Binomial distributions with different values of n and θ are plotted in Figures 3.2 and 3.3. For a binomial(n, θ) random variable,

$$\text{E}[Y|\theta] = n\theta;$$
$$\text{Var}[Y|\theta] = n\theta(1-\theta).$$

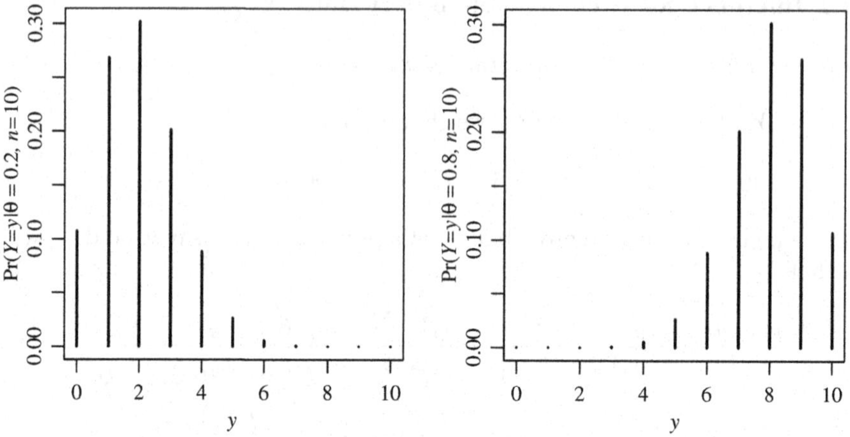

Fig. 3.2. Binomial distributions with $n = 10$ and $\theta \in \{0.2, 0.8\}$.

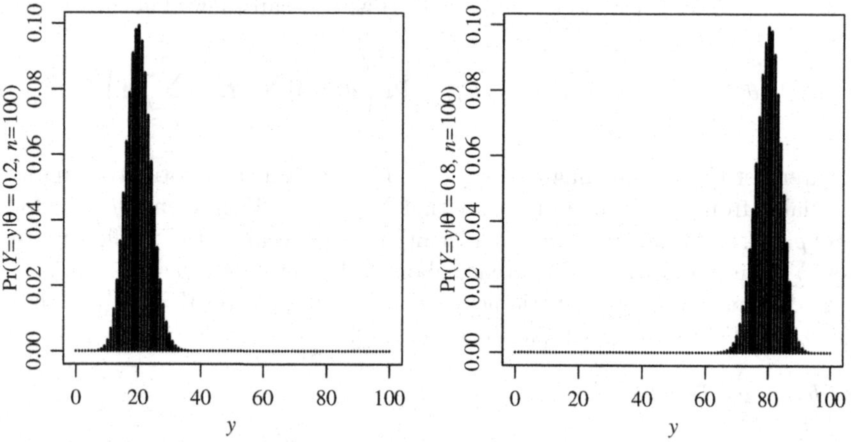

Fig. 3.3. Binomial distributions with $n = 100$ and $\theta \in \{0.2, 0.8\}$.

Posterior inference under a uniform prior distribution

Having observed $Y = y$ our task is to obtain the posterior distribution of θ:

$$\begin{aligned} p(\theta|y) &= \frac{p(y|\theta)p(\theta)}{p(y)} \\ &= \frac{\binom{n}{y}\theta^y(1-\theta)^{n-y}p(\theta)}{p(y)} \\ &= c(y)\theta^y(1-\theta)^{n-y}p(\theta) \end{aligned}$$

where $c(y)$ is a function of y and not of θ. For the uniform distribution with $p(\theta) = 1$, we can find out what $c(y)$ is using our calculus trick:

$$\begin{aligned} 1 &= \int_0^1 c(y)\theta^y(1-\theta)^{n-y}\, d\theta \\ 1 &= c(y)\int_0^1 \theta^y(1-\theta)^{n-y}\, d\theta \\ 1 &= c(y)\frac{\Gamma(y+1)\Gamma(n-y+1)}{\Gamma(n+2)} . \end{aligned}$$

The normalizing constant $c(y)$ is therefore equal to $\Gamma(n+2)/\{\Gamma(y+1)\Gamma(n-y+1)\}$, and we have

$$\begin{aligned} p(\theta|y) &= \frac{\Gamma(n+2)}{\Gamma(y+1)\Gamma(n-y+1)}\theta^y(1-\theta)^{n-y} \\ &= \frac{\Gamma(n+2)}{\Gamma(y+1)\Gamma(n-y+1)}\theta^{(y+1)-1}(1-\theta)^{(n-y+1)-1} \\ &= \text{beta}(y+1, n-y+1). \end{aligned}$$

Recall the happiness example, where we observed that $Y \equiv \sum Y_i = 118$:

$$n = 129, Y \equiv \sum Y_i = 118 \quad \Rightarrow \quad \theta|\{Y = 118\} \sim \text{beta}(119, 12).$$

This confirms the sufficiency result for this model and prior distribution, by showing that if $\sum y_i = y = 118$,

$$p(\theta|y_1, \ldots, y_n) = p(\theta|y) = \text{beta}(119, 12).$$

In other words, the information contained in $\{Y_1 = y_1, \ldots, Y_n = y_n\}$ is the same as the information contained in $\{Y = y\}$, where $Y = \sum Y_i$ and $y = \sum y_i$.

Posterior distributions under beta prior distributions

The uniform prior distribution has $p(\theta) = 1$ for all $\theta \in [0, 1]$. This distribution can be thought of as a beta prior distribution with parameters $a = 1, b = 1$:

$$p(\theta) = \frac{\Gamma(2)}{\Gamma(1)\Gamma(1)}\theta^{1-1}(1-\theta)^{1-1} = \frac{1}{1 \times 1} 1 \times 1 = 1 .$$

Note that $\Gamma(x+1) = x! = x \times (x-1) \cdots \times 1$ if x is a positive integer, and $\Gamma(1) = 1$ by convention. In the previous paragraph, we saw that

$$\text{if } \left\{ \begin{array}{c} \theta \sim \text{beta}(1,1) \text{ (uniform)} \\ Y \sim \text{ binomial}(n, \theta) \end{array} \right\}, \text{ then } \{\theta|Y = y\} \sim \text{beta}(1+y, 1+n-y),$$

and so to get the posterior distribution when our prior distribution is beta($a = 1, b = 1$), we can simply add the number of 1's to the a parameter and the number of 0's to the b parameter. Does this result hold for arbitrary beta priors? Let's find out: Suppose $\theta \sim \text{beta}(a,b)$ and $Y|\theta \sim \text{binomial}(n,\theta)$. Having observed $Y = y$,

$$\begin{aligned}
p(\theta|y) &= \frac{p(\theta)p(y|\theta)}{p(y)} \\
&= \frac{1}{p(y)} \times \frac{\Gamma(a+b)}{\Gamma(a)\Gamma(b)}\theta^{a-1}(1-\theta)^{b-1} \times \binom{n}{y}\theta^{y}(1-\theta)^{n-y} \\
&= c(n,y,a,b) \times \theta^{a+y-1}(1-\theta)^{b+n-y-1} \\
&= \text{dbeta}(\theta, a+y, b+n-y)\,.
\end{aligned}$$

It is important to understand the last two lines above: The second to last line says that $p(\theta|y)$ is, as a function of θ, proportional to $\theta^{a+y-1} \times (1-\theta)^{b+n-y-1}$. This means that it has the same *shape* as the beta density $\text{dbeta}(\theta, a+y, b+n-y)$. But we also know that $p(\theta|y)$ and the beta density must both integrate to 1, and therefore they also share the same *scale*. These two things together mean that $p(\theta|y)$ and the beta density are in fact the same function. Throughout the book we will use this trick to identify posterior distributions: We will recognize that the posterior distribution is proportional to a known probability density, and therefore must equal that density.

Conjugacy

We have shown that a beta prior distribution and a binomial sampling model lead to a beta posterior distribution. To reflect this, we say that the class of beta priors is *conjugate* for the binomial sampling model.

Definition 4 (Conjugate) *A class $\mathcal{P}$ of prior distributions for θ is called* conjugate *for a sampling model $p(y|\theta)$ if*

$$p(\theta) \in \mathcal{P} \Rightarrow p(\theta|y) \in \mathcal{P}.$$

Conjugate priors make posterior calculations easy, but might not actually represent our prior information. However, mixtures of conjugate prior distributions are very flexible and are computationally tractable (see Exercises 3.4 and 3.5).

Combining information

If $\theta|\{Y = y\} \sim \text{beta}(a+y, b+n-y)$, then

$$\text{E}[\theta|y] = \frac{a+y}{a+b+n},\ \text{mode}[\theta|y] = \frac{a+y-1}{a+b+n-2},\ \text{Var}[\theta|y] = \frac{\text{E}[\theta|y]\text{E}[1-\theta|y]}{a+b+n+1}\,.$$

The *posterior expectation* $\text{E}[\theta|y]$ is easily recognized as a combination of prior and data information:

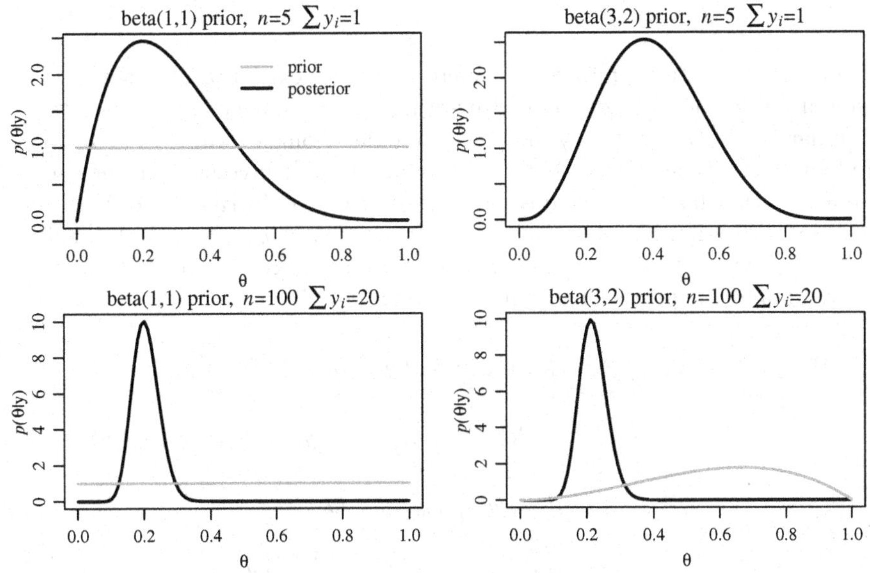

Fig. 3.4. Beta posterior distributions under two different sample sizes and two different prior distributions. Look across a row to see the effect of the prior distribution, and down a column to see the effect of the sample size.

$$
\begin{aligned}
\mathrm{E}[\theta|y] &= \frac{a+y}{a+b+n} \\
&= \frac{a+b}{a+b+n}\frac{a}{a+b} + \frac{n}{a+b+n}\frac{y}{n} \\
&= \frac{a+b}{a+b+n} \times \text{prior expectation} + \frac{n}{a+b+n} \times \text{data average}\,.
\end{aligned}
$$

For this model and prior distribution, the posterior expectation (also known as the posterior mean) is a weighted average of the prior expectation and the sample average, with weights proportional to $a+b$ and n respectively. This leads to the interpretation of a and b as "prior data":

$a \approx$ "prior number of 1's,"
$b \approx$ "prior number of 0's,"
$a+b \approx$ "prior sample size."

If our sample size n is larger than our prior sample size $a+b$, then it seems reasonable that a majority of our information about θ should be coming from the data as opposed to the prior distribution. This is indeed the case: For example, if $n >> a+b$, then

$$
\frac{a+b}{a+b+n} \approx 0\,,\ \mathrm{E}[\theta|y] \approx \frac{y}{n}\,,\ \mathrm{Var}[\theta|y] \approx \frac{1}{n}\frac{y}{n}\left(1-\frac{y}{n}\right).
$$

Prediction

An important feature of Bayesian inference is the existence of a predictive distribution for new observations. Reverting for the moment to our notation for binary data, let $y_1, \ldots, y_n$ be the outcomes from a sample of n binary random variables, and let $\tilde{Y} \in \{0, 1\}$ be an additional outcome from the same population that has yet to be observed. The *predictive distribution* of $\tilde{Y}$ is the conditional distribution of $\tilde{Y}$ given $\{Y_1 = y_1, \ldots, Y_n = y_n\}$. For conditionally i.i.d. binary variables this distribution can be derived from the distribution of $\tilde{Y}$ given θ and the posterior distribution of θ:

$$\begin{aligned}
\Pr(\tilde{Y} = 1|y_1, \ldots, y_n) &= \int \Pr(\tilde{Y} = 1, \theta|y_1, \ldots, y_n)\, d\theta \\
&= \int \Pr(\tilde{Y} = 1|\theta, y_1, \ldots, y_n) p(\theta|y_1, \ldots, y_n)\, d\theta \\
&= \int \theta p(\theta|y_1, \ldots, y_n)\, d\theta \\
&= \mathrm{E}[\theta|y_1, \ldots, y_n] = \frac{a + \sum_{i=1}^n y_i}{a + b + n} \\
\Pr(\tilde{Y} = 0|y_1, \ldots, y_n) &= 1 - \mathrm{E}[\theta|y_1, \ldots, y_n] = \frac{b + \sum_{i=1}^n (1 - y_i)}{a + b + n}.
\end{aligned}$$

You should notice two important things about the predictive distribution:

1. The predictive distribution does not depend on any unknown quantities. If it did, we would not be able to use it to make predictions.
2. The predictive distribution depends on our observed data. In this distribution, $\tilde{Y}$ is not independent of $Y_1, \ldots, Y_n$ (recall Section 2.7). This is because observing $Y_1, \ldots, Y_n$ gives information about θ, which in turn gives information about $\tilde{Y}$. It would be bad if $\tilde{Y}$ were independent of $Y_1, \ldots, Y_n$ - it would mean that we could never infer anything about the unsampled population from the sample cases.

Example

The uniform prior distribution, or beta(1,1) prior, can be thought of as equivalent to the information in a prior dataset consisting of a single "1" and a single "0". Under this prior distribution,

$$\begin{aligned}
\Pr(\tilde{Y} = 1|Y = y) = \mathrm{E}[\theta|Y = y] &= \frac{2}{2+n}\frac{1}{2} + \frac{n}{2+n}\frac{y}{n}, \\
\mathrm{mode}(\theta|Y = y) &= \frac{y}{n},
\end{aligned}$$

where $Y = \sum_{i=1}^n Y_i$. Does the discrepancy between these two posterior summaries of our information make sense? Consider the case in which $Y = 0$, for which $\mathrm{mode}(\theta|Y = 0) = 0$ but $\Pr(\tilde{Y} = 1|Y = 0) = 1/(2 + n)$.

3.1.2 Confidence regions

It is often desirable to identify regions of the parameter space that are likely to contain the true value of the parameter. To do this, after observing the data $Y = y$ we can construct an interval $(l(y), u(y))$ such that the probability that $l(y) < \theta < u(y)$ is large.

Definition 5 (Bayesian coverage) *An interval* $(l(y), u(y))$*, based on the observed data* $Y = y$*, has 95% Bayesian coverage for* θ *if*

$$\Pr(l(y) < \theta < u(y)|Y = y) = .95.$$

The interpretation of this interval is that it describes your information about the location of the true value of θ after you have observed $Y = y$. This is different from the frequentist interpretation of coverage probability, which describes the probability that the interval will cover the true value *before* the data are observed:

Definition 6 (frequentist coverage) *A random interval* $(l(Y), u(Y))$ *has 95% frequentist coverage for* θ *if, before the data are gathered,*

$$\Pr(l(Y) < \theta < u(Y)|\theta) = .95.$$

In a sense, the frequentist and Bayesian notions of coverage describe pre- and post-experimental coverage, respectively.

You may recall your introductory statistics instructor belaboring the following point: Once you observe $Y = y$ and you plug this data into your confidence interval formula $(l(y), u(y))$, then

$$\Pr(l(y) < \theta < u(y)|\theta) = \begin{cases} 0 & \text{if } \theta \notin (l(y), u(y)); \\ 1 & \text{if } \theta \in (l(y), u(y)). \end{cases}$$

This highlights the lack of a post-experimental interpretation of frequentist coverage. Although this may make the frequentist interpretation seem somewhat lacking, it is still useful in many situations. Suppose you are running a large number of unrelated experiments and are creating a confidence interval for each one of them. If your intervals each have 95% frequentist coverage probability, you can expect that 95% of your intervals contain the correct parameter value.

Can a confidence interval have the same Bayesian and frequentist coverage probability? Hartigan (1966) showed that, for the types of intervals we will construct in this book, an interval that has 95% Bayesian coverage additionally has the property that

$$\Pr(l(Y) < \theta < u(Y)|\theta) = .95 + \epsilon_n$$

where $|\epsilon_n| < \frac{a}{n}$ for some constant a. This means that a confidence interval procedure that gives 95% Bayesian coverage will have approximately 95% frequentist coverage as well, at least asymptotically. It is important to keep in mind that most non-Bayesian methods of constructing 95% confidence intervals also only achieve this coverage rate asymptotically. For more discussion of the similarities between intervals constructed by Bayesian and non-Bayesian methods, see Severini (1991) and Sweeting (2001).

Quantile-based interval

Perhaps the easiest way to obtain a confidence interval is to use posterior quantiles. To make a $100 \times (1-\alpha)\%$ quantile-based confidence interval, find numbers $\theta_{\alpha/2} < \theta_{1-\alpha/2}$ such that

1. $\Pr(\theta < \theta_{\alpha}|Y = y) = \alpha/2$;
2. $\Pr(\theta > \theta_{1-\alpha/2}|Y = y) = \alpha/2$.

The numbers $\theta_{\alpha/2}, \theta_{1-\alpha/2}$ are the $\alpha/2$ and $1-\alpha/2$ posterior quantiles of θ, and so

$$\begin{aligned}\Pr(\theta \in (\theta_{\alpha/2}, \theta_{1-\alpha/2})|Y = y) &= 1 - \Pr(\theta \notin (\theta_{\alpha/2}, \theta_{1-\alpha/2})|Y = y) \\ &= 1 - [\Pr(\theta < \theta_{\alpha/2}|Y = y) + \Pr(\theta > \theta_{1-\alpha/2}|Y = y)] \\ &= 1 - \alpha\,.\end{aligned}$$

Example: Binomial sampling and uniform prior

Suppose out of $n = 10$ conditionally independent draws of a binary random variable we observe $Y = 2$ ones. Using a uniform prior distribution for θ, the posterior distribution is $\theta|\{Y = 2\} \sim \text{beta}(1 + 2, 1 + 8)$. A 95% posterior confidence interval can be obtained from the .025 and .975 quantiles of this beta distribution. These quantiles are 0.06 and 0.52 respectively, and so the posterior probability that $\theta \in (0.06, 0.52)$ is 95%.

```
> a<-1  ; b<-1   #prior
> n<-10 ; y<-2   #data

> qbeta( c(.025,.975), a+y,b+n-y)

[1] 0.06021773 0.51775585
```

Highest posterior density (HPD) region

Figure 3.5 shows the posterior distribution and a 95% confidence interval for θ from the previous example. Notice that there are θ-values *outside* the quantile-based interval that have higher probability (density) than some points *inside* the interval. This suggests a more restrictive type of interval:

Definition 7 (HPD region) *A* $100 \times (1-\alpha)\%$ *HPD region consists of a subset of the parameter space,* $s(y) \subset \Theta$ *such that*

1. $\Pr(\theta \in s(y)|Y = y) = 1 - \alpha$;

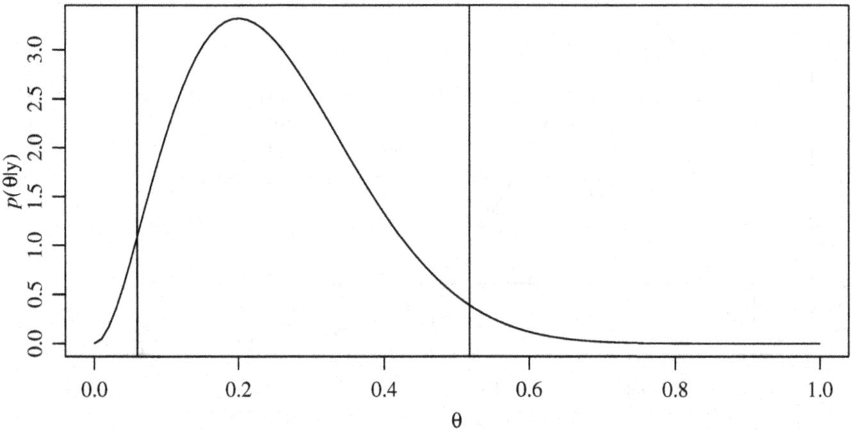

Fig. 3.5. A beta posterior distribution, with vertical bars indicating a 95% quantile-based confidence interval.

2. If $\theta_a \in s(y)$, *and* $\theta_b \notin s(y)$, *then* $p(\theta_a|Y=y) > p(\theta_b|Y=y)$.

All points in an HPD region have a higher posterior density than points outside the region. However, an HPD region might not be an interval if the posterior density is multimodal (having multiple peaks). Figure 3.6 gives the basic idea behind the construction of an HPD region: Gradually move a horizontal line down across the density, including in the HPD region all θ-values having a density above the horizontal line. Stop moving the line down when the posterior probability of the θ-values in the region reaches $(1-\alpha)$. For the binomial example above, the 95% HPD region is $(0.04, 0.48)$, which is narrower (more precise) than the quantile-based interval, yet both contain 95% of the posterior probability.

3.2 The Poisson model

Some measurements, such as a person's number of children or number of friends, have values that are whole numbers. In these cases our sample space is $\mathcal{Y} = \{0, 1, 2, \ldots\}$. Perhaps the simplest probability model on $\mathcal{Y}$ is the Poisson model.

Poisson distribution

Recall from Chapter 2 that a random variable Y has a Poisson distribution with mean θ if

$$\Pr(Y=y|\theta) = \text{dpois}(y,\theta) = \theta^y e^{-\theta}/y! \;\; \text{for } y \in \{0,1,2,\ldots\} \,.$$

For such a random variable,

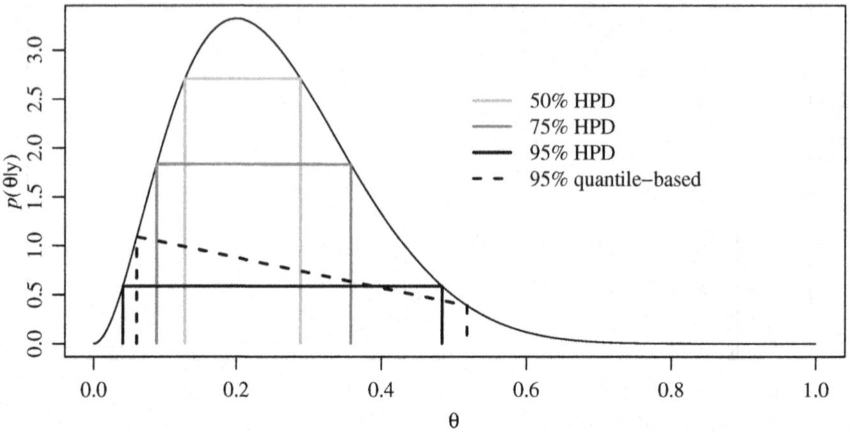

Fig. 3.6. Highest posterior density regions of varying probability content. The dashed line is the 95% quantile-based interval.

- $\text{E}[Y|\theta] = \theta$;
- $\text{Var}[Y|\theta] = \theta$.

People sometimes say that the Poisson family of distributions has a "mean-variance relationship" because if one Poisson distribution has a larger mean than another, it will have a larger variance as well.

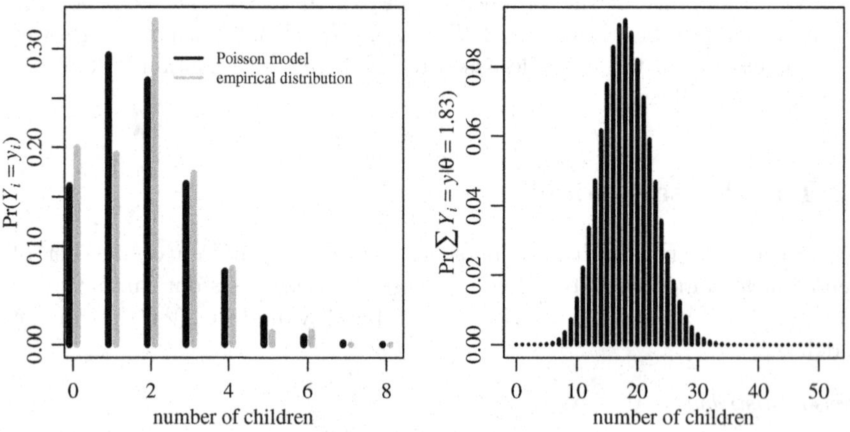

Fig. 3.7. Poisson distributions. The first panel shows a Poisson distribution with mean of 1.83, along with the empirical distribution of the number of children of women of age 40 from the GSS during the 1990s. The second panel shows the distribution of the sum of 10 i.i.d. Poisson random variables with mean 1.83. This is the same as a Poisson distribution with mean 18.3

3.2.1 Posterior inference

If we model $Y_1, \ldots, Y_n$ as i.i.d. Poisson with mean θ, then the joint pdf of our sample data is as follows:

$$\begin{aligned}
\Pr(Y_1 = y_1, \ldots, Y_n = y_n|\theta) &= \prod_{i=1}^{n} p(y_i|\theta) \\
&= \prod_{i=1}^{n} \frac{1}{y_i!}\theta^{y_i} e^{-\theta} \\
&= c(y_1, \ldots, y_n)\theta^{\sum y_i} e^{-n\theta} .
\end{aligned}$$

Comparing two values of θ *a posteriori*, we have

$$\begin{aligned}
\frac{p(\theta_a|y_1, \ldots, y_n)}{p(\theta_b|y_1, \ldots, y_n)} &= \frac{c(y_1, \ldots, y_n)\, e^{-n\theta_a}\, \theta_a^{\sum y_i}\, p(\theta_a)}{c(y_1, \ldots, y_n)\, e^{-n\theta_b}\, \theta_b^{\sum y_i}\, p(\theta_b)} \\
&= \frac{e^{-n\theta_a}}{e^{-n\theta_b}} \frac{\theta_a^{\sum y_i}}{\theta_b^{\sum y_i}} \frac{p(\theta_a)}{p(\theta_b)} .
\end{aligned}$$

As in the case of the i.i.d. binary model, $\sum_{i=1}^n Y_i$ contains all the information about θ that is available in the data, and again we say that $\sum_{i=1}^n Y_i$ is a sufficient statistic. Furthermore, $\{\sum_{i=1}^n Y_i|\theta\} \sim$ Poisson$(n\theta)$.

Conjugate prior

For now we will work with a class of conjugate prior distributions that will make posterior calculations simple. Recall that a class of prior densities is conjugate for a sampling model $p(y_1, \ldots, y_n|\theta)$ if the posterior distribution is also in the class. For the Poisson sampling model, our posterior distribution for θ has the following form:

$$\begin{aligned}
p(\theta|y_1, \ldots, y_n) &\propto p(\theta) \times p(y_1, \ldots, y_n|\theta) \\
&\propto p(\theta) \times \theta^{\sum y_i} e^{-n\theta} .
\end{aligned}$$

This means that whatever our conjugate class of densities is, it will have to include terms like $\theta^{c_1} e^{-c_2\theta}$ for numbers c_1 and c_2. The simplest class of such densities includes only these terms, and their corresponding probability distributions are known as the family of gamma distributions.

Gamma distribution

An uncertain positive quantity θ has a gamma(a, b) distribution if

$$p(\theta) = \text{dgamma}(\theta, a, b) = \frac{b^a}{\Gamma(a)}\theta^{a-1}e^{-b\theta}, \quad \text{for } \theta,\ a,\ b > 0.$$

For such a random variable,

- $\mathrm{E}[\theta] = a/b$;
- $\mathrm{Var}[\theta] = a/b^2$;
- $\mathrm{mode}[\theta] = \begin{cases} (a-1)/b & \text{if } a > 1 \\ 0 & \text{if } a \leq 1 \end{cases}$.

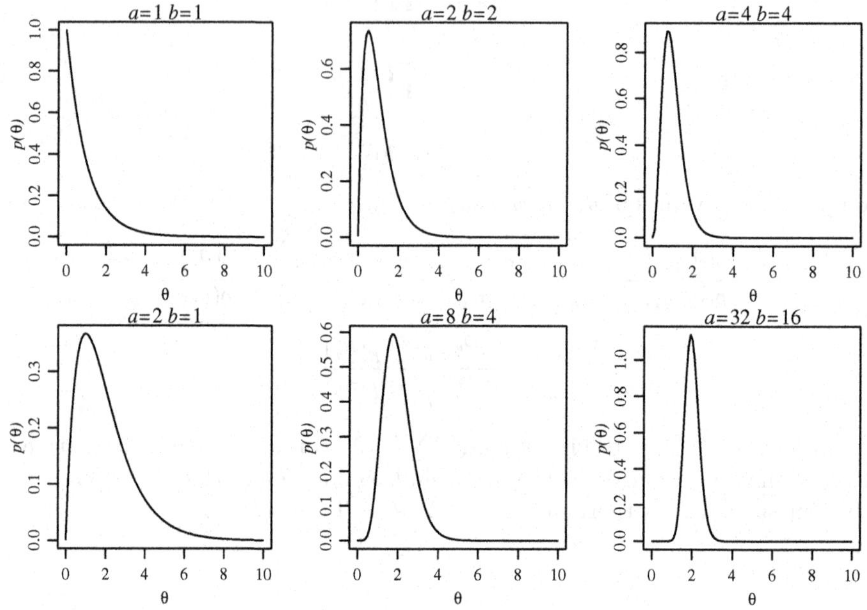

Fig. 3.8. Gamma densities.

Posterior distribution of θ

Suppose $Y_1, \ldots, Y_n|\theta \sim$ i.i.d. Poisson(θ) and $p(\theta)$=dgamma(θ, a, b). Then

$$\begin{aligned} p(\theta|y_1, \ldots, y_n) &= p(\theta) \times p(y_1, \ldots, y_n|\theta)/p(y_1, \ldots, y_n) \\ &= \left\{\theta^{a-1}e^{-b\theta}\right\} \times \left\{\theta^{\sum y_i}e^{-n\theta}\right\} \times c(y_1, \ldots, y_n, a, b) \\ &= \left\{\theta^{a+\sum y_i - 1}e^{-(b+n)\theta}\right\} \times c(y_1, \ldots, y_n, a, b). \end{aligned}$$

This is evidently a gamma distribution, and we have confirmed the conjugacy of the gamma family for the Poisson sampling model:

$$\left.\begin{array}{c} \theta \sim \mathrm{gamma}(a, b) \\ Y_1, \ldots, Y_n|\theta \sim \mathrm{Poisson}(\theta) \end{array}\right\} \Rightarrow \{\theta|Y_1, \ldots, Y_n\} \sim \mathrm{gamma}(a + \sum_{i=1}^n Y_i, b + n)\,.$$

Estimation and prediction proceed in a manner similar to that in the binomial model. The posterior expectation of θ is a convex combination of the prior expectation and the sample average:

$$\begin{aligned} \mathrm{E}[\theta|y_1,\ldots,y_n] &= \frac{a+\sum y_i}{b+n} \\ &= \frac{b}{b+n}\frac{a}{b} + \frac{n}{b+n}\frac{\sum y_i}{n} \end{aligned}$$

- b is interpreted as the number of prior observations;
- a is interpreted as the sum of counts from b prior observations.

For large n, the information from the data dominates the prior information:

$$n >> b \Rightarrow \mathrm{E}[\theta|y_1,\ldots,y_n] \approx \bar{y},\ \mathrm{Var}[\theta|y_1,\ldots,y_n] \approx \bar{y}/n\,.$$

Predictions about additional data can be obtained with the posterior predictive distribution:

$$\begin{aligned} p(\tilde{y}|y_1,\ldots,y_n) &= \int_0^\infty p(\tilde{y}|\theta,y_1,\ldots,y_n)p(\theta|y_1,\ldots,y_n)\,d\theta \\ &= \int p(\tilde{y}|\theta)p(\theta|y_1,\ldots,y_n)\,d\theta \\ &= \int \mathrm{dpois}(\tilde{y},\theta)\mathrm{dgamma}(\theta,a+\textstyle\sum y_i,b+n)\,d\theta \\ &= \int \left\{\frac{1}{\tilde{y}!}\theta^{\tilde{y}}e^{-\theta}\right\}\left\{\frac{(b+n)^{a+\sum y_i}}{\Gamma(a+\sum y_i)}\theta^{a+\sum y_i-1}e^{-(b+n)\theta}\right\}\,d\theta \\ &= \frac{(b+n)^{a+\sum y_i}}{\Gamma(\tilde{y}+1)\Gamma(a+\sum y_i)}\int_0^\infty \theta^{a+\sum y_i+\tilde{y}-1}e^{-(b+n+1)\theta}\,d\theta\,. \end{aligned}$$

Evaluation of this complicated integral looks daunting, but it turns out that it can be done without any additional calculus. Let's use what we know about the gamma density:

$$1 = \int_0^\infty \frac{b^a}{\Gamma(a)}\theta^{a-1}e^{-b\theta}\,d\theta \quad \text{for any values } a,b>0\,.$$

This means that

$$\int_0^\infty \theta^{a-1}e^{-b\theta}\,d\theta = \frac{\Gamma(a)}{b^a} \quad \text{for any values } a,b>0\,.$$

Now substitute in $a+\sum y_i+\tilde{y}$ instead of a and $b+n+1$ instead of b to get

$$\int_0^\infty \theta^{a+\sum y_i+\tilde{y}-1}e^{-(b+n+1)\theta}\,d\theta = \frac{\Gamma(a+\sum y_i+\tilde{y})}{(b+n+1)^{a+\sum y_i+\tilde{y}}}\,.$$

After simplifying some of the algebra, this gives

$$p(\tilde{y}|y_1,\ldots,y_n) = \frac{\Gamma(a+\sum y_i+\tilde{y})}{\Gamma(\tilde{y}+1)\Gamma(a+\sum y_i)}\left(\frac{b+n}{b+n+1}\right)^{a+\sum y_i}\left(\frac{1}{b+n+1}\right)^{\tilde{y}}$$

for $\tilde{y} \in \{0, 1, 2, \ldots\}$. This is a negative binomial distribution with parameters $(a + \sum y_i, b + n)$, for which

$$\mathrm{E}[\tilde{Y}|y_1, \ldots, y_n] = \frac{a + \sum y_i}{b + n} = \mathrm{E}[\theta|y_1, \ldots, y_n];$$

$$\mathrm{Var}[\tilde{Y}|y_1, \ldots, y_n] = \frac{a + \sum y_i}{b + n} \frac{b + n + 1}{b + n} = \mathrm{Var}[\theta|y_1, \ldots y_n] \times (b + n + 1)$$

$$= \mathrm{E}[\theta|y_1, \ldots, y_n] \times \frac{b + n + 1}{b + n}.$$

Let's try to obtain a deeper understanding of this formula for the predictive variance. Recall, the predictive variance is to some extent a measure of our posterior uncertainty about a new sample $\tilde{Y}$ from the population. Uncertainty about $\tilde{Y}$ stems from uncertainty about the population and the variability in sampling from the population. For large n, uncertainty about θ is small ($(b + n + 1)/(b + n) \approx 1$) and uncertainty about $\tilde{Y}$ stems primarily from sampling variability, which for the Poisson model is equal to θ. For small n, uncertainty in $\tilde{Y}$ also includes the uncertainty in θ, and so the total uncertainty is larger than just the sampling variability ($(b + n + 1)/(b + n) > 1$).

3.2.2 Example: Birth rates

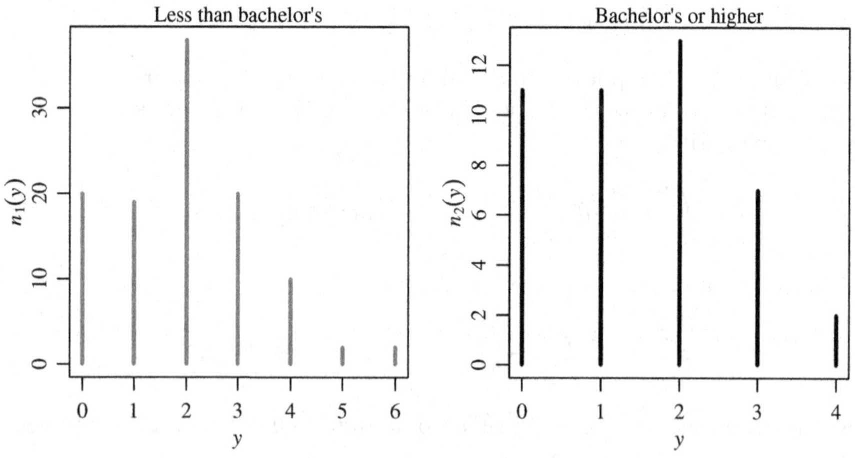

Fig. 3.9. Numbers of children for the two groups.

Over the course of the 1990s the General Social Survey gathered data on the educational attainment and number of children of 155 women who were 40 years of age at the time of their participation in the survey. These women were in their 20s during the 1970s, a period of historically low fertility rates in the

United States. In this example we will compare the women with college degrees to those without in terms of their numbers of children. Let $Y_{1,1} \ldots, Y_{n_1,1}$ denote the numbers of children for the n_1 women without college degrees and $Y_{1,2} \ldots, Y_{n_2,2}$ be the data for women with degrees. For this example, we will use the following sampling models:

$$\begin{aligned} Y_{1,1} \ldots, Y_{n_1,1}|\theta_1 &\sim \text{i.i.d. Poisson}(\theta_1) \\ Y_{1,2} \ldots, Y_{n_2,2}|\theta_2 &\sim \text{i.i.d. Poisson}(\theta_2) \end{aligned}$$

The appropriateness of the Poisson model for these data will be examined in the next chapter.

Empirical distributions for the data are displayed in Figure 3.9, and group sums and means are as follows:

Less than bachelor's: $n_1 = 111,\ \sum_{i=1}^{n_1} Y_{i,1} = 217,\ \bar{Y}_1 = 1.95$
Bachelor's or higher: $n_2 = 44,\ \sum_{i=1}^{n_2} Y_{i,2} = 66,\ \bar{Y}_2 = 1.50$

In the case where $\{\theta_1, \theta_2\} \sim$ i.i.d. gamma$(a = 2, b = 1)$, we have the following posterior distributions:

$$\theta_1|\{n_1 = 111, \textstyle\sum Y_{i,1} = 217\} \sim \text{gamma}(2 + 217, 1 + 111) = \text{gamma}(219, 112)$$
$$\theta_2|\{n_2 = 44, \textstyle\sum Y_{i,2} = 66\} \sim \text{gamma}(2 + 66, 1 + 44) = \text{gamma}(68, 45)$$

Posterior means, modes and 95% quantile-based confidence intervals for θ_1 and θ_2 can be obtained from their gamma posterior distributions:

```
> a<-2 ; b<-1                 # prior parameters
> n1<-111 ; sy1<-217          # data in group 1
> n2<-44  ; sy2<-66           # data in group 2

> (a+sy1)/(b+n1)              # posterior mean
[1] 1.955357
> (a+sy1-1)/(b+n1)            # posterior mode
[1] 1.946429
> qgamma( c(.025,.975),a+sy1,b+n1)    # posterior 95% CI
[1] 1.704943 2.222679

> (a+sy2)/(b+n2)
[1] 1.511111
> (a+sy2-1)/(b+n2)
[1] 1.488889
> qgamma( c(.025,.975),a+sy2,b+n2)
[1] 1.173437 1.890836
```

Posterior densities for the population means of the two groups are shown in the first panel of Figure 3.10. The posterior indicates substantial evidence that $\theta_1 > \theta_2$. For example, $\Pr(\theta_1 > \theta_2|\sum Y_{i,1} = 217, \sum Y_{i,2} = 66) = 0.97$. Now consider two randomly sampled individuals, one from each of the two

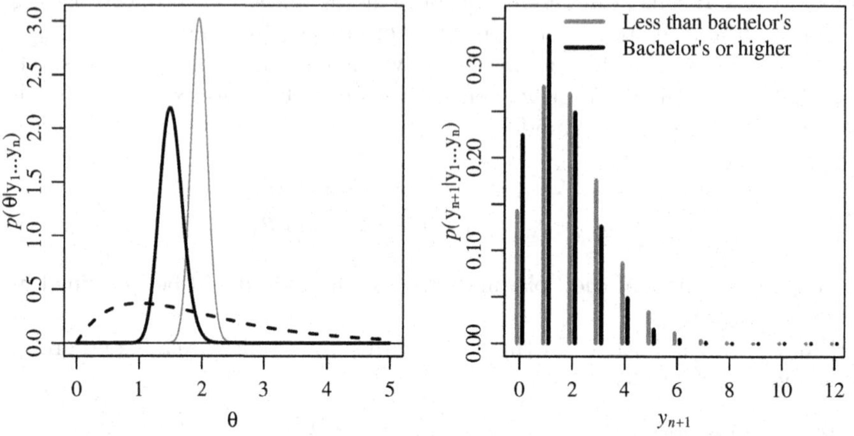

Fig. 3.10. Posterior distributions of mean birth rates (with the common prior distribution given by the dashed line), and posterior predictive distributions for number of children.

populations. To what extent do we expect the one without the bachelor's degree to have more children than the other? We can calculate the relevant probabilities exactly: The posterior predictive distributions for $\tilde{Y}_1$ and $\tilde{Y}_2$ are both negative binomial distributions and are plotted in the second panel of Figure 3.10.

```
> y<- 0:10

> dnbinom(y, size=(a+sy1), mu=(a+sy1)/(b+n1))
 [1] 1.427473e-01 2.766518e-01 2.693071e-01 1.755660e-01
 [5] 8.622930e-02 3.403387e-02 1.124423e-02 3.198421e-03
 [9] 7.996053e-04 1.784763e-04 3.601115e-05

> dnbinom(y, size=(a+sy2), mu=(a+sy2)/(b+n2))
 [1] 2.243460e-01 3.316420e-01 2.487315e-01 1.261681e-01
 [5] 4.868444e-02 1.524035e-02 4.030961e-03 9.263700e-04
 [9] 1.887982e-04 3.465861e-05 5.801551e-06
```

Notice that there is much more overlap between these two distributions than between the posterior distributions of θ_1 and θ_2. For example, $\Pr(\tilde{Y}_1 > \tilde{Y}_2|\sum Y_{i,1} = 217, \sum Y_{i,2} = 66) = .48$ and $\Pr(\tilde{Y}_1 = \tilde{Y}_2|\sum Y_{i,1} = 217, \sum Y_{i,2} = 66) = .22$. The distinction between the events $\{\theta_1 > \theta_2\}$ and $\{\tilde{Y}_1 > \tilde{Y}_2\}$ is extremely important: Strong evidence of a difference between two populations does not mean that the difference itself is large.

3.3 Exponential families and conjugate priors

The binomial and Poisson models discussed in this chapter are both instances of one-parameter *exponential family models*. A one-parameter exponential family model is any model whose densities can be expressed as $p(y|\phi) = h(y)c(\phi)e^{\phi t(y)}$, where ϕ is the unknown parameter and $t(y)$ is the sufficient statistic. Diaconis and Ylvisaker (1979) study conjugate prior distributions for general exponential family models, and in particular prior distributions of the form $p(\phi|n_0, t_0) = \kappa(n_0, t_0)c(\phi)^{n_0}e^{n_0 t_0 \phi}$. Combining such prior information with information from $Y_1, \ldots, Y_n \sim$ i.i.d. $p(y|\phi)$ gives the following posterior distribution:

$$
\begin{aligned}
p(\phi|y_1, \ldots, y_n) &\propto p(\phi)p(y_1, \ldots, y_n|\phi) \\
&\propto c(\phi)^{n_0+n} \exp\left\{\phi \times \left[n_0 t_0 + \sum_{i=1}^{n} t(y_i)\right]\right\} \\
&\propto p(\phi|n_0 + n, n_0 t_0 + n\bar{t}(\boldsymbol{y})),
\end{aligned}
$$

where $\bar{t}(\boldsymbol{y}) = \sum t(y_i)/n$. The similarity between the posterior and prior distributions suggests that n_0 can be interpreted as a "prior sample size" and t_0 as a "prior guess" of $t(Y)$. This interpretation can be made a bit more precise: Diaconis and Ylvisaker (1979) show that

$$
\begin{aligned}
\mathrm{E}[t(Y)] &= \mathrm{E}[\ \mathrm{E}[t(Y)|\phi]\ ] \\
&= \mathrm{E}[-c'(\phi)/c(\phi)] = t_0
\end{aligned}
$$

(see also Exercise 3.6), so t_0 represents the prior expected value of $t(Y)$. The parameter n_0 is a measure of how informative the prior is. There are a variety of ways of quantifying this, but perhaps the simplest is to note that, as a function of ϕ, $p(\phi|n_0, t_0)$ has the same shape as a likelihood $p(\tilde{y}_1, \ldots, \tilde{y}_{n_0}|\phi)$ based on n_0 "prior observations" $\tilde{y}_1, \ldots, \tilde{y}_{n_0}$ for which $\sum t(\tilde{y}_i)/n_0 = t_0$. In this sense the prior distribution $p(\phi|n_0, t_0)$ contains the same amount of information that would be obtained from n_0 independent samples from the population.

Example: Binomial model

The exponential family representation of the binomial(θ) model can be obtained from the density function for a single binary random variable:

$$
\begin{aligned}
p(y|\theta) &= \theta^y(1-\theta)^{1-y} \\
&= \left(\frac{\theta}{1-\theta}\right)^y (1-\theta) \\
&= e^{\phi y}(1+e^{\phi})^{-1},
\end{aligned}
$$

where $\phi = \log[\theta/(1-\theta)]$ is the log-odds. The conjugate prior for ϕ is thus given by $p(\phi|n_0, t_0) \propto (1+e^{\phi})^{-n_0}e^{n_0 t_0 \phi}$, where t_0 represents the prior expectation of $t(y) = y$, or equivalently, t_0 represents our prior probability that $Y = 1$. Using the change of variables formula (Exercise 3.10), this translates into a prior distribution for θ such that $p(\theta|n_0, t_0) \propto \theta^{n_0 t_0 - 1}(1-\theta)^{n_0(1-t_0)-1}$, which is a beta$(n_0 t_0, n_0(1-t_0))$ distribution. A weakly informative prior distribution can be obtained by setting t_0 equal to our prior expectation and $n_0 = 1$. If our prior expectation is 1/2, the resulting prior is a beta(1/2,1/2) distribution, which is equivalent to Jeffreys' prior distribution (Exercise 3.11) for the binomial sampling model. Under the weakly informative beta$(t_0, (1-t_0))$ prior distribution, the posterior would be $\{\theta|y_1, \ldots, y_n\} \sim$ beta$(t_0 + \sum y_i, (1-t_0) + \sum(1-y_i))$.

Example: Poisson model

The Poisson(θ) model can be shown to be an exponential family model with

- $t(y) = y$;
- $\phi = \log\theta$;
- $c(\phi) = \exp(e^{-\phi})$.

The conjugate prior distribution for ϕ is thus $p(\phi|n_0, t_0) = \exp(n_0 e^{-\phi})e^{n_0 t_0 \phi}$ where t_0 is the prior expectation of the population mean of Y. This translates into a prior density for θ of the form $p(\theta|n_0, t_0) \propto \theta^{n_0 t_0 - 1}e^{-n_0\theta}$, which is a gamma$(n_0 t_0, n_0)$ density. A weakly informative prior distribution can be obtained with t_0 set to the prior expectation of Y and $n_0 = 1$, giving a gamma$(t_0, 1)$ prior distribution. The posterior distribution under such a prior would be $\{\theta|y_1, \ldots, y_n\} \sim$ gamma$(t_0 + \sum y_i, 1 + n)$.

3.4 Discussion and further references

The notion of conjugacy for classes of prior distributions was developed in Raiffa and Schlaifer (1961). Important results on conjugacy for exponential families appear in Diaconis and Ylvisaker (1979) and Diaconis and Ylvisaker (1985). The latter shows that any prior distribution may be approximated by a mixture of conjugate priors.

Most authors refer to intervals of high posterior probability as "credible intervals" as opposed to confidence intervals. Doing so fails to recognize that Bayesian intervals do have frequentist coverage probabilities, often being very close to the specified Bayesian coverage level (Welch and Peers, 1963; Hartigan, 1966; Severini, 1991). Some authors suggest that accurate frequentist coverage can be a guide for the construction of prior distributions (Tibshirani, 1989; Sweeting, 1999, 2001). See also Kass and Wasserman (1996) for a review of formal methods for selecting prior distributions.

4

Monte Carlo approximation

In the last chapter we saw examples in which a conjugate prior distribution for an unknown parameter θ led to a posterior distribution for which there were simple formulae for posterior means and variances. However, often we will want to summarize other aspects of a posterior distribution. For example, we may want to calculate $\Pr(\theta \in A|y_1, \ldots, y_n)$ for arbitrary sets A. Alternatively, we may be interested in means and standard deviations of some function of θ, or the predictive distribution of missing or unobserved data. When comparing two or more populations we may be interested in the posterior distribution of $|\theta_1 - \theta_2|$, θ_1/θ_2, or $\max\{\theta_1, \ldots, \theta_m\}$, all of which are functions of more than one parameter. Obtaining exact values for these posterior quantities can be difficult or impossible, but if we can generate random sample values of the parameters from their posterior distributions, then all of these posterior quantities of interest can be approximated to an arbitrary degree of precision using the Monte Carlo method.

4.1 The Monte Carlo method

In the last chapter we obtained the following posterior distributions for birthrates of women without and with bachelor's degrees, respectively:

$$p(\theta_1|\sum_{i=1}^{111} Y_{i,1} = 217) = \text{dgamma}(\theta_1, 219, 112)$$

$$p(\theta_2|\sum_{i=1}^{44} Y_{i,2} = 66) = \text{dgamma}(\theta_2, 68, 45)$$

Additionally, we modeled θ_1 and θ_2 as conditionally independent given the data. It was claimed that $\Pr(\theta_1 > \theta_2|\sum Y_{i,1} = 217, \sum Y_{i,2} = 66) = 0.97$. How was this probability calculated? From Chapter 2, we have

P.D. Hoff, *A First Course in Bayesian Statistical Methods*,
Springer Texts in Statistics, DOI 10.1007/978-0-387-92407-6_4,

$$\begin{aligned}
&\Pr(\theta_1 > \theta_2 | y_{1,1}, \ldots, y_{n_2,2}) \\
&= \int_0^\infty \int_0^{\theta_1} p(\theta_1, \theta_2 | y_{1,1}, \ldots, y_{n_2,2}) \, d\theta_2 d\theta_1 \\
&= \int_0^\infty \int_0^{\theta_1} \text{dgamma}(\theta_1, 219, 112) \times \text{dgamma}(\theta_2, 68, 45) \, d\theta_2 d\theta_1 \\
&= \frac{112^{219} 45^{68}}{\Gamma(219)\Gamma(68)} \int_0^\infty \int_0^{\theta_1} \theta_1^{218} \theta_2^{67} e^{-112\theta_1 - 45\theta_2} \, d\theta_2 d\theta_1 .
\end{aligned}$$

There are a variety of ways to calculate this integral. It can be done with pencil and paper using results from calculus, and it can be calculated numerically in many mathematical software packages. However, the feasibility of these integration methods depends heavily on the particular details of this model, prior distribution and the probability statement that we are trying to calculate. As an alternative, in this text we will use an integration method for which the general principles and procedures remain relatively constant across a broad class of problems. The method, known as *Monte Carlo approximation*, is based on random sampling and its implementation does not require a deep knowledge of calculus or numerical analysis.

Let θ be a parameter of interest and let $y_1, \ldots, y_n$ be the numerical values of a sample from a distribution $p(y_1, \ldots, y_n | \theta)$. Suppose we could sample some number S of independent, random θ-values from the posterior distribution $p(\theta | y_1, \ldots, y_n)$:

$$\theta^{(1)}, \ldots, \theta^{(S)} \sim \text{i.i.d } p(\theta | y_1, \ldots, y_n).$$

Then the empirical distribution of the samples $\{\theta^{(1)}, \ldots, \theta^{(S)}\}$ would approximate $p(\theta | y_1, \ldots, y_n)$, with the approximation improving with increasing S. The empirical distribution of $\{\theta^{(1)}, \ldots, \theta^{(S)}\}$ is known as a *Monte Carlo approximation* to $p(\theta | y_1, \ldots, y_n)$. Many computer languages and computing environments have procedures for simulating this sampling process. For example, R has built-in functions to simulate i.i.d. samples from most of the distributions we will use in this book.

Figure 4.1 shows successive Monte Carlo approximations to the density of the gamma(68,45) distribution, along with the true density function for comparison. As we see, the empirical distribution of the Monte Carlo samples provides an increasingly close approximation to the true density as S gets larger. Additionally, let $g(\theta)$ be (just about) any function. The law of large numbers says that if $\theta^{(1)}, \ldots, \theta^{(S)}$ are i.i.d. samples from $p(\theta | y_1, \ldots, y_n)$, then

$$\frac{1}{S} \sum_{s=1}^{S} g(\theta^{(s)}) \to \text{E}[g(\theta) | y_1, \ldots, y_n] = \int g(\theta) p(\theta | y_1, \ldots, y_n) \, d\theta \text{ as } S \to \infty .$$

This implies that as $S \to \infty$,

- $\bar{\theta} = \sum_{s=1}^{S} \theta^{(s)} / S \to \text{E}[\theta | y_1, \ldots, y_n]$;
- $\sum_{s=1}^{S} (\theta^{(s)} - \bar{\theta})^2 / (S-1) \to \text{Var}[\theta | y_1, \ldots, y_n]$;

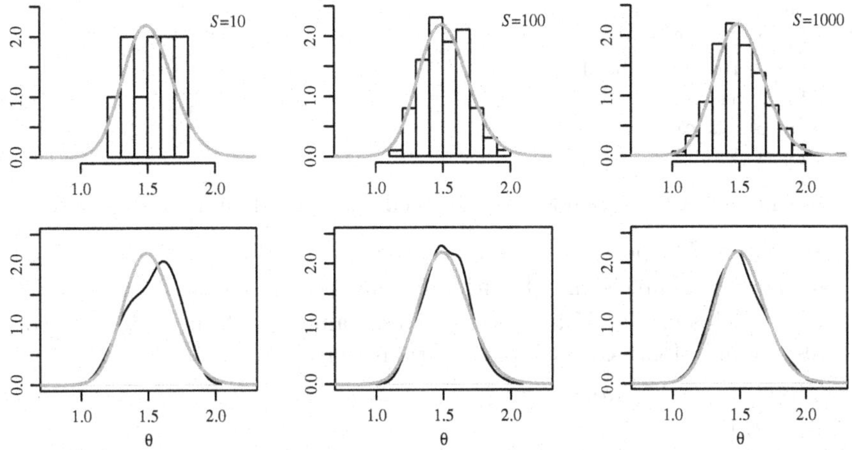

Fig. 4.1. Histograms and kernel density estimates of Monte Carlo approximations to the gamma(68,45) distribution, with the true density in gray.

- $\#(\theta^{(s)} \leq c)/S \rightarrow \Pr(\theta \leq c|y_1,\ldots,y_n)$;
- the empirical distribution of $\{\theta^{(1)},\ldots,\theta^{(S)}\} \rightarrow p(\theta|y_1,\ldots,y_n)$;
- the median of $\{\theta^{(1)},\ldots,\theta^{(S)}\} \rightarrow \theta_{1/2}$;
- the α-percentile of $\{\theta^{(1)},\ldots,\theta^{(S)}\} \rightarrow \theta_\alpha$.

Just about any aspect of a posterior distribution we may be interested in can be approximated arbitrarily exactly with a large enough Monte Carlo sample.

Numerical evaluation

We will first gain some familiarity and confidence with the Monte Carlo procedure by comparing its approximations to a few posterior quantities that we can compute exactly (or nearly so) by other methods. Suppose we model $Y_1,\ldots,Y_n|\theta$ as i.i.d. Poisson(θ), and have a gamma(a,b) prior distribution for θ. Having observed $Y_1 = y_1,\ldots,Y_n = y_n$, the posterior distribution is gamma $(a+\sum y_i, b+n)$. For the college-educated population in the birthrate example, $(a = 2, b = 1)$ and $(\sum y_i = 66, n = 44)$.

Expectation: The posterior mean is $(a+\sum y_i)/(b+n) = 68/45 = 1.51$. Monte Carlo approximations to this for $S \in \{10, 100, 1000\}$ can be obtained in R as follows:

```
a<-2  ; b<-1
sy<-66 ; n<-44

theta.mc10<-rgamma(10,a+sy,b+n)
theta.mc100<-rgamma(100,a+sy,b+n)
theta.mc1000<-rgamma(1000,a+sy,b+n)
```

```
> mean(theta.mc10)
[1] 1.532794
> mean(theta.mc100)
[1] 1.513947
> mean(theta.mc1000)
[1] 1.501015
```

Results will vary depending on the seed of the random number generator.

Probabilities: The posterior probability that $\{\theta < 1.75\}$ can be obtained to a high degree of precision in R with the command `pgamma(1.75,a+sy,b+n)`, which yields 0.8998. Using the simulated values of θ from above, the corresponding Monte Carlo approximations were:

```
> mean(theta.mc10<1.75)
[1] 0.9
> mean(theta.mc100<1.75)
[1] 0.94
> mean(theta.mc1000<1.75)
[1] 0.899
```

Quantiles: A 95% quantile-based confidence region can be obtained with `qgamma(c(.025,.975),a+sy,b+n)`, giving an interval of (1.173,1.891). Approximate 95% confidence regions can also be obtained from the Monte Carlo samples:

```
> quantile(theta.mc10, c(.025,.975))
    2.5%    97.5%
1.260291 1.750068
> quantile(theta.mc100, c(.025,.975))
    2.5%    97.5%
1.231646 1.813752
> quantile(theta.mc1000, c(.025,.975))
    2.5%    97.5%
1.180194 1.892473
```

Figure 4.2 shows the convergence of the Monte Carlo estimates to the correct values graphically, based on cumulative estimates from a sequence of $S = 1000$ samples from the gamma(68,45) distribution. Such plots can help indicate when enough Monte Carlo samples have been made. Additionally, Monte Carlo standard errors can be obtained to assess the accuracy of approximations to posterior means: Letting $\bar{\theta} = \sum_{s=1}^{S} \theta^{(s)}/S$ be the sample mean of the Monte Carlo samples, the Central Limit Theorem says that $\bar{\theta}$ is approximately normally distributed with expectation $\mathrm{E}[\theta|y_1,\ldots,y_n]$ and standard deviation equal to $\sqrt{\mathrm{Var}[\theta|y_1,\ldots y_n]/S}$. The Monte Carlo standard error is the approximation to this standard deviation: Letting $\hat{\sigma}^2 = \sum(\theta^{(s)} - \bar{\theta})^2/(S-1)$ be the Monte Carlo estimate of $\mathrm{Var}[\theta|y_1,\ldots,y_n]$, the Monte Carlo standard error is $\sqrt{\hat{\sigma}^2/S}$. An approximate 95% Monte Carlo confidence interval for the

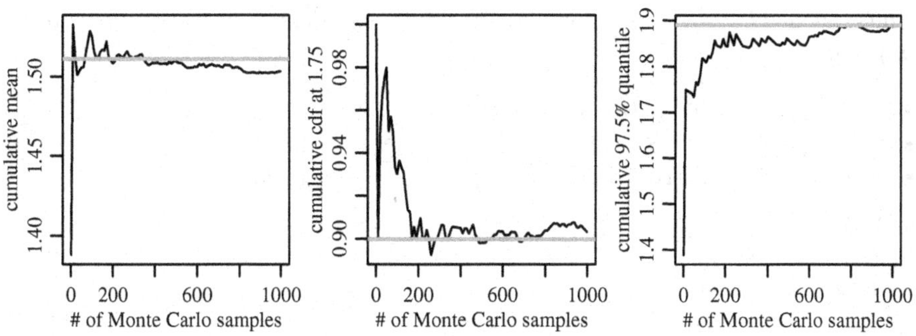

Fig. 4.2. Estimates of the posterior mean, $\Pr(\theta < 1.75|y_1, \ldots, y_n)$ and the 97.5% posterior quantile as a function of the number of Monte Carlo samples. Horizontal gray lines are the true values.

posterior mean of θ is $\hat{\theta} \pm 2\sqrt{\hat{\sigma}^2/S}$. Standard practice is to choose S to be large enough so that the Monte Carlo standard error is less than the precision to which you want to report $\mathrm{E}[\theta|y_1, \ldots, y_n]$. For example, suppose you had generated a Monte Carlo sample of size $S = 100$ for which the estimate of $\mathrm{Var}[\theta|y_1, \ldots, y_n]$ was 0.024. The approximate Monte Carlo standard error would then be $\sqrt{0.024/100} = 0.015$. If you wanted the difference between $\mathrm{E}[\theta|y_1, \ldots, y_n]$ and its Monte Carlo estimate to be less than 0.01 with high probability, you would need to increase your Monte Carlo sample size so that $2\sqrt{0.024/S} < 0.01$, i.e. $S > 960$.

4.2 Posterior inference for arbitrary functions

Suppose we are interested in the posterior distribution of some computable function $g(\theta)$ of θ. In the binomial model, for example, we are sometimes interested in the log odds:

$$\log \text{odds}(\theta) = \log \frac{\theta}{1-\theta} = \gamma\,.$$

The law of large numbers says that if we generate a sequence $\{\theta^{(1)}, \theta^{(2)}, \ldots\}$ from the posterior distribution of θ, then the average value of $\log \frac{\theta^{(s)}}{1-\theta^{(s)}}$ converges to $\mathrm{E}[\log \frac{\theta}{1-\theta}|y_1, \ldots, y_n]$. However, we may also be interested in other aspects of the posterior distribution of $\gamma = \log \frac{\theta}{1-\theta}$. Fortunately, these too can be computed using a Monte Carlo approach:

$$\left.\begin{array}{l}
\text{sample } \theta^{(1)} \sim p(\theta|y_1, \ldots, y_n), \;\; \text{compute } \gamma^{(1)} = g(\theta^{(1)}) \\
\text{sample } \theta^{(2)} \sim p(\theta|y_1, \ldots, y_n), \;\; \text{compute } \gamma^{(2)} = g(\theta^{(2)}) \\
\quad\vdots \\
\text{sample } \theta^{(S)} \sim p(\theta|y_1, \ldots, y_n), \;\; \text{compute } \gamma^{(S)} = g(\theta^{(S)})
\end{array}\right\} \text{ independently}\,.$$

The sequence $\{\gamma^{(1)},\ldots,\gamma^{(S)}\}$ constitutes S independent samples from $p(\gamma|y_1,\ldots,y_n)$, and so as $S\to\infty$

- $\bar{\gamma}=\sum_{s=1}^{S}\gamma^{(s)}/S\to \mathrm{E}[\gamma|y_1,\ldots,y_n]$,
- $\sum_{s=1}^{S}(\gamma^{(s)}-\bar{\gamma})^2/(S-1)\to \mathrm{Var}[\gamma|y_1,\ldots,y_n]$,
- the empirical distribution of $\{\gamma^{(1)},\ldots,\gamma^{(S)}\}\to p(\gamma|y_1,\ldots,y_n)$,

as before.

Example: Log-odds

Fifty-four percent of the respondents in the 1998 General Social Survey reported their religious preference as Protestant, leaving non-Protestants in the minority. Respondents were also asked if they agreed with a Supreme Court ruling that prohibited state or local governments from requiring the reading of religious texts in public schools. Of the $n=860$ individuals in the religious minority (non-Protestant), $y=441$ (51%) said they agreed with the Supreme Court ruling, whereas 353 of the 1011 Protestants (35%) agreed with the ruling.

Let θ be the population proportion agreeing with the ruling in the minority population. Using a binomial sampling model and a uniform prior distribution, the posterior distribution of θ is beta(442, 420). Using the Monte Carlo algorithm described above, we can obtain samples of the log-odds $\gamma=\log[\theta/(1-\theta)]$ from both the prior distribution and the posterior distribution of γ. In R, the Monte Carlo algorithm involves only a few commands:

```
a<-1 ; b<-1
theta.prior.mc<-rbeta(10000,a,b)
gamma.prior.mc<- log( theta.prior.mc/(1-theta.prior.mc) )

n0<-860-441 ; n1<-441
theta.post.mc<-rbeta(10000,a+n1,b+n0)
gamma.post.mc<- log( theta.post.mc/(1-theta.post.mc) )
```

Using the `density()` function in R , we can plot smooth kernel density approximations to these distributions, as shown in Figure 4.3.

Example: Functions of two parameters

Based on the prior distributions and the data in the birthrate example, the posterior distributions for the two educational groups are

$$\begin{aligned}\{\theta_1|y_{1,1},\ldots,y_{n_1,1}\}&\sim \text{gamma}(219,112)\ \text{(women without bachelor's degrees)}\\ \{\theta_2|y_{1,2},\ldots,y_{n_2,2}\}&\sim \text{gamma}(68,45)\quad \text{(women with bachelor's degrees)}.\end{aligned}$$

There are a variety of ways to describe our knowledge about the difference between θ_1 and θ_2. For example, we may be interested in the numerical value

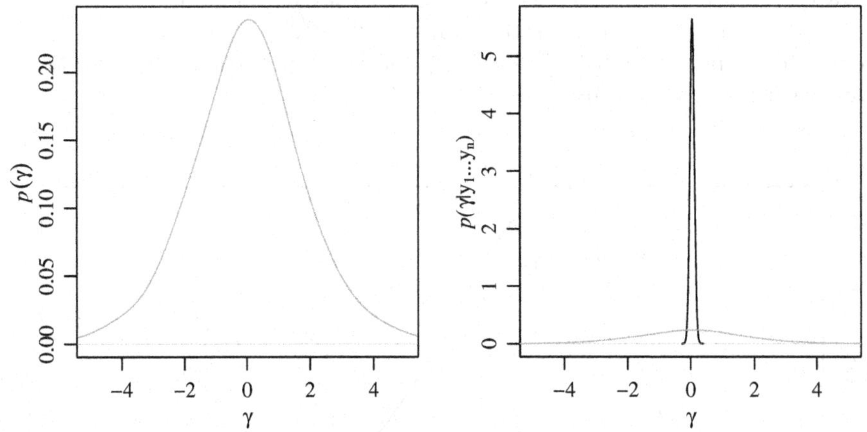

Fig. 4.3. Monte Carlo approximations to the prior and posterior distributions of the log-odds.

of $\Pr(\theta_1 > \theta_2 | Y_{1,1} = y_{1,1}, \ldots, Y_{n_2,2} = y_{n_2,2})$, or in the posterior distribution of θ_1/θ_2. Both of these quantities can be obtained with Monte Carlo sampling:

$$\begin{array}{ll} \text{sample } \theta_1^{(1)} \sim p(\theta_1 | \sum_{i=1}^{111} Y_{i,1} = 217), & \text{sample } \theta_2^{(1)} \sim p(\theta_2 | \sum_{i=1}^{44} Y_{i,2} = 66) \\ \text{sample } \theta_1^{(2)} \sim p(\theta_1 | \sum_{i=1}^{111} Y_{i,1} = 217), & \text{sample } \theta_2^{(2)} \sim p(\theta_2 | \sum_{i=1}^{44} Y_{i,2} = 66) \\ \vdots & \vdots \\ \text{sample } \theta_1^{(S)} \sim p(\theta_1 | \sum_{i=1}^{111} Y_{i,1} = 217), & \text{sample } \theta_2^{(S)} \sim p(\theta_2 | \sum_{i=1}^{44} Y_{i,2} = 66) \,. \end{array}$$

The sequence $\{(\theta_1^{(1)}, \theta_2^{(1)}), \ldots, (\theta_1^{(S)}, \theta_2^{(S)})\}$ consists of S independent samples from the joint posterior distribution of θ_1 and θ_2, and can be used to make Monte Carlo approximations to posterior quantities of interest. For example, $\Pr(\theta_1 > \theta_2 | \sum_{i=1}^{111} Y_{i,1} = 217, \sum_{i=1}^{44} Y_{i,2} = 66)$ is approximated by $\frac{1}{S}\sum_{s=1}^{S} 1(\theta_1^{(s)} > \theta_2^{(s)})$, where $1(x > y)$ is the indicator function which is 1 if $x > y$ and zero otherwise. The approximation can be calculated in R with the following commands:

```
> a<-2 ; b<-1
> sy1<-217 ;  n1<-111
> sy2<-66  ;  n2<-44

> theta1.mc<-rgamma(10000,a+sy1, b+n1)
> theta2.mc<-rgamma(10000,a+sy2, b+n2)

> mean(theta1.mc>theta2.mc)

[1] 0.9708
```

Additionally, if we were interested in the ratio of the means of the two groups, we could use the empirical distribution of $\{\theta_1^{(1)}/\theta_2^{(1)}, \ldots, \theta_1^{(S)}/\theta_2^{(S)}\}$ to approximate the posterior distribution of θ_1/θ_2. A Monte Carlo estimate of this posterior density is given in Figure 4.4.

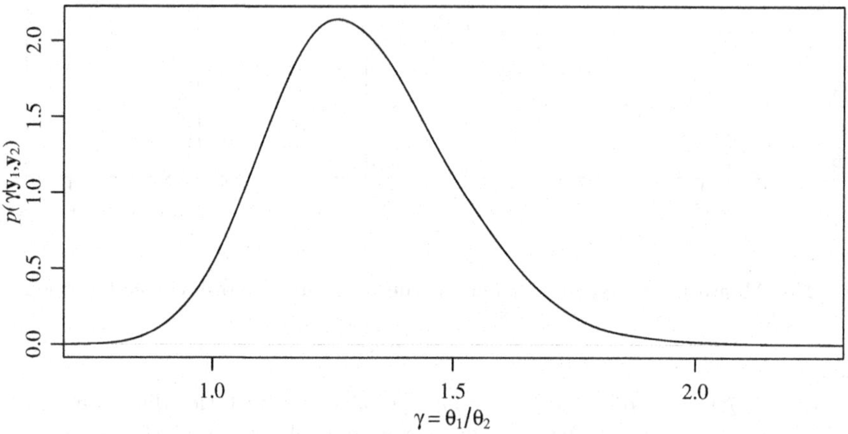

Fig. 4.4. Monte Carlo estimate to the posterior predictive distribution of $\gamma = \theta_1/\theta_2$.

4.3 Sampling from predictive distributions

As described in Section 3.1, the predictive distribution of a random variable $\tilde{Y}$ is a probability distribution for $\tilde{Y}$ such that

- known quantities have been conditioned on;
- unknown quantities have been integrated out.

For example, let $\tilde{Y}$ be the number of children of a person who is sampled from the population of women aged 40 with a college degree. If we knew the true mean birthrate θ of this population, we might describe our uncertainty about $\tilde{Y}$ with a Poisson(θ) distribution:

Sampling model: $\Pr(\tilde{Y} = \tilde{y}|\theta) = p(\tilde{y}|\theta) = \theta^{\tilde{y}} e^{-\theta}/\tilde{y}!$

We cannot make predictions from this model, however, because we do not actually know θ. If we did not have any sample data from the population, our predictive distribution would be obtained by integrating out θ:

Predictive model: $\Pr(\tilde{Y} = \tilde{y}) = \int p(\tilde{y}|\theta)p(\theta)d\theta$

In the case where $\theta \sim \text{gamma}(a, b)$, we showed in the last chapter that this predictive distribution is the negative binomial(a, b) distribution. A predictive distribution that integrates over unknown parameters but is not conditional on observed data is called a *prior predictive distribution.* Such a distribution can be useful in evaluating if a prior distribution for θ actually translates into reasonable prior beliefs for observable data $\tilde{Y}$ (see Exercise 7.4). After we have observed a sample $Y_1, \ldots, Y_n$ from the population, the relevant predictive distribution for a new observation becomes

$$\begin{aligned} \Pr(\tilde{Y} = \tilde{y}|Y_1 = y_1, \ldots, Y_n = y_n) &= \int p(\tilde{y}|\theta, y_1, \ldots, y_n) p(\theta|y_1, \ldots, y_n)\, d\theta \\ &= \int p(\tilde{y}|\theta) p(\theta|y_1, \ldots, y_n)\, d\theta. \end{aligned}$$

This is called a *posterior predictive distribution*, because it conditions on an observed dataset. In the case of a Poisson model with a gamma prior distribution, we showed in Chapter 3 that the posterior predictive distribution is negative binomial$(a + \sum y_i, b + n)$.

In many modeling situations, we will be able to sample from $p(\theta|y_1, \ldots, y_n)$ and $p(y|\theta)$, but $p(\tilde{y}|y_1, \ldots, y_n)$ will be too complicated to sample from directly. In this situation we can sample from the posterior predictive distribution indirectly using a Monte Carlo procedure. Since $p(\tilde{y}|y_1, \ldots, y_n) = \int p(\tilde{y}|\theta) p(\theta|y_1, \ldots, y_n)\, d\theta$, we see that $p(\tilde{y}|y_1, \ldots, y_n)$ is the posterior expectation of $p(\tilde{y}|\theta)$. To obtain the posterior predictive probability that $\tilde{Y}$ is equal to some specific value $\tilde{y}$, we could just apply the Monte Carlo method of the previous section: Sample $\theta^{(1)}, \ldots, \theta^{(S)} \sim$ i.i.d. $p(\theta|y_1, \ldots, y_n)$, and then approximate $p(\tilde{y}|y_1, \ldots, y_n)$ with $\sum_{s=1}^{S} p(\tilde{y}|\theta^{(s)})/S$. This procedure will work well if $p(y|\theta)$ is discrete and we are interested in quantities that are easily computed from $p(y|\theta)$. However, it will generally be useful to have a set of samples of $\tilde{Y}$ from its posterior predictive distribution. Obtaining these samples can be done quite easily as follows:

$$\begin{array}{ll} \text{sample } \theta^{(1)} \sim p(\theta|y_1, \ldots, y_n), & \text{sample } \tilde{y}^{(1)} \sim p(\tilde{y}|\theta^{(1)}) \\ \text{sample } \theta^{(2)} \sim p(\theta|y_1, \ldots, y_n), & \text{sample } \tilde{y}^{(2)} \sim p(\tilde{y}|\theta^{(2)}) \\ \vdots & \\ \text{sample } \theta^{(S)} \sim p(\theta|y_1, \ldots, y_n), & \text{sample } \tilde{y}^{(S)} \sim p(\tilde{y}|\theta^{(S)})\,. \end{array}$$

The sequence $\{(\theta, \tilde{y})^{(1)}, \ldots, (\theta, \tilde{y})^{(S)}\}$ constitutes S independent samples from the joint posterior distribution of $(\theta, \tilde{Y})$, and the sequence $\{\tilde{y}^{(1)}, \ldots, \tilde{y}^{(S)}\}$ constitutes S independent samples from the *marginal* posterior distribution of $\tilde{Y}$, which is the posterior predictive distribution.

Example: Poisson model

At the end of Chapter 3 it was reported that the predictive probability that an age-40 woman without a college degree would have more children than an age-40 woman with a degree was 0.48. To arrive at this answer exactly we would have to do the following doubly infinite sum:

$$\Pr(\tilde{Y}_1 > \tilde{Y}_2 | \sum Y_{i,1} = 217, \sum Y_{i,2} = 66) =$$
$$\sum_{\tilde{y}_2=0}^{\infty} \sum_{\tilde{y}_1=\tilde{y}_2+1}^{\infty} \text{dnbinom}(\tilde{y}_1, 219, 112) \times \text{dnbinom}(\tilde{y}_2, 68, 45)\,.$$

Alternatively, this sum can be approximated with Monte Carlo sampling. Since $\tilde{Y}_1$ and $\tilde{Y}_2$ are *a posteriori* independent, samples from their joint posterior distribution can be made by sampling values of each variable separately from their individual posterior distributions. Posterior predictive samples from the conjugate Poisson model can be generated as follows:

$$\begin{array}{ll}
\text{sample } \theta^{(1)} \sim \text{gamma}(a + \sum y_i, b+n), & \text{sample } \tilde{y}^{(1)} \sim \text{Poisson}(\theta^{(1)}) \\
\text{sample } \theta^{(2)} \sim \text{gamma}(a + \sum y_i, b+n), & \text{sample } \tilde{y}^{(2)} \sim \text{Poisson}(\theta^{(2)}) \\
\vdots & \\
\text{sample } \theta^{(S)} \sim \text{gamma}(a + \sum y_i, b+n) & \text{sample } \tilde{y}^{(S)} \sim \text{Poisson}(\theta^{(S)})\,.
\end{array}$$

Monte Carlo samples from the posterior predictive distributions of our two educational groups can be obtained with just a few commands in R:

```
> a<-2 ; b<-1
> sy1<-217 ;   n1<-111
> sy2<-66  ;   n2<-44

> theta1.mc<-rgamma(10000,a+sy1, b+n1)
> theta2.mc<-rgamma(10000,a+sy2, b+n2)
> y1.mc<-rpois(10000,theta1.mc)
> y2.mc<-rpois(10000,theta2.mc)

> mean(y1.mc>y2.mc)
[1] 0.4823
```

Once we have generated these Monte Carlo samples from the posterior predictive distribution, we can use them again to calculate other posterior quantities of interest. For example, Figure 4.5 shows the Monte Carlo approximation to the posterior distribution of $D = (\tilde{Y}_1 - \tilde{Y}_2)$, the difference in number of children between two individuals, one sampled from each of the two groups.

4.4 Posterior predictive model checking

Let's consider for the moment the sample of 40-year-old women without a college degree. The empirical distribution of the number of children of these women, along with the corresponding posterior predictive distribution, is shown in the first panel of Figure 4.6. In this sample of $n = 111$ women, the number of women with exactly two children is 38, which is twice the number of women in the sample with one child. In contrast, this group's posterior predictive distribution, shown in gray, suggests that the probability of

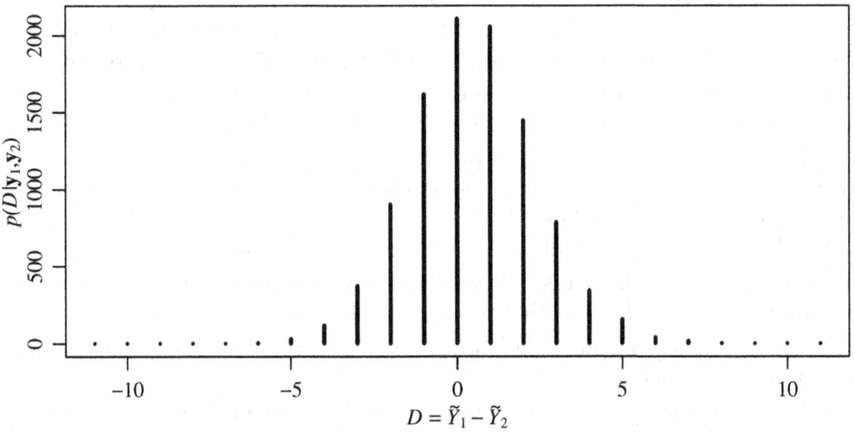

Fig. 4.5. The posterior predictive distribution of $D = \tilde{Y}_1 - \tilde{Y}_2$, the difference in the number of children of two randomly sampled women, one from each of the two educational populations.

sampling a woman with two children is slightly less probable than sampling a woman with one (probabilities of 0.27 and 0.28, respectively). These two distributions seem to be in conflict. If the observed data have twice as many women with two children than one, why should we be predicting otherwise?

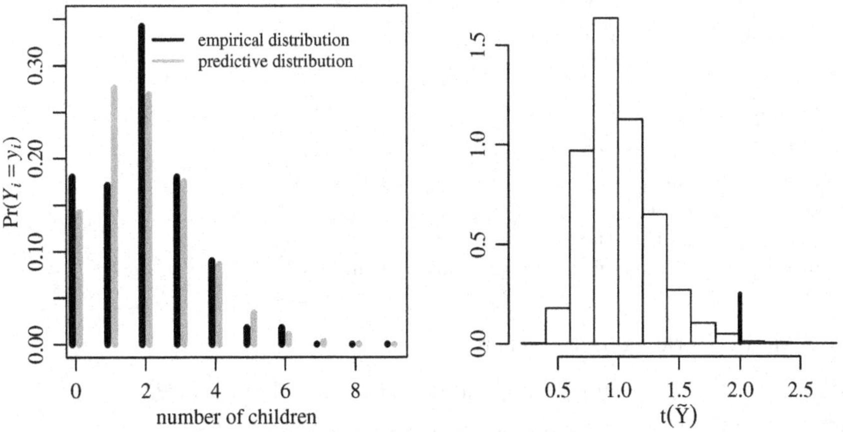

Fig. 4.6. Evaluation of model fit. The first panel shows the empirical and posterior predictive distributions of the number of children of women without a bachelor's degree. The second panel shows the posterior predictive distribution of the empirical odds of having two children versus one child in a dataset of size $n = 111$. The observed odds are given in the short vertical line.

One explanation for the large number of women in the sample with two children is that it is a result of sampling variability: The empirical distribution of sampled data does not generally match exactly the distribution of the population from which the data were sampled, and in fact may look quite different if the sample size is small. A smooth population distribution can produce sample empirical distributions that are quite bumpy. In such cases, having a predictive distribution that smoothes over the bumps of the empirical distribution may be desirable.

An alternative explanation for the large number of women in the sample with two children is that this is indeed a feature of the population, and the data are correctly reflecting this feature. In contrast, the Poisson model is unable to represent this feature of the population because there is no Poisson distribution that has such a sharp peak at $y = 2$.

These explanations for the discrepancy between the empirical and predictive distributions can be assessed numerically with Monte Carlo simulation. For every vector $\boldsymbol{y}$ of length $n = 111$, let $t(\boldsymbol{y})$ be the ratio of the number of 2's in $\boldsymbol{y}$ to the number of 1's, so for our observed data $\boldsymbol{y}_{\text{obs}}$, $t(\boldsymbol{y}_{\text{obs}}) = 2$. Now suppose we were to sample a different set of 111 women, obtaining a data vector $\tilde{\boldsymbol{Y}}$ of length 111 recording their number of children. What sort of values of $t(\tilde{\boldsymbol{Y}})$ would we expect? Monte Carlo samples from the posterior predictive distribution of $t(\tilde{\boldsymbol{Y}})$ can be obtained with the following procedure and R-code:

For each $s \in \{1, \ldots, S\}$,
1. sample $\theta^{(s)} \sim p(\theta|\boldsymbol{Y} = \boldsymbol{y}_{\text{obs}})$
2. sample $\tilde{\boldsymbol{Y}}^{(s)} = (\tilde{y}_1^{(s)}, \ldots, \tilde{y}_n^{(s)}) \sim$ i.i.d. $p(y|\theta^{(s)})$
3. compute $t^{(s)} = t(\tilde{\boldsymbol{Y}}^{(s)})$.

```
a<-2 ; b<-1
t.mc<-NULL

for(s in 1:10000) {
  theta1<-rgamma(1, a+sy1, b+n1)
  y1.mc<-rpois(n1, theta1)
  t.mc<-c(t.mc,sum(y1.mc==2)/sum(y1.mc==1))
                    }
```

In this Monte Carlo sampling scheme,

$\{\theta^{(1)}, \ldots, \theta^{(S)}\}$ are samples from the posterior distribution of θ;
$\{\tilde{\boldsymbol{Y}}^{(1)}, \ldots, \tilde{\boldsymbol{Y}}^{(S)}\}$ are posterior predictive *datasets*, each of size n;
$\{t^{(1)}, \ldots, t^{(S)}\}$ are samples from the posterior predictive distribution of $t(\tilde{\boldsymbol{Y}})$.

A Monte Carlo approximation to the distribution of $t(\tilde{\boldsymbol{Y}})$ is shown in the second panel of Figure 4.6, with the observed value $t(\mathbf{y}_{obs})$ indicated with a short vertical line. Out of 10,000 Monte Carlo datasets, only about a half of a percent had values of $t(\boldsymbol{y})$ that equaled or exceeded $t(\boldsymbol{y}_{\text{obs}})$. This indicates that our Poisson model is flawed: It predicts that we would hardly ever see a dataset that resembled our observed one in terms of $t(\boldsymbol{y})$. If we were interested in making inference on the true probability distribution $p_{\text{true}}(y)$ for each value of y, then the Poisson model would be inadequate, and we would have to consider a more complicated model (for example, a multinomial sampling model). However, a simple Poisson model may suffice if we are interested only in certain aspects of p_{true}. For example, the predictive distribution generated by the Poisson model will have a mean that approximates the true population mean, even though p_{true} may not be a Poisson distribution. Additionally, for these data the sample mean and variance are similar, being 1.95 and 1.90 respectively, suggesting that the Poisson model can represent both the mean and variance of the population.

In terms of data description, we should at least make sure that our model generates predictive datasets $\tilde{\boldsymbol{Y}}$ that resemble the observed dataset in terms of features that are of interest. If this condition is not met, we may want to consider using a more complex model. However, an incorrect model can still provide correct inference for some aspects of the true population (White, 1982; Bunke and Milhaud, 1998; Kleijn and van der Vaart, 2006). For example, the Poisson model provides consistent estimation of the population mean, as well as accurate confidence intervals if the population mean is approximately equal to the variance.

4.5 Discussion and further references

The use of Monte Carlo methods is widespread in statistics and science in general. Rubinstein and Kroese (2008) cover Monte Carlo methods for a wide variety of statistical problems, and Robert and Casella (2004) include more coverage of Bayesian applications (and cover Markov chain Monte Carlo methods as well).

Using the posterior predictive distribution to assess model fit was suggested by Guttman (1967) and Rubin (1984), and is now common practice. In some problems, it is useful to evaluate goodness-of-fit using functions that depend on parameters as well as predicted data. This is discussed in Gelman et al (1996) and more recently in Johnson (2007). These types of posterior predictive checks have given rise to a notion of posterior predictive p-values, which despite their name, do not generally share the same frequentist properties as p-values based on classical goodness-of-fit tests. This distinction is discussed in Bayarri and Berger (2000), who also consider alternative types of Bayesian goodness of fit probabilities to serve as a replacement for frequentist p-values.

5

The normal model

Perhaps the most useful (or utilized) probability model for data analysis is the normal distribution. There are several reasons for this, one being the central limit theorem, and another being that the normal model is a simple model with separate parameters for the population mean and variance - two quantities that are often of primary interest. In this chapter we discuss some of the properties of the normal distribution, and show how to make posterior inference on the population mean and variance parameters. We also compare the sampling properties of the standard Bayesian estimator of the population mean to those of the unbiased sample mean. Lastly, we discuss the appropriateness of the normal model when the underlying data are not normally distributed.

5.1 The normal model

A random variable Y is said to be normally distributed with mean θ and variance $\sigma^2 > 0$ if the density of Y is given by

$$p(y|\theta, \sigma^2) = \frac{1}{\sqrt{2\pi\sigma^2}} e^{-\frac{1}{2}\left(\frac{y-\theta}{\sigma}\right)^2}, \quad -\infty < y < \infty.$$

Figure 5.1 shows normal density curves for a few values of θ and σ^2. Some important things to remember about this distribution include that

- the distribution is symmetric about θ, and the mode, median and mean are all equal to θ;
- about 95% of the population lies within two standard deviations of the mean (more precisely, 1.96 standard deviations);
- if $X \sim \text{normal}(\mu, \tau^2)$, $Y \sim \text{normal}(\theta, \sigma^2)$ and X and Y are independent, then $aX + bY \sim \text{normal}(a\mu + b\theta, a^2\tau^2 + b^2\sigma^2)$;
- the `dnorm, rnorm, pnorm`, and `qnorm` commands in R take the standard deviation σ as their argument, not the variance σ^2. Be very careful about this when using R - confusing σ with σ^2 can drastically change your results.

P.D. Hoff, *A First Course in Bayesian Statistical Methods*,
Springer Texts in Statistics, DOI 10.1007/978-0-387-92407-6_5,

```
> dnorm
function (x, mean = 0, sd = 1, log = FALSE)
.Internal(dnorm(x, mean, sd, log))
```

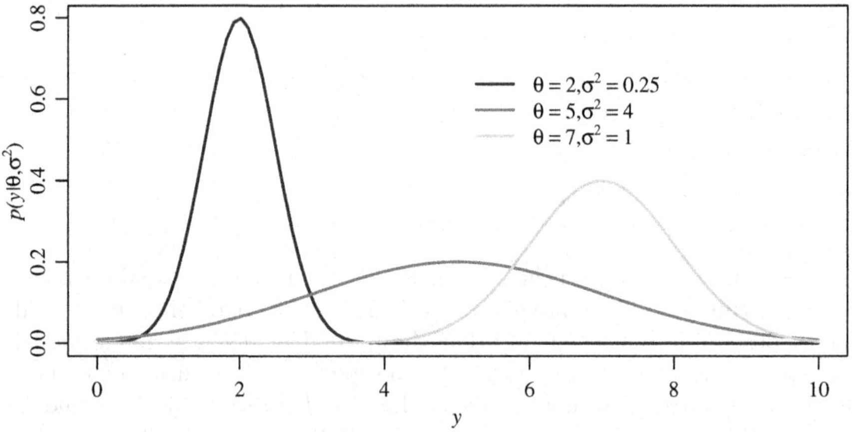

Fig. 5.1. Some normal densities.

The importance of the normal distribution stems primarily from the central limit theorem, which says that under very general conditions, the sum (or mean) of a set of random variables is approximately normally distributed. In practice, this means that the normal sampling model will be appropriate for data that result from the additive effects of a large number of factors.

Example: women's height

A study of 1,100 English families from 1893 to 1898 gathered height data on $n = 1375$ women over the age of 18. A histogram of these data is shown in Figure 5.2. The sample mean of these data is $\bar{y} = 63.75$ and the sample standard deviation is $s = 2.62$ inches. One explanation for the variability in heights among these women is that the women were heterogeneous in terms of a number of factors controlling human growth, such as genetics, diet, disease, stress and so on. Variability in these factors among the women results in variability in their heights. Letting y_i be the height in inches of woman i, a simple additive model for height might be

$$\begin{aligned}
y_1 &= a + b \times \text{gene}_1 + c \times \text{diet}_1 + d \times \text{disease}_1 + \cdots \\
y_2 &= a + b \times \text{gene}_2 + c \times \text{diet}_2 + d \times \text{disease}_2 + \cdots \\
&\vdots \\
y_n &= a + b \times \text{gene}_n + c \times \text{diet}_n + d \times \text{disease}_n + \cdots
\end{aligned}$$

where gene_i might denote the presence of a particular height-promoting gene, diet_i might measure some aspect of woman i's diet, and disease_i might indicate if woman i had ever had a particular disease. Of course, there may be a large number of genes, diseases, dietary and other factors that contribute to a woman's height. If the effects of these factors are approximately additive, then each height measurement y_i is roughly equal to a linear combination of a large number of terms. For such situations, the central limit theorem says that the empirical distribution of $y_1, \ldots, y_n$ will look like a normal distribution, and so the normal model provides an appropriate sampling model for the data.

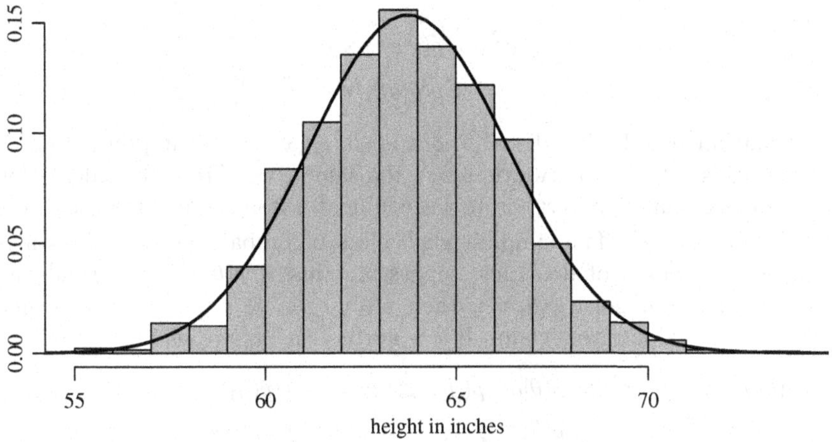

Fig. 5.2. Height data and a normal density with $\theta = 63.75$ and $\sigma = 2.62$.

5.2 Inference for the mean, conditional on the variance

Suppose our model is $\{Y_1, \ldots, Y_n | \theta, \sigma^2\} \sim$ i.i.d. normal (θ, σ^2). Then the joint sampling density is given by

$$\begin{aligned} p(y_1, \ldots, y_n | \theta, \sigma^2) &= \prod_{i=1}^{n} p(y_i | \theta, \sigma^2) \\ &= \prod_{i=1}^{n} \frac{1}{\sqrt{2\pi\sigma^2}} e^{-\frac{1}{2}\left(\frac{y_i - \theta}{\sigma}\right)^2} \\ &= (2\pi\sigma^2)^{-n/2} \exp\left\{-\frac{1}{2}\sum\left(\frac{y_i - \theta}{\sigma}\right)^2\right\}. \end{aligned}$$

Expanding the quadratic term in the exponent, we see that $p(y_1, \ldots, y_n | \theta, \sigma^2)$ depends on $y_1, \ldots, y_n$ through

$$\sum_{i=1}^{n}\left(\frac{y_i-\theta}{\sigma}\right)^2 = \frac{1}{\sigma^2}\sum y_i^2 - 2\frac{\theta}{\sigma^2}\sum y_i + n\frac{\theta^2}{\sigma^2}.$$

From this you can show that $\{\sum y_i^2, \sum y_i\}$ make up a two-dimensional sufficient statistic. Knowing the values of these quantities is equivalent to knowing the values of $\bar{y} = \sum y_i/n$ and $s^2 = \sum(y_i - \bar{y})^2/(n-1)$, and so $\{\bar{y}, s^2\}$ are also a sufficient statistic.

Inference for this two-parameter model can be broken down into two one-parameter problems. We will begin with the problem of making inference for θ when σ^2 is known, and use a conjugate prior distribution for θ. For any (conditional) prior distribution $p(\theta|\sigma^2)$, the posterior distribution will satisfy

$$\begin{aligned} p(\theta|y_1,\ldots,y_n,\sigma^2) &\propto p(\theta|\sigma^2) \times e^{-\frac{1}{2\sigma^2}\sum(y_i-\theta)^2} \\ &\propto p(\theta|\sigma^2) \times e^{c_1(\theta-c_2)^2}. \end{aligned}$$

Recall that a class of prior distributions is conjugate for a sampling model if the resulting posterior distribution is in the same class. From the calculation above, we see that if $p(\theta|\sigma^2)$ is to be conjugate, it must include quadratic terms like $e^{c_1(\theta-c_2)^2}$. The simplest such class of probability densities on $\mathbb{R}$ is the normal family of densities, suggesting that if $p(\theta|\sigma^2)$ is normal and $y_1,\ldots,y_n$ are i.i.d. normal(θ,σ^2), then $p(\theta|y_1,\ldots,y_n,\sigma^2)$ is also a normal density. Let's evaluate this claim: If $\theta \sim$ normal (μ_0, τ_0^2), then

$$\begin{aligned} p(\theta|y_1,\ldots,y_n,\sigma^2) &= p(\theta|\sigma^2)p(y_1,\ldots,y_n|\theta,\sigma^2)/p(y_1,\ldots,y_n|\sigma^2) \\ &\propto p(\theta|\sigma^2)p(y_1,\ldots,y_n|\theta,\sigma^2) \\ &\propto \exp\{-\frac{1}{2\tau_0^2}(\theta-\mu_0)^2\}\exp\{-\frac{1}{2\sigma^2}\sum(y_i-\theta)^2\}. \end{aligned}$$

Adding the terms in the exponents and ignoring the -1/2 for the moment, we have

$$\frac{1}{\tau_0^2}(\theta^2 - 2\theta\mu_0 + \mu_0^2) + \frac{1}{\sigma^2}(\sum y_i^2 - 2\theta\sum y_i + n\theta^2) = a\theta^2 - 2b\theta + c, \text{ where}$$

$$a = \frac{1}{\tau_0^2} + \frac{n}{\sigma^2}, \quad b = \frac{\mu_0}{\tau_0^2} + \frac{\sum y_i}{\sigma^2}, \quad \text{and } c = c(\mu_0, \tau_0^2, \sigma^2, y_1, \ldots, y_n).$$

Now let's see if $p(\theta|\sigma^2, y_1,\ldots,y_n)$ takes the form of a normal density:

$$\begin{aligned} p(\theta|\sigma^2,y_1,\ldots,y_n) &\propto \exp\{-\frac{1}{2}(a\theta^2 - 2b\theta)\} \\ &= \exp\{-\frac{1}{2}a(\theta^2 - 2b\theta/a + b^2/a^2) + \frac{1}{2}b^2/a\} \\ &\propto \exp\{-\frac{1}{2}a(\theta - b/a)^2\} \\ &= \exp\left\{-\frac{1}{2}\left(\frac{\theta - b/a}{1/\sqrt{a}}\right)^2\right\}. \end{aligned}$$

This function has exactly the same shape as a normal density curve, with $1/\sqrt{a}$ playing the role of the standard deviation and b/a playing the role of the mean. Since probability distributions are determined by their shape, this means that $p(\theta|\sigma^2, y_1, \ldots, y_n)$ is indeed a normal density. We refer to the mean and variance of this density as μ_n and τ_n^2, where

$$\tau_n^2 = \frac{1}{a} = \frac{1}{\frac{1}{\tau_0^2} + \frac{n}{\sigma^2}} \quad \text{and} \quad \mu_n = \frac{b}{a} = \frac{\frac{1}{\tau_0^2}\mu_0 + \frac{n}{\sigma^2}\bar{y}}{\frac{1}{\tau_0^2} + \frac{n}{\sigma^2}}.$$

Combining information

The (conditional) posterior parameters τ_n^2 and μ_n combine the prior parameters τ_0^2 and μ_0 with terms from the data.

- Posterior variance and precision: The formula for $1/\tau_n^2$ is

$$\frac{1}{\tau_n^2} = \frac{1}{\tau_0^2} + \frac{n}{\sigma^2}, \tag{5.1}$$

  and so the prior *inverse* variance is combined with the inverse of the data variance. Inverse variance is often referred to as the *precision*. For the normal model let,
  $\tilde{\sigma}^2 = 1/\sigma^2$ = sampling precision, i.e. how close the y_i's are to θ;
  $\tilde{\tau}_0^2 = 1/\tau_0^2$ = prior precision;
  $\tilde{\tau}_n^2 = 1/\tau_n^2$ = posterior precision.
  It is convenient to think about precision as the quantity of information on an additive scale. For the normal model, Equation 5.1 implies that

$$\tilde{\tau}_n^2 = \tilde{\tau}_0^2 + n\tilde{\sigma}^2,$$

  and so posterior information = prior information + data information.
- Posterior mean: Notice that

$$\mu_n = \frac{\tilde{\tau}_0^2}{\tilde{\tau}_0^2 + n\tilde{\sigma}^2}\mu_0 + \frac{n\tilde{\sigma}^2}{\tilde{\tau}_0^2 + n\tilde{\sigma}^2}\bar{y},$$

  and so the posterior mean is a weighted average of the prior mean and the sample mean. The weight on the sample mean is n/σ^2, the sampling precision of the sample mean. The weight on the prior mean is $1/\tau_0^2$, the prior precision. If the prior mean were based on κ_0 prior observations from the same (or similar) population as $Y_1, \ldots, Y_n$, then we might want to set $\tau_0^2 = \sigma^2/\kappa_0$, the variance of the *mean* of the prior observations. In this case, the formula for the posterior mean reduces to

$$\mu_n = \frac{\kappa_0}{\kappa_0 + n}\mu_0 + \frac{n}{\kappa_0 + n}\bar{y}.$$

Prediction

Consider predicting a new observation $\tilde{Y}$ from the population after having observed $(Y_1 = y_1, \dots, Y_n = y_n)$. To find the predictive distribution, let's use the following fact:

$$\{\tilde{Y}|\theta, \sigma^2\} \sim \text{normal}(\theta, \sigma^2) \Leftrightarrow \tilde{Y} = \theta + \tilde{\epsilon}, \quad \{\tilde{\epsilon}|\theta, \sigma^2\} \sim \text{normal}(0, \sigma^2)\,.$$

In other words, saying that $\tilde{Y}$ is normal with mean θ is the same as saying $\tilde{Y}$ is equal to θ plus some mean-zero normally distributed noise. Using this result, let's first compute the posterior mean and variance of $\tilde{Y}$:

$$\begin{aligned} \mathrm{E}[\tilde{Y}|y_1, \dots, y_n, \sigma^2] &= \mathrm{E}[\theta + \tilde{\epsilon}|y_1, \dots, y_n, \sigma^2] \\ &= \mathrm{E}[\theta|y_1, \dots, y_n, \sigma^2] + \mathrm{E}[\tilde{\epsilon}|y_1, \dots, y_n, \sigma^2] \\ &= \mu_n + 0 = \mu_n \end{aligned}$$

$$\begin{aligned} \mathrm{Var}[\tilde{Y}|y_1, \dots, y_n, \sigma^2] &= \mathrm{Var}[\theta + \tilde{\epsilon}|y_1, \dots, y_n, \sigma^2] \\ &= \mathrm{Var}[\theta|y_1, \dots, y_n, \sigma^2] + \mathrm{Var}[\tilde{\epsilon}|y_1, \dots, y_n, \sigma^2] \\ &= \tau_n^2 + \sigma^2\,. \end{aligned}$$

Recall from the beginning of the chapter that the sum of independent normal random variables is also normal. Therefore, since both θ and $\tilde{\epsilon}$, conditional on $y_1, \dots, y_n$ and σ^2, are normally distributed, so is $\tilde{Y} = \theta + \tilde{\epsilon}$. The predictive distribution is therefore

$$\tilde{Y}|\sigma^2, y_1, \dots, y_n \sim \text{normal}(\mu_n, \tau_n^2 + \sigma^2)\,.$$

It is worthwhile to have some intuition about the form of the variance of $\tilde{Y}$: In general, our uncertainty about a new sample $\tilde{Y}$ is a function of our uncertainty about the center of the population (τ_n^2) as well as how variable the population is (σ^2). As $n \to \infty$ we become more and more certain about where θ is, and the posterior variance τ_n^2 of θ goes to zero. But certainty about θ does not reduce the sampling variability σ^2, and so our uncertainty about $\tilde{Y}$ never goes below σ^2.

Example: Midge wing length

Grogan and Wirth (1981) provide data on the wing length in millimeters of nine members of a species of midge (small, two-winged flies). From these nine measurements we wish to make inference on the population mean θ. Studies from other populations suggest that wing lengths are typically around 1.9 mm, and so we set $\mu_0 = 1.9$. We also know that lengths must be positive, implying that $\theta > 0$. Therefore, ideally we would use a prior distribution for θ that has mass only on $\theta > 0$. We can approximate this restriction with a normal prior distribution for θ as follows: Since for any normal distribution most of the probability is within two standard deviations of the mean, we choose τ_0^2 so that $\mu_0 - 2 \times \tau_0 > 0$, or equivalently $\tau_0 < 1.9/2 = 0.95$. For now, we take $\tau_0 = 0.95$, which somewhat overstates our prior uncertainty about θ.

The observations in order of increasing magnitude are (1.64, 1.70, 1.72, 1.74, 1.82, 1.82, 1.82, 1.90, 2.08), giving $\bar{y} = 1.804$. Using the formulae above for μ_n and τ_n^2, we have $\{\theta|y_1, \ldots, y_9, \sigma^2\} \sim$ normal (μ_n, τ_n^2), where

$$\mu_n = \frac{\frac{1}{\tau_0^2}\mu_0 + \frac{n}{\sigma^2}\bar{y}}{\frac{1}{\tau_0^2} + \frac{n}{\sigma^2}} = \frac{1.11 \times 1.9 + \frac{9}{\sigma^2}1.804}{1.11 + \frac{9}{\sigma^2}}$$

$$\tau_n^2 = \frac{1}{\frac{1}{\tau_0^2} + \frac{n}{\sigma^2}} = \frac{1}{1.11 + \frac{9}{\sigma^2}}.$$

If $\sigma^2 = s^2 = 0.017$, then $\{\theta|y_1, \ldots, y_9, \sigma^2 = 0.017\} \sim$ normal $(1.805, 0.002)$. A 95% quantile-based confidence interval for θ based on this distribution is $(1.72, 1.89)$. However, this interval assumes that we are certain that $\sigma^2 = s^2$, when in fact s^2 is only a rough estimate of σ^2 based on only nine observations. To get a more accurate representation of our information we need to account for the fact that σ^2 is not known.

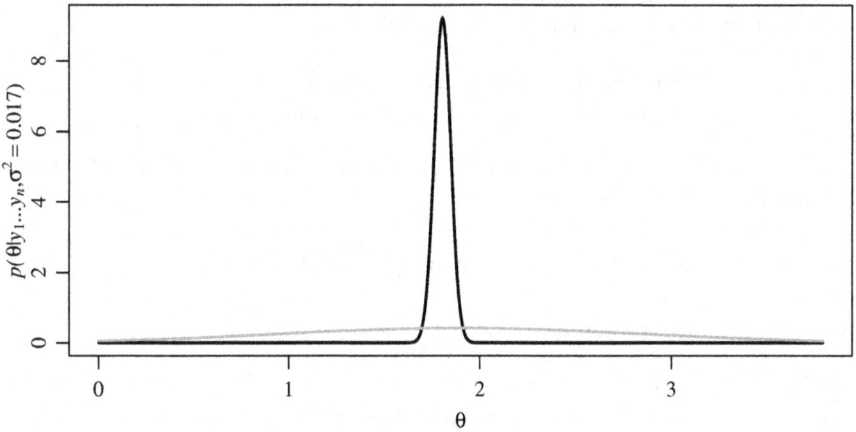

Fig. 5.3. Prior and conditional posterior distributions for the population mean wing length in the midge example.

5.3 Joint inference for the mean and variance

Bayesian inference for two or more unknown parameters is not conceptually different from the one-parameter case. For any joint prior distribution $p(\theta, \sigma^2)$ for θ and σ^2, posterior inference proceeds using Bayes' rule:

$$p(\theta, \sigma^2 | y_1, \dots, y_n) = p(y_1, \dots, y_n | \theta, \sigma^2) p(\theta, \sigma^2) / p(y_1, \dots, y_n).$$

As before, we will begin by developing a simple conjugate class of prior distributions which make posterior calculations easy.

Recall from our axioms of probability that a joint distribution for two quantities can be expressed as the product of a conditional probability and a marginal probability:

$$p(\theta, \sigma^2) = p(\theta | \sigma^2) p(\sigma^2).$$

In the last section, we saw that if σ^2 were known, then a conjugate prior distribution for θ was normal(μ_0, τ_0^2). Let's consider the particular case in which $\tau_0^2 = \sigma^2 / \kappa_0$:

$$p(\theta, \sigma^2) = p(\theta | \sigma^2) p(\sigma^2) = \text{dnorm}(\theta, \mu_0, \tau_0 = \sigma / \sqrt{\kappa_0}) \times p(\sigma^2).$$

In this case, the parameters μ_0 and κ_0 can be interpreted as the mean and sample size from a set of prior observations.

For σ^2 we need a family of prior distributions that has support on $(0, \infty)$. One such family of distributions is the gamma family, as we used for the Poisson sampling model. Unfortunately, this family is not conjugate for the normal variance. However, the gamma family does turn out to be a conjugate class of densities for $1/\sigma^2$ (the precision). When using such a prior distribution we say that σ^2 has an *inverse-gamma* distribution:

$$\begin{aligned} \text{precision} = 1/\sigma^2 &\sim \text{gamma}(a, b) \\ \text{variance} = \sigma^2 &\sim \text{inverse-gamma}(a, b) \end{aligned}$$

For interpretability later on, instead of using a and b we will parameterize this prior distribution as

$$1/\sigma^2 \sim \text{gamma}\left(\frac{\nu_0}{2}, \frac{\nu_0}{2} \sigma_0^2\right).$$

Under this parameterization,

- $\text{E}[\sigma^2] = \sigma_0^2 \frac{\nu_0/2}{\nu_0/2 - 1}$;
- $\text{mode}[\sigma^2] = \sigma_0^2 \frac{\nu_0/2}{\nu_0/2 + 1}$, so $\text{mode}[\sigma^2] < \sigma_0^2 < \text{E}[\sigma^2]$;
- $\text{Var}[\sigma^2]$ is decreasing in ν_0.

As we will see in a moment, we can interpret the prior parameters (σ_0^2, ν_0) as the sample variance and sample size of prior observations.

Posterior inference

Suppose our prior distributions and sampling model are as follows:

$$\begin{aligned} 1/\sigma^2 &\sim \text{gamma}(\nu_0/2, \nu_0 \sigma_0^2 / 2) \\ \theta | \sigma^2 &\sim \text{normal}(\mu_0, \sigma^2 / \kappa_0) \\ Y_1, \dots, Y_n | \theta, \sigma^2 &\sim \text{i.i.d. normal}\ (\theta, \sigma^2). \end{aligned}$$

Just as the prior distribution for θ and σ^2 can be decomposed as $p(\theta,\sigma^2) = p(\theta|\sigma^2)p(\sigma^2)$, the posterior distribution can be similarly decomposed:

$$p(\theta,\sigma^2|y_1,\ldots,y_n) = p(\theta|\sigma^2,y_1,\ldots,y_n)p(\sigma^2|y_1,\ldots,y_n)\,.$$

The conditional distribution of θ given the data and σ^2 can be obtained using the results of the previous section: Plugging in σ^2/κ_0 for τ_0^2, we have

$$\{\theta|y_1,\ldots,y_n,\sigma^2\} \sim \text{normal}(\mu_n,\sigma^2/\kappa_n), \text{ where}$$
$$\kappa_n = \kappa_0 + n \text{ and } \mu_n = \frac{(\kappa_0/\sigma^2)\mu_0 + (n/\sigma^2)\bar{y}}{\kappa_0/\sigma^2 + n/\sigma^2} = \frac{\kappa_0\mu_0 + n\bar{y}}{\kappa_n}.$$

Therefore, if μ_0 is the mean of κ_0 prior observations, then $\text{E}[\theta|y_1,\ldots,y_n,\sigma^2]$ is the sample mean of the current and prior observations, and $\text{Var}[\theta|y_1,\ldots,y_n,\sigma^2]$ is σ^2 divided by the total number of observations, both prior and current.

The posterior distribution of σ^2 can be obtained by performing an integration over the unknown value of θ:

$$\begin{aligned} p(\sigma^2|y_1,\ldots,y_n) &\propto p(\sigma^2)p(y_1,\ldots,y_n|\sigma^2) \\ &= p(\sigma^2)\int p(y_1,\ldots,y_n|\theta,\sigma^2)p(\theta|\sigma^2)\,d\theta\,. \end{aligned}$$

This integral can be done without much knowledge of calculus, but it is somewhat tedious and is left as an exercise (Exercise 5.3). The result is that

$$\begin{aligned} \{1/\sigma^2|y_1,\ldots,y_n\} &\sim \text{gamma}(\nu_n/2,\nu_n\sigma_n^2/2), \text{ where} \\ \nu_n &= \nu_0 + n \\ \sigma_n^2 &= \frac{1}{\nu_n}[\nu_0\sigma_0^2 + (n-1)s^2 + \frac{\kappa_0 n}{\kappa_n}(\bar{y}-\mu_0)^2]. \end{aligned}$$

These formulae suggest an interpretation of ν_0 as a prior sample size, from which a prior sample variance of σ_0^2 has been obtained. Recall that $s^2 = \sum_{i=1}^n (y_i - \bar{y})^2/(n-1)$ is the sample variance, and $(n-1)s^2$ is the sum of squared observations from the sample mean, which is often called the "sum of squares." Similarly, we can think of $\nu_0\sigma_0^2$ and $\nu_n\sigma_n^2$ as prior and posterior sums of squares, respectively. Multiplying both sides of the last equation by ν_n almost gives us "posterior sum of squares equals prior sum of squares plus data sum of squares." However, the third term in the last equation is a bit harder to understand - it says that a large value of $(\bar{y}-\mu_0)^2$ increases the posterior probability of a large σ^2. This makes sense for our particular joint prior distribution for θ and σ^2: If we want to think of μ_0 as the sample mean of κ_0 prior observations with variance σ^2, then $\frac{\kappa_0 n}{\kappa_0+n}(\bar{y}-\mu_0)^2$ is an estimate of σ^2 and so we want to use the information that this term provides. For situations in which μ_0 should not be thought of as the mean of prior observations, we will develop an alternative prior distribution in the next chapter.

Example

Returning to the midge data, studies of other populations suggest that the true mean and standard deviation of our population under study should not be too far from 1.9 mm and 0.1 mm respectively, suggesting $\mu_0 = 1.9$ and $\sigma_0^2 = 0.01$. However, this population may be different from the others in terms of wing length, and so we choose $\kappa_0 = \nu_0 = 1$ so that our prior distributions are only weakly centered around these estimates from other populations.

The sample mean and variance of our observed data are $\bar{y} = 1.804$ and $s^2 = 0.0169$ ($s = 0.130$). From these values and the prior parameters, we compute μ_n and σ_n^2:

$$\mu_n = \frac{\kappa_0\mu_0 + n\bar{y}}{\kappa_n} = \frac{1.9 + 9 \times 1.804}{1+9} = 1.814$$

$$\sigma_n^2 = \frac{1}{\nu_n}[\nu_0\sigma_0^2 + (n-1)s^2 + \frac{\kappa_0 n}{\kappa_n}(\bar{y}-\mu_0)^2]$$

$$= \frac{0.010 + 0.135 + 0.008}{10} = 0.015\,.$$

These calculations can be done with the following commands in R:

```
# prior
mu0<-1.9  ; k0<-1
s20<-.010 ; nu0<-1

# data
y<-c(1.64,1.70,1.72,1.74,1.82,1.82,1.82,1.90,2.08)
n<-length(y) ; ybar<-mean(y) ; s2<-var(y)

# posterior inference
kn<-k0+n ; nun<-nu0+n
mun<- (k0*mu0 + n*ybar)/kn
s2n<- (nu0*s20 +(n-1)*s2 +k0*n*(ybar-mu0)^2/(kn))/(nun)

> mun
[1] 1.814
> s2n
[1] 0.015324
> sqrt(s2n)
[1] 0.1237901
```

Our joint posterior distribution is completely determined by the values $\mu_n = 1.814$, $\kappa_n = 10$, $\sigma_n^2 = 0.015$, $\nu_n = 10$, and can be expressed as

$$\{\theta|y_1,\ldots,y_n,\sigma^2\} \sim \text{normal}(1.814, \sigma^2/10),$$
$$\{1/\sigma^2|y_1,\ldots,y_n\} \sim \text{gamma}(10/2, 10 \times 0.015/2).$$

Letting $\tilde{\sigma}^2 = 1/\sigma^2$, a contour plot of the bivariate posterior density of $(\theta, \tilde{\sigma}^2)$ appears in the first panel of Figure 5.4. This plot was obtained by computing dnorm($\theta_k, \mu_n, 1/\sqrt{10\tilde{\sigma}_l^2}$) $\times$ dgamma($\tilde{\sigma}_l^2, 10/2, 10\sigma_n^2/2$) for each pair of values $(\theta_k, \tilde{\sigma}_l^2)$ on a grid. Similarly, the second panel plots the joint posterior density of (θ, σ^2). Notice that the contours are more peaked as a function of θ for low values of σ^2 than high values.

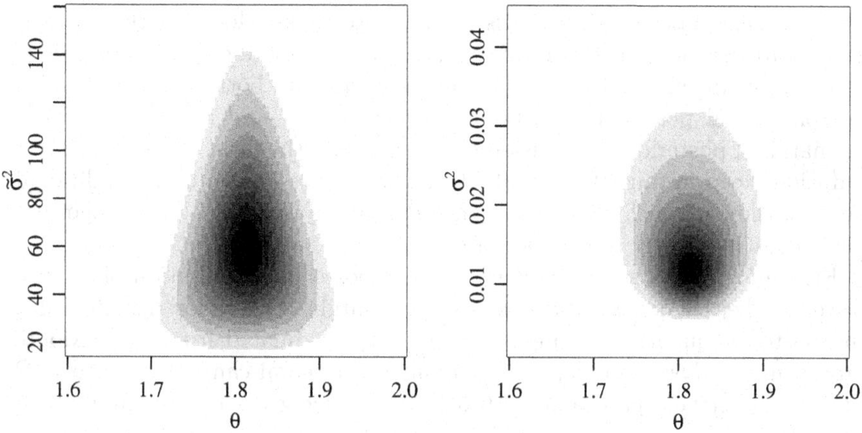

Fig. 5.4. Joint posterior distributions of $(\theta, \tilde{\sigma}^2)$ and (θ, σ^2).

Monte Carlo sampling

For many data analyses, interest primarily lies in estimating the population mean θ, and so we would like to calculate quantities like $\mathrm{E}[\theta|y_1, \ldots, y_n]$, $\mathrm{sd}[\theta|y_1, \ldots, y_n]$, $\Pr(\theta_1 < \theta_2|\ y_{1,1}, \ldots,\ y_{n_2,2})$, and so on. These quantities are all determined by the *marginal* posterior distribution of θ given the data. But all we know (so far) is that the *conditional* distribution of θ given the data and σ^2 is normal, and that σ^2 given the data is inverse-gamma. If we could generate *marginal* samples of θ, from $p(\theta|y_1, \ldots, y_n)$, then we could use the Monte Carlo method to approximate the above quantities of interest. It turns out that this is quite easy to do by generating samples of θ and σ^2 from their joint posterior distribution. Consider simulating parameter values using the following Monte Carlo procedure:

$$
\begin{aligned}
\sigma^{2(1)} &\sim \text{inverse gamma}(\nu_n/2, \sigma_n^2\nu_n/2), & \theta^{(1)} &\sim \text{normal}(\mu_n, \sigma^{2(1)}/\kappa_n) \\
&\vdots & &\vdots \\
\sigma^{2(S)} &\sim \text{inverse gamma}(\nu_n/2, \sigma_n^2\nu_n/2), & \theta^{(S)} &\sim \text{normal}(\mu_n, \sigma^{2(S)}/\kappa_n)\,.
\end{aligned}
$$

Note that each $\theta^{(s)}$ is sampled from its conditional distribution given the data and $\sigma^2 = \sigma^{2(s)}$. This Monte Carlo procedure can be implemented in R with only two lines of code:

```
s2.postsample <- 1/rgamma(10000, nun/2, s2n*nun/2 )
theta.postsample <- rnorm(10000, mun, sqrt(s2.postsample/kn))
```

A sequence of pairs $\{(\sigma^{2(1)}, \theta^{(1)}), \dots, (\sigma^{2(S)}, \theta^{(S)})\}$ simulated using this procedure are independent samples from the joint posterior distribution of $p(\theta, \sigma^2|y_1, \dots, y_n)$. Additionally, the simulated sequence $\{\theta^{(1)}, \dots, \theta^{(S)}\}$ can be seen as independent samples from the *marginal* posterior distribution of $p(\theta|y_1, \dots, y_n)$, and so we use this sequence to make Monte Carlo approximations to functions involving $p(\theta|y_1, \dots, y_n)$, as described in Chapter 4. It may seem confusing that each $\theta^{(s)}$-value is referred to both as a sample from the conditional posterior distribution of θ given σ^2 *and* as a sample from the marginal posterior distribution of θ given only the data. To alleviate this confusion, keep in mind that while $\theta^{(1)}, \dots, \theta^{(S)}$ are indeed each conditional samples, they are each conditional on *different* values of σ^2. Taken together, they constitute marginal samples of θ.

Figure 5.5 shows samples from the joint posterior distribution of (θ, σ^2), as well as kernel density estimates of the marginal posterior distributions. Any posterior quantities of interest can be approximated from these Monte Carlo samples. For example, a 95% confidence interval can be obtained in R with `quantile(theta.postsample,c(.025,.975))`, which gives an interval of (1.73, 1.90). This is extremely close to (1.70, 1.90), a frequentist 95% confidence interval obtained from the t-test. There is a reason for this: It turns out that $p(\theta|y_1, \dots, y_n)$, the marginal posterior distribution of θ, can be obtained in a closed form. From this form, it can be shown that the posterior distribution of $t(\theta) = \frac{(\theta - \mu_n)}{\sigma_n/\sqrt{\kappa_n}}$, given $\bar{y}$ and s^2, has a t-distribution with $\nu_0 + n$ degrees of freedom. If κ_0 and ν_0 are small, then the posterior distribution of $t(\theta)$ will be very close to the t_{n-1} distribution. How small can κ_0 and ν_0 be?

Improper priors

What if you want to "be Bayesian" so you can talk about things like $\Pr(\theta < c|y_1, \dots, y_n)$ but want to "be objective" by not using any prior information? Since we have referred to κ_0 and ν_0 as prior sample sizes, it might seem that the smaller these parameters are, the more objective the estimates will be. So it is natural to wonder what happens to the posterior distribution as κ_0 and ν_0 get smaller and smaller. The formulae for μ_n and σ_n^2 are

$$
\begin{aligned}
\mu_n &= \frac{\kappa_0 \mu_0 + n\bar{y}}{\kappa_0 + n} \\
\sigma_n^2 &= \frac{1}{\nu_0 + n}[\nu_0 \sigma_0^2 + (n-1)s^2 + \frac{\kappa_0 n}{\kappa_0 + n}(\bar{y} - \mu_0)^2],
\end{aligned}
$$

and so as $\kappa_0, \nu_0 \to 0$,

$$
\begin{aligned}
\mu_n &\to \bar{y}, \text{ and} \\
\sigma_n^2 &\to \frac{n-1}{n} s^2 = \frac{1}{n} \sum (y_i - \bar{y})^2 .
\end{aligned}
$$

This has led some to suggest the following "posterior distribution":

$$\{1/\sigma^2|y_1,\ldots,y_n\} \sim \text{gamma}(\frac{n}{2},\frac{n}{2}\frac{1}{n}\sum(y_i-\bar{y})^2)$$
$$\{\theta|\sigma^2,y_1,\ldots,y_n\} \sim \text{normal}(\bar{y},\frac{\sigma^2}{n}).$$

Somewhat more formally, if we let $\tilde{p}(\theta,\sigma^2)=1/\sigma^2$ (which is not a probability density) and set $p(\theta,\sigma^2|\boldsymbol{y}) \propto p(\boldsymbol{y}|\theta,\sigma^2)\times\tilde{p}(\theta,\sigma^2)$, we get the same "conditional distribution" for θ but a gamma$(\frac{n-1}{2},\frac{1}{2}\sum(y_i-\bar{y})^2)$ distribution for $1/\sigma^2$ (Gelman et al (2004), Chapter 3). You can integrate this latter joint distribution over σ^2 to show that

$$\frac{\theta-\bar{y}}{s/\sqrt{n}}|y_1,\ldots,y_n \sim t_{n-1}.$$

It is interesting to compare this result to the sampling distribution of the t-statistic, conditional on θ but unconditional on the data:

$$\frac{\bar{Y}-\theta}{s/\sqrt{n}}|\theta \sim t_{n-1}.$$

The second statement says that, *before* you sample the data, your uncertainty about the scaled deviation of the sample mean $\bar{Y}$ from the population mean θ is represented with a t_{n-1} distribution. The first statement says that *after* you sample your data, your uncertainty is still represented with a t_{n-1} distribution. The difference is that before you sample your data, both $\bar{Y}$ and θ are unknown. After you sample your data, then $\bar{Y}=\bar{y}$ is known and this provides us with information about θ.

There are no proper prior probability distributions on (θ,σ^2) that will lead to the above t_{n-1} posterior distribution for θ, and so inference based on this posterior distribution is not formally Bayesian. However, sometimes taking limits like this leads to sensible answers: Theoretical results in Stein (1955) show that from a decision-theoretic point of view, any reasonable estimator is a Bayesian estimator or a limit of a sequence of Bayesian estimators, and that any Bayesian estimator is reasonable (the technical term here is *admissible*; see also Berger (1980)).

5.4 Bias, variance and mean squared error

A *point estimator* of an unknown parameter θ is a function that converts your data into a single element of the parameter space Θ. For example, in the case of a normal sampling model and conjugate prior distribution of the last section, the posterior mean estimator of θ is

$$\hat{\theta}_b(y_1,\ldots,y_n)=\mathrm{E}[\theta|y_1,\ldots,y_n]=\frac{n}{\kappa_0+n}\bar{y}+\frac{\kappa_0}{\kappa_0+n}\mu_0=w\bar{y}+(1-w)\mu_0.$$

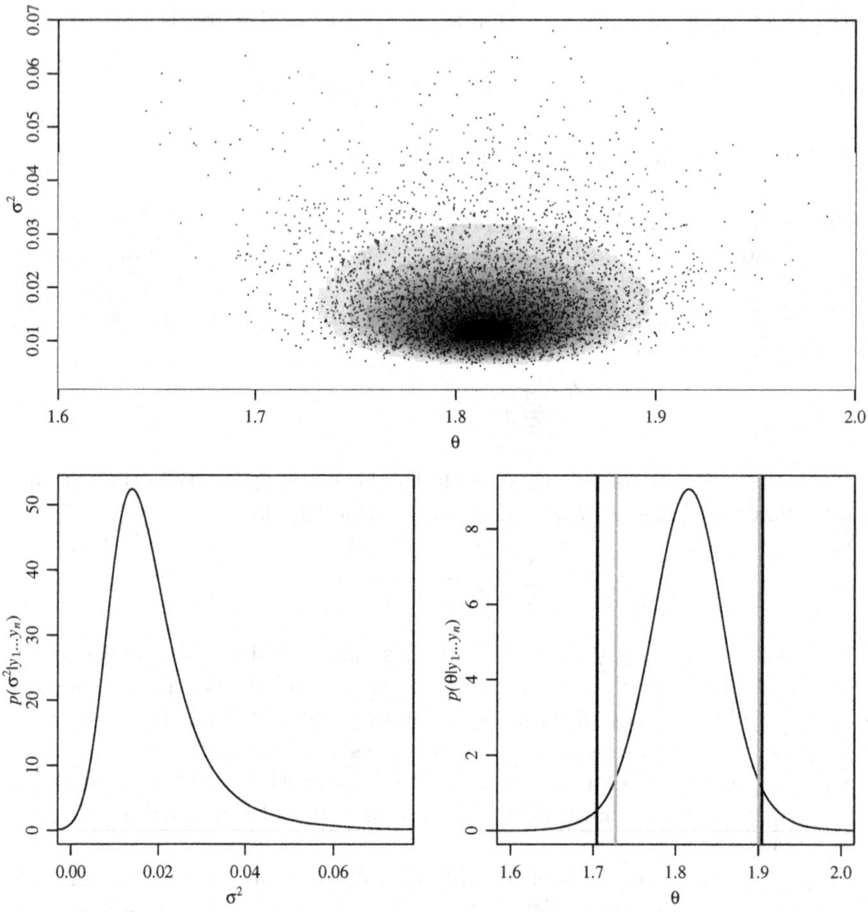

Fig. 5.5. Monte Carlo samples from and estimates of the joint and marginal distributions of the population mean and variance. The vertical lines in the third plot give a 95% quantile-based posterior interval for θ (gray), as well as the 95% confidence interval based on the t-statistic (black).

The *sampling properties* of an estimator such as $\hat{\theta}_b$ refer to its behavior under hypothetically repeatable surveys or experiments. Let's compare the sampling properties of $\hat{\theta}_b$ to $\hat{\theta}_e(y_1, \ldots, y_n) = \bar{y}$, the sample mean, when the true value of the population mean is θ_0:

$$\mathrm{E}[\hat{\theta}_e | \theta = \theta_0] = \theta_0, \quad \text{and we say that } \hat{\theta}_e \text{ is "unbiased,"}$$
$$\mathrm{E}[\hat{\theta}_b | \theta = \theta_0] = w\theta_0 + (1-w)\mu_0, \quad \text{and if } \mu_0 \neq \theta_0 \text{ we say that } \hat{\theta}_b \text{ is "biased."}$$

Bias refers to how close the center of mass of the sampling distribution of an estimator is to the true value. An unbiased estimator is an estimator with zero bias, which sounds desirable. However, bias does not tell us how far away an estimate might be from the true value. For example, y_1 is an unbiased estimator of the population mean θ_0, but will generally be farther away from θ_0 than $\bar{y}$. To evaluate how close an estimator $\hat{\theta}$ is likely to be to the true value θ_0, we might use the *mean squared error* (MSE). Letting $m = \mathrm{E}[\hat{\theta}|\theta_0]$, the MSE is

$$\begin{aligned}
\mathrm{MSE}[\hat{\theta}|\theta_0] &= \mathrm{E}[(\hat{\theta} - \theta_0)^2|\theta_0] \\
&= \mathrm{E}[(\hat{\theta} - m + m - \theta_0)^2|\theta_0] \\
&= \mathrm{E}[(\hat{\theta} - m)^2|\theta_0] + 2\mathrm{E}[(\hat{\theta} - m)(m - \theta_0)|\theta_0] + \mathrm{E}[(m - \theta_0)^2|\theta_0].
\end{aligned}$$

Since $m = \mathrm{E}[\hat{\theta}|\theta_0]$ it follows that $\mathrm{E}[\hat{\theta} - m|\theta_0] = 0$ and so the second term is zero. The first term is the variance of $\hat{\theta}$ and the third term is the square of the bias and so

$$\mathrm{MSE}[\hat{\theta}|\theta_0] = \mathrm{Var}[\hat{\theta}|\theta_0] + \mathrm{Bias}^2[\hat{\theta}|\theta_0].$$

This means that, before the data are gathered, the expected distance from the estimator to the true value depends on how close θ_0 is to the center of the distribution of $\hat{\theta}$ (the bias), as well as how spread out the distribution is (the variance). Getting back to our comparison of $\hat{\theta}_b$ to $\hat{\theta}_e$, the bias of $\hat{\theta}_e$ is zero, but

$$\begin{aligned}
\mathrm{Var}[\hat{\theta}_e|\theta = \theta_0, \sigma^2] &= \frac{\sigma^2}{n}, \text{ whereas} \\
\mathrm{Var}[\hat{\theta}_b|\theta = \theta_0, \sigma^2] &= w^2 \times \frac{\sigma^2}{n} < \frac{\sigma^2}{n},
\end{aligned}$$

and so $\hat{\theta}_b$ has lower variability. Which one is better in terms of MSE?

$$\begin{aligned}
\mathrm{MSE}[\hat{\theta}_e|\theta_0] &= E[(\hat{\theta}_e - \theta_0)^2|\theta_0] = \frac{\sigma^2}{n} \\
\mathrm{MSE}[\hat{\theta}_b|\theta_0] &= E[(\hat{\theta}_b - \theta_0)^2|\theta_0] = E[\{w(\bar{y} - \theta_0) + (1 - w)(\mu_0 - \theta_0)\}^2|\theta_0] \\
&= w^2 \times \frac{\sigma^2}{n} + (1 - w)^2(\mu_0 - \theta_0)^2
\end{aligned}$$

With some algebra, you can show that $\mathrm{MSE}[\hat{\theta}_b|\theta_0] < \mathrm{MSE}[\hat{\theta}_e|\theta_0]$ if

$$\begin{aligned}
(\mu_0 - \theta_0)^2 &< \frac{\sigma^2}{n}\frac{1 + w}{1 - w} \\
&= \sigma^2\left(\frac{1}{n} + \frac{2}{\kappa_0}\right).
\end{aligned}$$

Some argue that if you know even just a little bit about the population you are about to sample from, you should be able to find values of μ_0 and κ_0 such that this inequality holds. In this case, you can construct a Bayesian estimator that will have a lower average squared distance to the truth than does the sample mean. For example, if you are pretty sure that your best prior guess μ_0 is within about one and a half (1.4) standard deviations of the true population mean, then if you pick $\kappa_0 = 1$ you can be pretty sure that the Bayes estimator has a lower MSE. To make some of this more clear, let's take a look at the sampling distributions of a few different estimators in the context of an example.

Example: IQ scores

Scoring on IQ tests is designed to produce a normal distribution with a mean of 100 and a standard deviation of 15 (a variance of 225) when applied to the general population. Now suppose we are to sample n individuals from a particular town in the United States and then estimate θ, the town-specific mean IQ score, based on the sample of size n. For Bayesian estimation, if we lack much information about the town in question, a natural choice of μ_0 would be $\mu_0 = 100$.

Suppose that unknown to us the people in this town are extremely exceptional and the true mean and standard deviation of IQ scores in the town are $\theta = 112$ and $\sigma = 13$ ($\sigma^2 = 169$). The MSEs of the estimators $\hat{\theta}_e$ and $\hat{\theta}_b$ are then

$$\text{MSE}[\hat{\theta}_e|\theta_0] = \text{Var}[\hat{\theta}_e] = \frac{\sigma^2}{n} = \frac{169}{n}$$
$$\text{MSE}[\hat{\theta}_b|\theta_0] = w^2 \frac{169}{n} + (1-w)^2 144,$$

where $w = n/(\kappa_0 + n)$. The ratio $\text{MSE}[\hat{\theta}_b|\theta_0]/\text{MSE}[\hat{\theta}_e|\theta_0]$ is plotted in the first panel of Figure 5.6 as a function of n, for $\kappa_0 = 1$, 2 and 3.

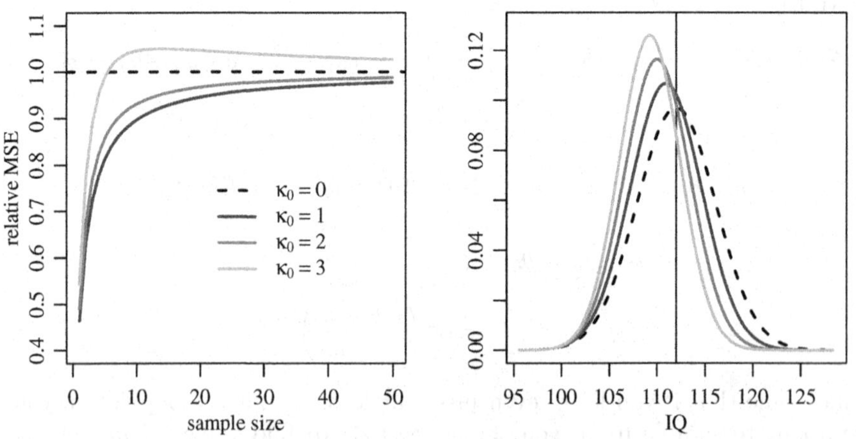

Fig. 5.6. Mean squared errors and sampling distributions of different estimators of the population mean IQ score.

Notice that when $\kappa_0 = 1$ or 2 the Bayes estimate has lower MSE than the sample mean, especially when the sample size is low. This is because even though the prior guess $\mu_0 = 100$ is seemingly way off, it is not actually that far off when considering the uncertainty in our sample data. A choice of $\kappa_0 = 3$ on the other hand puts more weight on the value of 100, and the corresponding estimator has a generally higher MSE than the sample mean. As n increases, the bias of each of the estimators shrinks to zero, and the MSEs converge to the common value of σ^2/n. The second panel of Figure 5.6 shows the sampling distributions of the sample mean when $n = 10$, as well as those of the three Bayes estimators corresponding to $\kappa_0 = 1$, 2 and 3. This plot highlights the relative contributions of the bias and variance to the MSE. The sampling distribution of the sample mean is centered around the true value of 112, but is more spread out than any of the other distributions. The distribution of the $\kappa_0 = 1$ estimator is not quite centered around the true mean, but its variance is low and so this estimator is closer on average to the truth than the sample mean.

5.5 Prior specification based on expectations

A p-dimensional exponential family model is a model whose densities can be written as $p(y|\boldsymbol{\phi}) = h(y)c(\boldsymbol{\phi})\exp\{\boldsymbol{\phi}^T\boldsymbol{t}(y)\}$, where $\boldsymbol{\phi}$ is the parameter to be estimated and $\boldsymbol{t}(y) = \{t_1(y), \ldots, t_p(y)\}$ is the sufficient statistic. The normal model is a two-dimensional exponential family model, where

- $\boldsymbol{t}(y) = (y, y^2)$,
- $\boldsymbol{\phi} = (\theta/\sigma^2, -(2\sigma^2)^{-1})$ and
- $c(\boldsymbol{\phi}) = |\phi_2|^{1/2}\exp\{\phi_1^2/(2\phi_2)\}$.

As was the case for one-parameter exponential family models in Section 3.3, a conjugate prior distribution can be written in terms of $\boldsymbol{\phi}$, giving $p(\boldsymbol{\phi}|n_0, \boldsymbol{t}_0) \propto c(\boldsymbol{\phi})^{n_0}\exp(n_0\boldsymbol{t}_0^T\boldsymbol{\phi})$, where $\boldsymbol{t}_0 = (t_{01}, t_{02}) = (\mathrm{E}[Y], \mathrm{E}[Y^2])$, the prior expectations of Y and Y^2. If we reparameterize in terms of (θ, σ^2), we get

$$p(\theta, \sigma^2|n_0, \boldsymbol{t}_0) \propto \left[(\sigma^2)^{-1/2}\exp\left\{\frac{-n_0(\theta - t_{01})^2}{2\sigma^2}\right\}\right] \times \\ \left[(\sigma^2)^{-(n_0+5)/2}\exp\left\{\frac{-n_0(t_{02} - t_{01}^2)}{2\sigma^2}\right\}\right].$$

The first term in the big braces is proportional to a normal$(t_{01}, \sigma^2/n_0)$ density, and the second is proportional to an inverse-gamma$((n_0+3)/2, n_0(t_2 - t_1^2)/2)$ density. To see how our prior parameters t_{01} and t_{02} should be determined, let's consider the case where we have a prior expectation μ_0 for the population mean (so $\mathrm{E}[Y] = \mathrm{E}[\mathrm{E}[Y|\theta]] = \mathrm{E}[\theta] = \mu_0$), and a prior expectation σ_0^2 for the population variance (so that $\mathrm{E}[\mathrm{Var}[Y|\theta,\sigma^2]] = \mathrm{E}[\sigma^2] = \sigma_0^2$). Then we would set t_{01} equal to μ_0, and determine t_{02} from

$$\begin{aligned} t_{02} = \mathrm{E}[Y^2] &= \mathrm{E}[\mathrm{E}[Y^2|\theta,\sigma^2]] \\ &= \mathrm{E}[\sigma^2 + \theta^2] \\ &= \sigma_0^2 + \sigma_0^2/n_0 + \mu_0^2 = \sigma_0^2(n_0+1)/n_0 + \mu_0^2, \end{aligned}$$

so $n_0(t_{02} - t_{01}^2) = (n_0+1)\sigma_0^2$. Thus our joint prior distribution for (θ,σ^2) would be

$$\begin{aligned} \theta|\sigma^2 &\sim \text{normal}(\mu_0, \sigma^2/n_0), \text{ and} \\ \sigma^2 &\sim \text{inverse-gamma}((n_0+3)/2, (n_0+1)\sigma_0^2/2). \end{aligned}$$

For example, if our prior information is weak we might set $n_0 = 1$, giving $\theta|\sigma^2 \sim \text{normal}(\mu_0, \sigma^2)$ and $1/\sigma^2 \sim \text{inverse-gamma}(2, \sigma_0^2)$. It is easy to check that under this prior distribution the prior expectation of Y is μ_0, and the prior expectation of $\mathrm{Var}[Y|\theta,\sigma^2]$ is σ_0^2, as desired. Given n i.i.d samples from the population, our posterior distribution under this prior would be

$$\begin{aligned} \{\theta|\sigma^2, y_1,\ldots,y_n\} &\sim \text{normal}\left(\frac{\mu_0/\sigma^2 + n\bar{y}}{1/\sigma^2 + n}, \frac{\sigma^2}{n+1}\right) \\ \{\sigma^2|y_1,\ldots,y_n\} &\sim \left(2 + n/2, \sigma_0^2 + (n-1)s^2 + \frac{n}{n+1}(\bar{y} - \mu_0)^2\right). \end{aligned}$$

5.6 The normal model for non-normal data

People use the normal model all the time in situations where the data are not even close to being normally distributed. The justification of this is generally that while the sampling distribution of a single data point is not normal, the sampling distribution of the sample mean is close to normal. Let's explore this distinction via a Monte Carlo sampling experiment: The 1998 General Social Survey (GSS) recorded the number of children for 921 women over the age of 40. Let's take these 921 women as our population, and consider estimating the mean number of children for this population (which is 2.42) based on random samples $Y_1,\ldots,Y_n$ of different sizes n.

The true population distribution is plotted in the first panel of Figure 5.7, and is clearly not normal. For example, the distribution is discrete, bounded and skewed, whereas a normal distribution is continuous, unbounded and symmetric. Now let's consider the sampling distribution of the sample mean $\bar{Y}_n = \frac{1}{n}\sum_{i=1}^n Y_i$ for $n \in \{5, 15, 45\}$. This can be done using a Monte Carlo approximation as follows: For each n and some large value of S, simulate $\{\bar{y}_n^{(1)},\ldots,\bar{y}_n^{(S)}\}$, where each $\bar{y}_n^{(s)}$ is the sample mean of n samples taken without replacement from the 921 values in the population. The second panel of Figure 5.7 shows the Monte Carlo approximations to the sampling distributions of

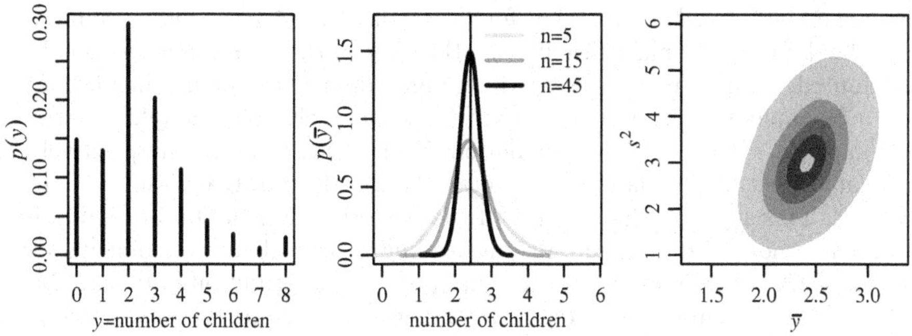

Fig. 5.7. A non-normal distribution and the distribution of its sample mean for $n \in \{5, 15, 45\}$. The third panel shows a contour plot of the joint sampling density of $\{\bar{y}, s^2\}$ for the case $n = 45$.

$\bar{Y}_5$,$\bar{Y}_{15}$ and $\bar{Y}_{45}$. While the distribution of $\bar{Y}_5$ looks a bit skewed, the sampling distributions of $\bar{Y}_{15}$ and $\bar{Y}_{45}$ are hard to distinguish from normal distributions. This should not be too much of a surprise, as the central limit theorem tells us that

$$p(\bar{y}|\theta, \sigma^2) \approx \text{dnorm}(\bar{y}, \theta, \sqrt{\sigma^2/n}),$$

with the approximation becoming increasingly good as n gets larger. If the population variance σ^2 were known, then an approximate posterior distribution of the population mean, conditional on the sample mean, could be obtained as

$$\begin{aligned} p(\theta|\bar{y}, \sigma^2) &\propto p(\theta) \times p(\bar{y}|\theta, \sigma^2) \\ &\approx p(\theta) \times \text{dnorm}(\bar{y}, \theta, \sqrt{\sigma^2/n}). \end{aligned}$$

Of course σ^2 is generally not known, but it is estimated by s^2. The approximate posterior distribution of (θ, σ^2) conditional on the estimates $(\bar{y}, s^2)$ is given by

$$\begin{aligned} p(\theta, \sigma^2|\bar{y}, s^2) &\propto p(\theta, \sigma^2) \times p(\bar{y}, s^2|\theta, \sigma^2) \\ &= p(\theta, \sigma^2) \times p(\bar{y}|\theta, \sigma^2) \times p(s^2|\bar{y}, \theta, \sigma^2) \\ &\approx p(\theta, \sigma^2) \times \text{dnorm}(\bar{y}, \theta, \sqrt{\sigma^2/n}) \times p(s^2|\bar{y}, \theta, \sigma^2). \qquad (5.2) \end{aligned}$$

Again, for large n, the approximation of $p(\bar{y}|\theta, \sigma^2)$ by the normal density is generally a good one even if the population is not normally distributed. However, it is not clear what to put for $p(s^2|\bar{y}, \theta, \sigma^2)$. If we knew that the data were actually sampled from a normal distribution, then results from statistical theory would say that

$$p(s^2|\bar{y}, \theta, \sigma^2) = \text{dgamma}(s^2, \frac{n-1}{2}, \frac{n-1}{2\sigma^2}).$$

Note that this result says that, for normal populations, $\bar{Y}$ and s^2 are independent. Using this sampling model for s^2 in Equation 5.2 results in exactly the same conditional distribution for (θ, σ^2) as $p(\theta, \sigma^2|y_1, \ldots, y_n)$ assuming that the data are normally distributed. However, if the data are not normally distributed, then s^2 is not necessarily gamma-distributed or independent of $\bar{y}$. For example, the third panel of Figure 5.7 shows the joint sampling distribution of $\{\bar{Y}, s^2\}$ for the GSS population. Notice that $\bar{Y}$ and s^2 are positively correlated for this population, as is often the case for positively skewed populations. This suggests that the use of the posterior distribution in Equation 5.2 for non-normal data could give misleading results about the joint distribution of $\{\theta, \sigma^2\}$. However, the marginal posterior distribution of θ based on 5.2 can be remarkably accurate, even for non-normal data. The reasoning is as follows: The central limit theorem says that for large n

$$\sqrt{n}\frac{\bar{Y} - \theta}{\sigma} \stackrel{\cdot}{\sim} \text{normal}(0, 1),$$

where $\stackrel{\cdot}{\sim}$ means "approximately distributed as." Additionally, if n is sufficiently large, then $s^2 \approx \sigma^2$ and so

$$\sqrt{n}\frac{\bar{Y} - \theta}{s} \stackrel{\cdot}{\sim} \text{normal}(0, 1).$$

This should seem familiar: Recall from introductory statistics that for normal data, $\sqrt{n}(\bar{Y} - \theta)/s$ has a t-distribution with $n - 1$ degrees of freedom. For large n, s^2 is very close to σ^2 and the t_{n-1} distribution is very close to a normal$(0, 1)$ distribution.

Even though the posterior distribution based on a normal model may provide good inference for the population mean, the normal model can provide misleading results for other sample quantities. For example, every normal density is symmetric and has a skew of zero, whereas our true population in the above example has a skew of $\text{E}[(Y - \theta)^3]/\sigma^3 = 0.89$. Normal-model inference for samples from this population will underestimate the number of people in the right tail of the distribution, and so will provide poor estimates of the percentage of people with large numbers of children. In general, using the normal model for non-normal data is reasonable if we are only interested in obtaining a posterior distribution for the population mean. For other population quantities the normal model can provide misleading results.

5.7 Discussion and further references

The normal sampling model can be justified in many different ways. For example, Lukacs (1942) shows that a characterizing feature of the normal distribution is that the sample mean and the sample variance are independent (see also Rao (1958)). From a subjective probability perspective, this suggests that if your beliefs about the sample mean are independent from those about the sample variance, then a normal model is appropriate. Also, among all distributions with a given mean θ and variance σ^2, the normal(θ, σ^2) distribution is the most diffuse in terms of a measure known as entropy (see Jaynes, 2003, Chap.7, Chap.11).

From a data analysis perspective, one justification of the normal sampling model is that, as described in Section 5.6, the sample mean will generally be approximately normally distributed due to the central limit theorem. Thus the normal model provides a reasonable sampling model for the sample mean, if not the sample data. Additionally, the normal model is a simple exponential family model with sufficient statistics equivalent to the sample mean and variance. As a result, it will provide consistent estimation of the population mean and variance even if the underlying population is not normal. Additionally, confidence intervals for the population mean based on the normal model will generally be asymptotically correct (these results can be derived from those in White (1982)). However, the normal model may give inaccurate inference for other population quantities.

6

Posterior approximation with the Gibbs sampler

For many multiparameter models the joint posterior distribution is non-standard and difficult to sample from directly. However, it is often the case that it is easy to sample from the full conditional distribution of each parameter. In such cases, posterior approximation can be made with the Gibbs sampler, an iterative algorithm that constructs a dependent sequence of parameter values whose distribution converges to the target joint posterior distribution. In this chapter we outline the Gibbs sampler in the context of the normal model with a semiconjugate prior distribution, and discuss how well the method is able to approximate the posterior distribution.

6.1 A semiconjugate prior distribution

In the previous chapter we modeled our uncertainty about θ as depending on σ^2:

$$p(\theta|\sigma^2) = \text{ dnorm } (\theta, \mu_0, \sigma/\sqrt{\kappa_0}) .$$

This prior distribution relates the prior variance of θ to the sampling variance of our data in such a way that μ_0 can be thought of as κ_0 prior samples from the population. In some situations this makes sense, but in others we may want to specify our uncertainty about θ as being independent of σ^2, so that $p(\theta, \sigma^2) = p(\theta) \times p(\sigma^2)$. One such joint distribution is the following "semiconjugate" prior distribution:

$$\begin{aligned} \theta &\sim \text{normal}(\mu_0, \tau_0^2) \\ 1/\sigma^2 &\sim \text{gamma}(\nu_0/2, \nu_0\sigma_0^2/2) . \end{aligned}$$

If $\{Y_1, \ldots, Y_n|\theta, \sigma^2\} \sim$ i.i.d. normal(θ, σ^2), we showed in Section 5.2 that $\{\theta|\sigma^2, y_1, \ldots, y_n\} \sim$ normal(μ_n, τ_n^2) with

$$\mu_n = \frac{\mu_0/\tau_0^2 + n\bar{y}/\sigma^2}{1/\tau_0^2 + n/\sigma^2} \quad \text{and} \quad \tau_n^2 = \left(\frac{1}{\tau_0^2} + \frac{n}{\sigma^2}\right)^{-1} .$$

P.D. Hoff, *A First Course in Bayesian Statistical Methods*,
Springer Texts in Statistics, DOI 10.1007/978-0-387-92407-6_6,

In the conjugate case where τ_0^2 was proportional to σ^2, we showed that $p(\sigma^2|y_1,\ldots,y_n)$ was an inverse-gamma distribution, and that a Monte Carlo sample of $\{\theta,\sigma^2\}$ from their joint posterior distribution could be obtained by sampling

1. a value $\sigma^{2(s)}$ from $p(\sigma^2|y_1,\ldots,y_n)$, an inverse-gamma distribution, then
2. a value $\theta^{(s)}$ from $p(\theta|\sigma^{2(s)},y_1,\ldots,y_n)$, a normal distribution.

However, in the case where τ_0^2 is not proportional to σ^2, the marginal density of $1/\sigma^2$ is *not* a gamma distribution, or any other standard distribution from which we can easily sample.

6.2 Discrete approximations

Letting $\tilde{\sigma}^2 = 1/\sigma^2$ be the precision, recall that the posterior distribution of $\{\theta,\tilde{\sigma}^2\}$ is equal to the *joint* distribution of $\{\theta,\sigma^2,y_1,\ldots,y_n\}$, divided by $p(y_1,\ldots,y_n)$, which does not depend on the parameters. Therefore the relative posterior probabilities of one set of parameter values $\{\theta_1,\tilde{\sigma}_1^2\}$ to another $\{\theta_2,\tilde{\sigma}_2^2\}$ is directly computable:

$$\begin{aligned}\frac{p(\theta_1,\tilde{\sigma}_1^2|y_1,\ldots,y_n)}{p(\theta_2,\tilde{\sigma}_2^2|y_1,\ldots,y_n)} &= \frac{p(\theta_1,\tilde{\sigma}_1^2,y_1,\ldots,y_n)/p(y_1,\ldots,y_n)}{p(\theta_2,\tilde{\sigma}_2^2,y_1,\ldots,y_n)/p(y_1,\ldots,y_n)}\\ &= \frac{p(\theta_1,\tilde{\sigma}_1^2,y_1,\ldots,y_n)}{p(\theta_2,\tilde{\sigma}_2^2,y_1,\ldots,y_n)}.\end{aligned}$$

The joint distribution is easy to compute as it was built out of standard prior and sampling distributions:

$$\begin{aligned}p(\theta,\tilde{\sigma}^2,y_1,\ldots,y_n) &= p(\theta,\tilde{\sigma}^2)\times p(y_1,\ldots,y_n|\theta,\tilde{\sigma}^2)\\ &= \text{dnorm}(\theta,\mu_0,\tau_0)\times \text{dgamma}(\tilde{\sigma}^2,\nu_0/2,\nu_0\sigma_0^2/2)\times\\ &\quad\prod_{i=1}^{n}\text{dnorm}(y_i,\theta,1/\sqrt{\tilde{\sigma}^2}).\end{aligned}$$

A discrete approximation to the posterior distribution makes use of these facts by constructing a posterior distribution over a grid of parameter values, based on relative posterior probabilities. This is done by evaluating $p(\theta,\tilde{\sigma}^2,y_1,\ldots,y_n)$ on a two-dimensional grid of values of $\{\theta,\tilde{\sigma}^2\}$. Letting $\{\theta_1,\ldots,\theta_G\}$ and $\{\tilde{\sigma}_1^2,\ldots,\tilde{\sigma}_H^2\}$ be sequences of evenly spaced parameter values, the discrete approximation to the posterior distribution assigns a posterior probability to each pair $\{\theta_k,\tilde{\sigma}_l^2\}$ on the grid, given by

$$\begin{aligned} p_D(\theta_k, \tilde{\sigma}_l^2 | y_1, \dots, y_n) &= \frac{p(\theta_k, \tilde{\sigma}_l^2 | y_1, \dots, y_n)}{\sum_{g=1}^G \sum_{h=1}^H p(\theta_g, \tilde{\sigma}_h^2 | y_1, \dots, y_n)} \\ &= \frac{p(\theta_k, \tilde{\sigma}_l^2, y_1, \dots, y_n)/p(y_1, \dots, y_n)}{\sum_{g=1}^G \sum_{h=1}^H p(\theta_g, \tilde{\sigma}_h^2, y_1, \dots, y_n)/p(y_1, \dots, y_n)} \\ &= \frac{p(\theta_k, \tilde{\sigma}_l^2, y_1, \dots, y_n)}{\sum_{g=1}^G \sum_{h=1}^H p(\theta_g, \tilde{\sigma}_h^2, y_1, \dots, y_n)}. \end{aligned}$$

This is a real joint probability distribution for $\theta \in \{\theta_1, \dots, \theta_G\}$ and $\tilde{\sigma}^2 \in \{\tilde{\sigma}_1^2, \dots, \tilde{\sigma}_H^2\}$, in the sense that it sums to 1. In fact, it is the actual posterior distribution of $\{\theta, \tilde{\sigma}^2\}$ if the joint prior distribution for these parameters is discrete on this grid.

Let's try the approximation for the midge data from the previous chapter. Recall that our data were $\{n = 9, \bar{y} = 1.804, s^2 = 0.017\}$. The conjugate prior distribution on θ and σ^2 of Chapter 5 required that the prior variance on θ be σ^2/κ_0, i.e. proportional to the sampling variance. A small value of the sampling variance then has the possibly undesirable effect of reducing the nominal prior uncertainty for θ. In contrast, the semiconjugate prior distribution frees us from this constraint. Recall that we first suggested that the prior mean and standard deviation of θ should be $\mu_0 = 1.9$ and $\tau_0 = .95$, as this would put most of the prior mass on $\theta > 0$, which we know to be true. For σ^2, let's use prior parameters of $\nu_0 = 1$ and $\sigma_0^2 = 0.01$.

The R-code below evaluates $p(\theta, \tilde{\sigma}^2 | y_1, \dots, y_n)$ on a 100×100 grid of evenly spaced parameter values, with $\theta \in \{1.505, 1.510, \dots, 1.995, 2.00\}$ and $\tilde{\sigma}^2 \in \{1.75, 3.5, \dots, 173.25, 175.0\}$. The first panel of Figure 6.1 gives the discrete approximation to the joint distribution of $\{\theta, \tilde{\sigma}^2\}$. Marginal and conditional posterior distributions for θ and $\tilde{\sigma}^2$ can be obtained from the approximation to the joint distribution with simple arithmetic. For example,

$$p_D(\theta_k | y_1, \dots, y_n) = \sum_{h=1}^H p_D(\theta_k, \tilde{\sigma}_h^2 | y_1, \dots, y_n).$$

The resulting discrete approximations to the marginal posterior distributions of θ and $\tilde{\sigma}^2$ are shown in the second and third panels of Figure 6.1.

```
mu0<-1.9  ; t20<-0.95^2 ; s20<-.01 ; nu0<-1
y<-c(1.64,1.70,1.72,1.74,1.82,1.82,1.82,1.90,2.08)

G<-100 ; H<-100

mean.grid<-seq(1.505,2.00,length=G)
prec.grid<-seq(1.75,175,length=H)
post.grid<-matrix(nrow=G,ncol=H)

for(g in 1:G) {
```

```
for(h in 1:H) {

  post.grid[g,h]<-
     dnorm(mean.grid[g], mu0, sqrt(t20)) *
     dgamma(prec.grid[h], nu0/2, s20*nu0/2 ) *
     prod(dnorm(y,mean.grid[g],1/sqrt(prec.grid[h])))

              }
              }

post.grid<-post.grid/sum(post.grid)
```

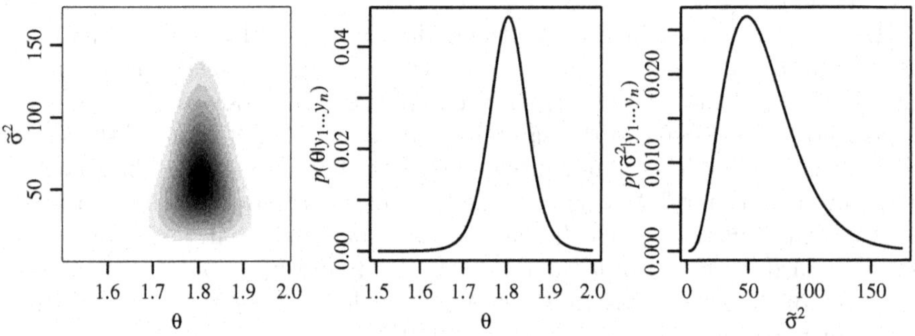

Fig. 6.1. Joint and marginal posterior distributions based on a discrete approximation.

Evaluation of this two-parameter posterior distribution at 100 values of each parameter required a grid of size $100 \times 100 = 100^2$. In general, to construct a similarly fine approximation for a p-dimensional posterior distribution we would need a p-dimensional grid containing 100^p posterior probabilities. This means that discrete approximations will only be feasible for densities having a small number of parameters.

6.3 Sampling from the conditional distributions

Suppose for the moment you knew the value of θ. The conditional distribution of $\tilde{\sigma}^2$ given θ and $\{y_1, \ldots, y_n\}$ is

$$\begin{aligned} p(\tilde{\sigma}^2|\theta, y_1, \ldots, y_n) &\propto p(y_1, \ldots, y_n, \theta, \tilde{\sigma}^2) \\ &= p(y_1, \ldots, y_n|\theta, \tilde{\sigma}^2) p(\theta|\tilde{\sigma}^2) p(\tilde{\sigma}^2) . \end{aligned}$$

If θ and $\tilde{\sigma}^2$ are independent in the prior distribution, then $p(\theta|\tilde{\sigma}^2) = p(\theta)$ and

$$
\begin{aligned}
p(\tilde{\sigma}^2|\theta, y_1, \ldots, y_n) &\propto p(y_1, \ldots, y_n|\theta, \tilde{\sigma}^2)p(\tilde{\sigma}^2) \\
&\propto \left((\tilde{\sigma}^2)^{n/2} \exp\{-\tilde{\sigma}^2 \sum_{i=1}^{n} (y_i - \theta)^2/2\} \right) \times \\
&\quad \left((\tilde{\sigma}^2)^{\nu_0/2-1} \exp\{-\tilde{\sigma}^2 \nu_0 \sigma_0^2/2\} \right) \\
&= (\tilde{\sigma}^2)^{(\nu_0+n)/2-1} \times \exp\{-\tilde{\sigma}^2 \times [\nu_0\sigma_0^2 + \sum (y_i - \theta)^2]/2\}.
\end{aligned}
$$

This is the form of a gamma density, and so evidently $\{\sigma^2|\theta, y_1, \ldots, y_n\} \sim$ inverse-gamma$(\nu_n/2, \nu_n\sigma_n^2(\theta)/2)$, where

$$
\nu_n = \nu_0 + n \ , \quad \sigma_n^2(\theta) = \frac{1}{\nu_n} \left[\nu_0\sigma_0^2 + ns_n^2(\theta) \right],
$$

and $s_n^2(\theta) = \sum(y_i - \theta)^2/n$, the unbiased estimate of σ^2 if θ were known. This means that we can easily sample directly from $p(\sigma^2|\theta, y_1, \ldots, y_n)$, as well as from $p(\theta|\sigma^2, y_1, \ldots, y_n)$ as shown at the beginning of the chapter. However, we do not yet have a way to sample directly from $p(\theta, \sigma^2|y_1, \ldots, y_n)$. Can we use the full conditional distributions to sample from the joint posterior distribution?

Suppose we were given $\sigma^{2(1)}$, a single sample from the marginal posterior distribution $p(\sigma^2|y_1, \ldots, y_n)$. Then we could sample

$$
\theta^{(1)} \sim p(\theta|\sigma^{2(1)}, y_1, \ldots, y_n)
$$

and $\{\theta^{(1)}, \sigma^{2(1)}\}$ would be a sample from the joint distribution of $\{\theta, \sigma^2\}$. Additionally, $\theta^{(1)}$ can be considered a sample from the marginal distribution $p(\theta|y_1, \ldots, y_n)$. From this θ-value, we can generate

$$
\sigma^{2(2)} \sim p(\sigma^2|\theta^{(1)}, y_1, \ldots, y_n).
$$

But since $\theta^{(1)}$ is a sample from the marginal distribution of θ, and $\sigma^{2(2)}$ is a sample from the conditional distribution of σ^2 given $\theta^{(1)}$, then $\{\theta^{(1)}, \sigma^{2(2)}\}$ is also a sample from the joint distribution of $\{\theta, \sigma^2\}$. This in turn means that $\sigma^{2(2)}$ is a sample from the marginal distribution $p(\sigma^2|y_1, \ldots, y_n)$, which then could be used to generate a new sample $\theta^{(2)}$, and so on. It seems that the two *conditional* distributions could be used to generate samples from the *joint* distribution, if only we had a $\sigma^{2(1)}$ from which to start.

6.4 Gibbs sampling

The distributions $p(\theta|\sigma^2, y_1, \ldots, y_n)$ and $p(\sigma^2|\theta, y_1, \ldots, y_n)$ are called the *full conditional distributions* of θ and σ^2 respectively, as they are each a conditional distribution of a parameter given everything else. Let's make the iterative sampling idea described in the previous paragraph more precise. Given a current state of the parameters $\phi^{(s)} = \{\theta^{(s)}, \tilde{\sigma}^{2(s)}\}$, we generate a new state as follows:

1. sample $\theta^{(s+1)} \sim p(\theta|\tilde{\sigma}^{2(s)}, y_1, \ldots, y_n)$;
2. sample $\tilde{\sigma}^{2(s+1)} \sim p(\tilde{\sigma}^2|\theta^{(s+1)}, y_1, \ldots, y_n)$;
3. let $\phi^{(s+1)} = \{\theta^{(s+1)}, \tilde{\sigma}^{2(s+1)}\}$.

This algorithm is called the *Gibbs sampler*, and generates a *dependent* sequence of our parameters $\{\phi^{(1)}, \phi^{(2)}, \ldots, \phi^{(S)}\}$. The R-code to perform this sampling scheme for the normal model with the semiconjugate prior distribution is as follows:

```
### data
mean.y<-mean(y) ; var.y<-var(y) ; n<-length(y)
###

### starting values
S<-1000
PHI<-matrix(nrow=S,ncol=2)
PHI[1,]<-phi<-c( mean.y, 1/var.y)
###

### Gibbs sampling
set.seed(1)
for(s in 2:S) {

# generate a new theta value from its full conditional
mun<- ( mu0/t20 + n*mean.y*phi[2] ) / ( 1/t20 + n*phi[2] )
t2n<- 1/( 1/t20 + n*phi[2] )
phi[1]<-rnorm(1, mun, sqrt(t2n) )

# generate a new 1/sigma^2 value from its full conditional
nun<- nu0+n
s2n<- (nu0*s20 + (n-1)*var.y + n*(mean.y-phi[1])^2 ) /nun
phi[2]<- rgamma(1, nun/2, nun*s2n/2)

PHI[s,]<-phi            }
###
```

In this code, we have used the identity

$$
\begin{aligned}
ns_n^2(\theta) = \sum_{i=1}^n (y_i - \theta)^2 &= \sum_{i=1}^n (y_i - \bar{y} + \bar{y} - \theta)^2 \\
&= \sum_{i=1}^n [(y_i - \bar{y})^2 + 2(y_i - \bar{y})(\bar{y} - \theta) + (\bar{y} - \theta)^2] \\
&= \sum_{i=1}^n (y_i - \bar{y})^2 + 0 + \sum_{i=1}^n (\bar{y} - \theta)^2 \\
&= (n-1)s^2 + n(\bar{y} - \theta)^2.
\end{aligned}
$$

The reason for writing the code this way is because s^2 and $\bar{y}$ do not change with each new θ-value, and computing $(n-1)s^2 + n(\bar{y}-\theta)^2$ is faster than having to recompute $\sum_{i=1}^{n}(y_i-\theta)^2$ at each iteration.

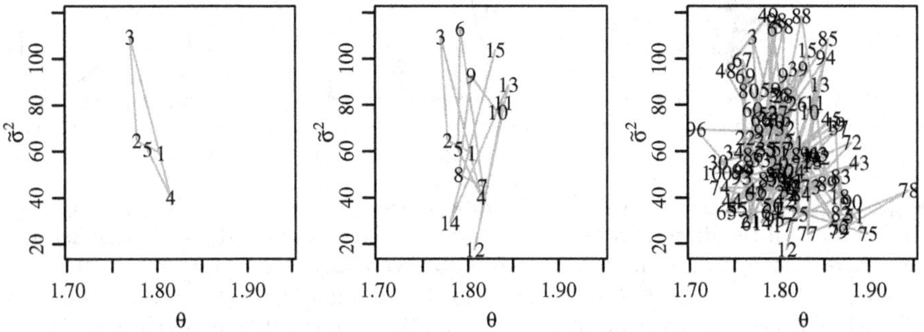

Fig. 6.2. The first 5, 15 and 100 iterations of a Gibbs sampler.

Using the midge data from the previous chapter and the prior distributions described above, a Gibbs sampler consisting of 1,000 iterations was constructed. Figure 6.2 plots the first 5, 15 and 100 simulated values, and the first panel of Figure 6.3 plots the 1,000 values over the contours of the discrete approximation to $p(\theta, \tilde{\sigma}^2|y_1, \ldots, y_n)$. The second and third panels of Figure 6.3 give density estimates of the distributions of the simulated values of θ and $\tilde{\sigma}^2$. Finally, let's find some empirical quantiles of our Gibbs samples:

```
### CI for population mean
> quantile(PHI[,1],c(.025,.5,.975))
    2.5%       50%     97.5%
1.707282 1.804348 1.901129

### CI for population precision
> quantile(PHI[,2],c(.025,.5, .975))
    2.5%       50%     97.5%
 17.48020  53.62511 129.20020


### CI for population standard deviation
> quantile(1/sqrt(PHI[,2]),c(.025,.5, .975))
      2.5%        50%      97.5%
0.08797701 0.13655763 0.23918408
```

The empirical distribution of these Gibbs samples very closely resembles the discrete approximation to their posterior distribution, as can be seen by comparing Figures 6.1 and 6.3. This gives some indication that the Gibbs sampling procedure is a valid method for approximating $p(\theta, \sigma^2|y_1, \ldots, y_n)$.

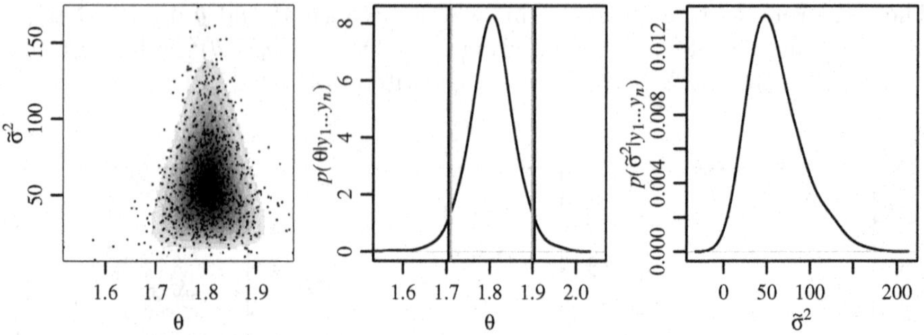

Fig. 6.3. The first panel shows 1,000 samples from the Gibbs sampler, plotted over the contours of the discrete approximation. The second and third panels give kernel density estimates to the distributions of Gibbs samples of θ and $\tilde{\sigma}^2$. Vertical gray bars on the second plot indicate 2.5% and 97.5% quantiles of the Gibbs samples of θ, while nearly identical black vertical bars indicate the 95% confidence interval based on the t-test.

6.5 General properties of the Gibbs sampler

Suppose you have a vector of parameters $\boldsymbol{\phi} = \{\phi_1, \ldots, \phi_p\}$, and your information about $\boldsymbol{\phi}$ is measured with $p(\boldsymbol{\phi}) = p(\phi_1, \ldots, \phi_p)$. For example, in the normal model $\boldsymbol{\phi} = \{\theta, \sigma^2\}$, and the probability measure of interest is $p(\theta, \sigma^2 | y_1, \ldots, y_n)$. Given a starting point $\boldsymbol{\phi}^{(0)} = \{\phi_1^{(0)}, \ldots, \phi_p^{(0)}\}$, the Gibbs sampler generates $\boldsymbol{\phi}^{(s)}$ from $\boldsymbol{\phi}^{(s-1)}$ as follows:

1. sample $\phi_1^{(s)} \sim p(\phi_1 | \phi_2^{(s-1)}, \phi_3^{(s-1)}, \ldots, \phi_p^{(s-1)})$
2. sample $\phi_2^{(s)} \sim p(\phi_2 | \phi_1^{(s)}, \phi_3^{(s-1)}, \ldots, \phi_p^{(s-1)})$

⋮

p. sample $\phi_p^{(s)} \sim p(\phi_p | \phi_1^{(s)}, \phi_2^{(s)}, \ldots, \phi_{p-1}^{(s)})$.

This algorithm generates a *dependent* sequence of vectors:

$$
\begin{aligned}
\boldsymbol{\phi}^{(1)} &= \{\phi_1^{(1)}, \ldots, \phi_p^{(1)}\} \\
\boldsymbol{\phi}^{(2)} &= \{\phi_1^{(2)}, \ldots, \phi_p^{(2)}\} \\
&\vdots \\
\boldsymbol{\phi}^{(S)} &= \{\phi_1^{(S)}, \ldots, \phi_p^{(S)}\} .
\end{aligned}
$$

In this sequence, $\boldsymbol{\phi}^{(s)}$ depends on $\boldsymbol{\phi}^{(0)}, \ldots, \boldsymbol{\phi}^{(s-1)}$ only through $\boldsymbol{\phi}^{(s-1)}$, i.e. $\boldsymbol{\phi}^{(s)}$ is conditionally independent of $\boldsymbol{\phi}^{(0)}, \ldots, \boldsymbol{\phi}^{(s-2)}$ given $\boldsymbol{\phi}^{(s-1)}$. This is called the Markov property, and so the sequence is called a *Markov chain*. Under some conditions that will be met for all of the models discussed in this text,

$$\Pr(\boldsymbol{\phi}^{(s)} \in A) \to \int_A p(\boldsymbol{\phi})\, d\boldsymbol{\phi} \quad \text{as } s \to \infty.$$

In words, the *sampling distribution* of $\boldsymbol{\phi}^{(s)}$ approaches the *target distribution* as $s \to \infty$, no matter what the starting value $\boldsymbol{\phi}^{(0)}$ is (although some starting values will get you to the target sooner than others). More importantly, for most functions g of interest,

$$\frac{1}{S}\sum_{s=1}^{S} g(\boldsymbol{\phi}^{(s)}) \to \mathrm{E}[g(\boldsymbol{\phi})] = \int g(\boldsymbol{\phi}) p(\boldsymbol{\phi})\, d\boldsymbol{\phi} \quad \text{as } S \to \infty. \tag{6.1}$$

This means we can approximate $\mathrm{E}[g(\boldsymbol{\phi})]$ with the sample average of $\{g(\boldsymbol{\phi}^{(1)}), \ldots, g(\boldsymbol{\phi}^{(S)})\}$, just as in Monte Carlo approximation. For this reason, we call such approximations *Markov chain Monte Carlo* (MCMC) approximations, and the procedure an MCMC algorithm. In the context of the semiconjugate normal model, Equation 6.1 implies that the joint distribution of $\{(\theta^{(1)}, \sigma^{2(1)}), \ldots, (\theta^{(1000)}, \sigma^{2(1000)})\}$ is approximately equal to $p(\theta, \sigma^2 | \, y_1, \ldots, y_n)$, and that

$$\mathrm{E}[\theta | y_1, \ldots, y_n] \approx \frac{1}{1000}\sum_{s=1}^{1000} \theta^{(s)} = 1.804, \text{ and}$$
$$\Pr(\theta \in [1.71, 1.90] | y_1, \ldots, y_n) \approx 0.95.$$

We will discuss practical aspects of MCMC in the context of specific models in the next section and in the next several chapters.

Distinguishing parameter estimation from posterior approximation

A Bayesian data analysis using Monte Carlo methods often involves a confusing array of sampling procedures and probability distributions. With this in mind it is helpful to distinguish the part of the data analysis which is statistical from that which is numerical approximation. Recall from Chapter 1 that the necessary ingredients of a Bayesian data analysis are

1. *Model specification:* a collection of probability distributions $\{p(\boldsymbol{y}|\phi), \phi \in \Phi\}$ which should represent the sampling distribution of your data for some value of $\phi \in \Phi$;
2. *Prior specification:* a probability distribution $p(\phi)$, ideally representing someone's prior information about which parameter values are likely to describe the sampling distribution.

Once these items are specified and the data have been gathered, the posterior $p(\phi|\boldsymbol{y})$ is completely determined. It is given by

$$p(\phi|\boldsymbol{y}) = \frac{p(\phi)p(\boldsymbol{y}|\phi)}{p(\boldsymbol{y})} = \frac{p(\phi)p(\boldsymbol{y}|\phi)}{\int p(\phi)p(\boldsymbol{y}|\phi)\, d\phi},$$

and so in a sense there is no more modeling or estimation. All that is left is

3. *Posterior summary*: a description of the posterior distribution $p(\phi|\boldsymbol{y})$, done in terms of particular quantities of interest such as posterior means, medians, modes, predictive probabilities and confidence regions.

For many models, $p(\phi|\boldsymbol{y})$ is complicated, hard to write down, and so on. In these cases, a useful way to "look at" $p(\phi|\boldsymbol{y})$ is by studying Monte Carlo samples from $p(\phi|\boldsymbol{y})$. Thus, Monte Carlo and MCMC sampling algorithms

- are not models,
- they do not generate "more information" than is in $\boldsymbol{y}$ and $p(\phi)$,
- they are simply "ways of looking at" $p(\phi|\boldsymbol{y})$.

For example, if we have Monte Carlo samples $\phi^{(1)}, \ldots, \phi^{(S)}$ that are approximate draws from $p(\phi|y)$, then these samples help describe $p(\phi|\boldsymbol{y})$:

$\frac{1}{S}\sum \phi^{(s)} \approx \int \phi p(\phi|\boldsymbol{y})\, d\phi$
$\frac{1}{S}\sum 1(\phi^{(s)} \leq c) \approx \Pr(\phi \leq c|\boldsymbol{y}) = \int_{-\infty}^{c} p(\phi|\boldsymbol{y})\, d\phi$.

and so on. To keep this distinction in mind, it is useful to reserve the word *estimation* to describe how we use $p(\phi|\boldsymbol{y})$ to make inference about ϕ, and to use the word *approximation* to describe the use of Monte Carlo procedures to approximate integrals.

6.6 Introduction to MCMC diagnostics

The purpose of Monte Carlo or Markov chain Monte Carlo approximation is to obtain a sequence of parameter values $\{\phi^{(1)}, \ldots, \phi^{(S)}\}$ such that

$$\frac{1}{S}\sum_{s=1}^{S} g(\phi^{(s)}) \approx \int g(\phi)p(\phi)\, d\phi,$$

for any functions g of interest. In other words, we want the empirical average of $\{g(\phi^{(1)}), \ldots, g(\phi^{(S)})\}$ to approximate the expected value of $g(\phi)$ under a target probability distribution $p(\phi)$ (in Bayesian inference, the target distribution is usually the posterior distribution). In order for this to be a good approximation for a wide range of functions g, we need the empirical distribution of the simulated sequence $\{\phi^{(1)}, \ldots, \phi^{(S)}\}$ to look like the target distribution $p(\phi)$. Monte Carlo and Markov chain Monte Carlo are two ways of generating such a sequence. Monte Carlo simulation, in which we generate independent samples from the target distribution, is in some sense the "gold standard." Independent MC samples automatically create a sequence that is representative of $p(\phi)$: The probability that $\phi^{(s)} \in A$ for any set A is $\int_A p(\phi)\, d\phi$. This is true for every $s \in \{1, \ldots, S\}$ and conditionally or unconditionally on the other values in the sequence. This is not true for MCMC samples, in which case all we are sure of is that

$$\lim_{s\to\infty} \Pr(\phi^{(s)} \in A) = \int_A p(\phi)\, d\phi.$$

Let's explore the differences between MC and MCMC with a simple example. Our target distribution will be the joint probability distribution of two variables: a discrete variable $\delta \in \{1,2,3\}$ and a continuous variable $\theta \in \mathbb{R}$. The target density for this example will be defined as $\{\Pr(\delta = 1), \Pr(\delta = 2), \Pr(\delta = 3)\} = (.45, .10, .45)$ and $p(\theta|\delta) = \text{dnorm}(\theta, \mu_\delta, \sigma_\delta)$, where $(\mu_1, \mu_2, \mu_3) = (-3, 0, 3)$ and $(\sigma_1^2, \sigma_2^2, \sigma_3^2) = (1/3, 1/3, 1/3)$. This is a mixture of three normal densities, where we might think of δ as being a group membership variable and $(\mu_\delta, \sigma_\delta^2)$ as the population mean and variance for group δ. A plot of the exact marginal density of θ, $p(\theta) = \sum p(\theta|\delta)p(\delta)$, appears in the black lines of Figure 6.4. Notice that there are three modes representing the three different group means.

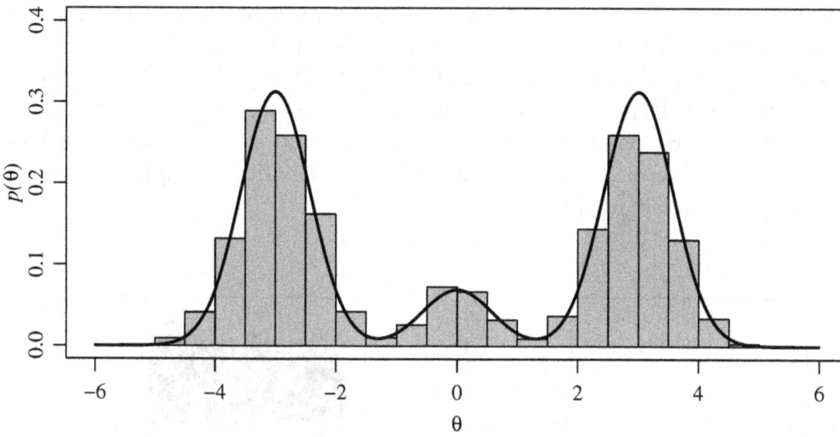

Fig. 6.4. A mixture of normal densities and a Monte Carlo approximation.

It is very easy to obtain independent Monte Carlo samples from the joint distribution of $\phi = (\delta, \theta)$. First, a value of δ is sampled from its marginal distribution, then the value is plugged into $p(\theta|\delta)$, from which a value of θ is sampled. The sampled pair (δ, θ) represents a sample from the joint distribution of $p(\delta, \theta) = p(\delta)p(\theta|\delta)$. The empirical distribution of the θ-samples provides an approximation to the marginal distribution $p(\theta) = \sum p(\theta|\delta)p(\delta)$. A histogram of 1,000 Monte Carlo θ-values generated in this way is shown in Figure 6.4. The empirical distribution of the Monte Carlo samples looks a lot like $p(\theta)$.

It is also straightforward to construct a Gibbs sampler for $\phi = (\delta, \theta)$. A Gibbs sampler would alternately sample values of θ and δ from their full conditional distributions. The full conditional distribution of θ is already provided, and using Bayes' rule we can show that the full conditional distribution of δ is given by

$$\Pr(\delta = d|\theta) = \frac{\Pr(\delta = d) \times \text{dnorm}(\theta, \mu_d, \sigma_d)}{\sum_{d=1}^{3} \Pr(\delta = d) \times \text{dnorm}(\theta, \mu_d, \sigma_d)}, \text{ for } d \in \{1, 2, 3\}.$$

The first panel of Figure 6.5 shows a histogram of 1,000 MCMC values of θ generated with the Gibbs sampler. Notice that the empirical distribution of the MCMC samples gives a poor approximation to $p(\theta)$. Values of θ near -3 are underrepresented, whereas values near zero and +3 are overrepresented. What went wrong? A plot of the θ-values versus iteration number in the second panel of the figure tells the story. The θ-values get "stuck" in certain regions, and rarely move among the three regions represented by the three values of μ. The technical term for this "stickiness" is *autocorrelation*, or correlation between consecutive values of the chain. In this Gibbs sampler, if we have a value of θ near 0 for example, then the next value of δ is likely to be 2. If δ is 2, then the next value of θ is likely to be near 0, resulting in a high degree of positive correlation between consecutive θ-values in the chain.

Isn't the Gibbs sampler guaranteed to eventually provide a good approximation to $p(\theta)$? It is, but "eventually" can be a very long time in some situations. The first panel of Figure 6.6 indicates that our approximation has greatly improved after using 10,000 iterations of the Gibbs sampler, although it is still somewhat inadequate.

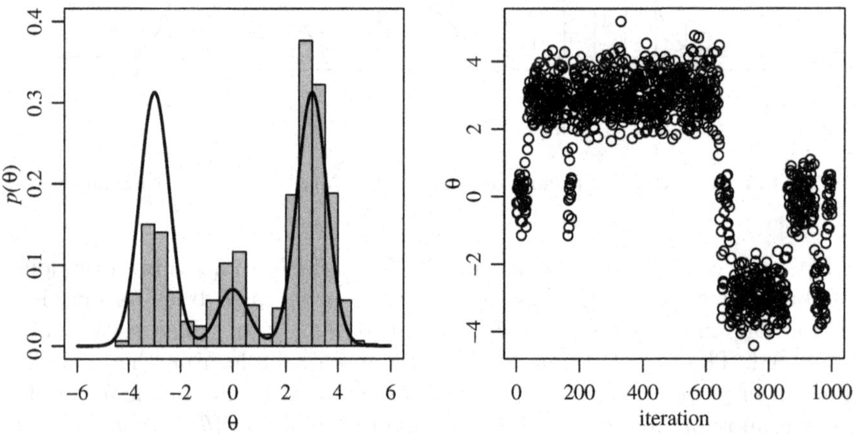

Fig. 6.5. Histogram and traceplot of 1,000 Gibbs samples.

In the case of a generic parameter ϕ and target distribution $p(\phi)$, it is helpful to think of the sequence $\{\phi^{(1)}, \ldots, \phi^{(S)}\}$ as the trajectory of a particle ϕ moving around the parameter space. In terms of MCMC integral approxi-

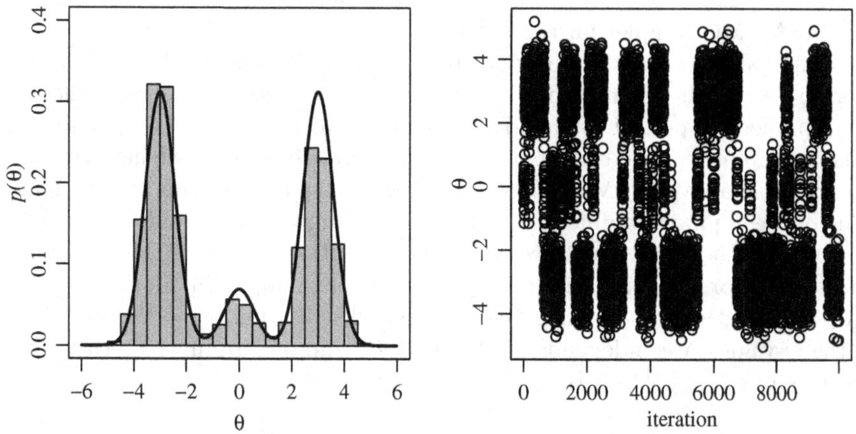

Fig. 6.6. Histogram and traceplot of 10,000 Gibbs samples.

mation, the critical thing is that the amount of time the particle spends in a given set A is proportional to the target probability $\int_A p(\phi)\, d\phi$.

Now suppose A_1, A_2 and A_3 are three disjoint subsets of the parameter space, with $\Pr(A_2) < \Pr(A_1) \approx \Pr(A_3)$ (these could be, for example, the regions near the three modes of the normal mixture distribution above). In terms of the integral approximation, this means that we want the particle to spend little time in A_2, and about the same amount of time in A_1 as in A_3. Since in general we do not know $p(\phi)$ (otherwise we would not be trying to approximate it), it is possible that we would accidentally start our Markov chain in A_2. In this case, it is critical that the number of iterations S is large enough so that the particle has a chance to

1. move out of A_2 and into higher probability regions, and
2. move between A_1 and A_3, and any other sets of high probability.

The technical term for attaining item 1 is to say that the chain has *achieved stationarity* or has *converged*. If your Markov chain starts off in a region of the parameter space that has high probability, then convergence generally is not a big issue. If you do not know if you are starting off in a good region, assessing convergence is fraught with epistemological problems. In general, you cannot know for sure if your chain has converged. But sometimes you can know if your chain has not converged, so we at least check for this latter possibility. One thing to check for is *stationarity*, or that samples taken in one part of the chain have a similar distribution to samples taken in other parts. For the normal model with semiconjugate prior distributions from the previous section, stationarity is achieved quite quickly and is not a big issue. However, for some highly parameterized models that we will see later on, the autocorrelation in the chain is high, good starting values can be hard to find and it can take a long time to get to stationarity. In these cases we need to run the MCMC sampler for a very long time.

Item 2 above relates to how quickly the particle moves around the parameter space, which is sometimes called the speed of *mixing*. An independent MC sampler has perfect mixing: It has zero autocorrelation and can jump between different regions of the parameter space in one step. As we have seen in the example above, an MCMC sampler might have poor mixing, take a long time between jumps to different parts of the parameter space and have a high degree of autocorrelation. How does the correlation of the MCMC samples affect posterior approximation? Suppose we want to approximate the integral $\mathrm{E}[\phi] = \int \phi p(\phi)\, d\phi = \phi_0$ using the empirical distribution of $\{\phi^{(1)}, \ldots, \phi^{(S)}\}$. If the ϕ-values are independent Monte Carlo samples from $p(\phi)$, then the variance of $\bar{\phi} = \sum \phi^{(s)}/S$ is

$$\mathrm{Var}_{\mathrm{MC}}[\bar{\phi}] = \mathrm{E}[(\bar{\phi} - \phi_0)^2] = \frac{\mathrm{Var}[\phi]}{S},$$

where $\mathrm{Var}[\phi] = \int \phi^2 p(\phi)\, d\phi - \phi_0^2$. Recall from Chapter 4 that the square root of $\mathrm{Var}_{\mathrm{MC}}[\bar{\phi}]$ is the Monte Carlo standard error, and is a measure of how well we expect $\bar{\phi}$ to approximate the integral $\int \phi p(\phi)\, d\phi$. If we were to rerun the MC approximation procedure many times, perhaps with different starting values or random number generators, we expect that ϕ_0, the true value of the integral, would be contained within the interval $\bar{\phi} \pm 2\sqrt{\mathrm{Var}_{\mathrm{MC}}[\bar{\phi}]}$ for roughly 95% of the MC approximations. The width of this interval is $4 \times \sqrt{\mathrm{Var}_{\mathrm{MC}}[\bar{\phi}]}$, and we can make this as small as we want by generating more MC samples.

What if we use an MCMC algorithm such as the Gibbs sampler? As can be seen in Figures 6.5 and 6.6, consecutive MCMC samples $\phi^{(s)}$ and $\phi^{(s+1)}$ can be positively correlated. Assuming stationarity has been achieved, the expected squared difference from the MCMC integral approximation $\bar{\phi}$ to the target $\phi_0 = \int \phi p(\phi)\, d\phi$ is the MCMC variance, and is given by

$$\begin{aligned}
\mathrm{Var}_{\mathrm{MCMC}}[\bar{\phi}] &= \mathrm{E}[(\bar{\phi} - \phi_0)^2] \\
&= \mathrm{E}[\{\frac{1}{S}\sum(\phi^{(s)} - \phi_0)\}^2] \\
&= \frac{1}{S^2}\mathrm{E}[\sum_{s=1}^{S}(\phi^{(s)} - \phi_0)^2 + \sum_{s \neq t}(\phi^{(s)} - \phi_0)(\phi^{(t)} - \phi_0)] \\
&= \frac{1}{S^2}\sum_{s=1}^{S}\mathrm{E}[(\phi^{(s)} - \phi_0)^2] + \frac{1}{S^2}\sum_{s \neq t}\mathrm{E}[(\phi^{(s)} - \phi_0)(\phi^{(t)} - \phi_0)] \\
&= \mathrm{Var}_{\mathrm{MC}}[\bar{\phi}] + \frac{1}{S^2}\sum_{s \neq t}\mathrm{E}[(\phi^{(s)} - \phi_0)(\phi^{(t)} - \phi_0)].
\end{aligned}$$

So the MCMC variance is equal to the MC variance plus a term that depends on the correlation of samples within the Markov chain. This term is generally positive and so the MCMC variance is higher than the MC variance, meaning that we expect the MCMC approximation to be further away from ϕ_0 than the MC approximation is. The higher the autocorrelation in the chain, the larger the MCMC variance and the worse the approximation is. To assess how much correlation there is in the chain we often compute the *sample autocorrelation function*. For a generic sequence of numbers $\{\phi_1, \ldots, \phi_S\}$, the lag-$t$ autocorrelation function estimates the correlation between elements of the sequence that are t steps apart:

$$\text{acf}_t(\phi) = \frac{\frac{1}{S-t}\sum_{s=1}^{S-t}(\phi_s - \bar{\phi})(\phi_{s+t} - \bar{\phi})}{\frac{1}{S-1}\sum_{s=1}^{S}(\phi_s - \bar{\phi})^2},$$

which is computed by the R-function `acf`. For the sequence of 10,000 θ-values plotted in Figure 6.6, the lag-10 autocorrelation is 0.93, and the lag-50 autocorrelation is 0.812. A Markov chain with such a high autocorrelation moves around the parameter space slowly, taking a long time to achieve the correct balance among the different regions of the parameter space. The higher the autocorrelation, the more MCMC samples we need to attain a given level of precision for our approximation. One way to measure this is to calculate the *effective sample size* for an MCMC sequence, using the R-command `effectiveSize` in the "coda" package. The effective sample size function estimates the value S_{eff} such that

$$\text{Var}_{\text{MCMC}}[\bar{\phi}] = \frac{\text{Var}[\phi]}{S_{\text{eff}}},$$

so that S_{eff} can be interpreted as the number of independent Monte Carlo samples necessary to give the same precision as the MCMC samples. For the normal mixture density example above, the effective sample size of the 10,000 Gibbs samples of θ is 18.42, indicating that the precision of the MCMC approximation to $\text{E}[\theta]$ is as good as the precision that would have been obtained by only about 18 independent samples of θ.

There is a large literature on the practical implementation and assessment of Gibbs sampling and MCMC approximation. Much insight can be gained by hands-on experience supplemented by reading books and articles. A good article to start with is "Practical Markov chain Monte Carlo" (Geyer, 1992), which includes a discussion by many researchers and a large variety of viewpoints on and techniques for MCMC approximation.

MCMC diagnostics for the semiconjugate normal analysis

We now assess the Markov chain of θ and σ^2 values generated by the Gibbs sampler in Section 6.4. Figure 6.7 plots the values of these two parameters in sequential order, and seems to indicate immediate convergence and a low degree of autocorrelation. The lag-1 autocorrelation for the sequence

$\{\theta^{(1)}, \ldots, \theta^{(1000)}\}$ is 0.031, which is essentially zero for approximation purposes. The effective sample size for this sequence is computed in R to be 1,000. The lag-1 autocorrelation for the σ^2-values is 0.147, with an effective sample size of 742. While not quite as good as an independently sampled sequence of parameter values, the Gibbs sampler for this model and prior distribution performs quite well.

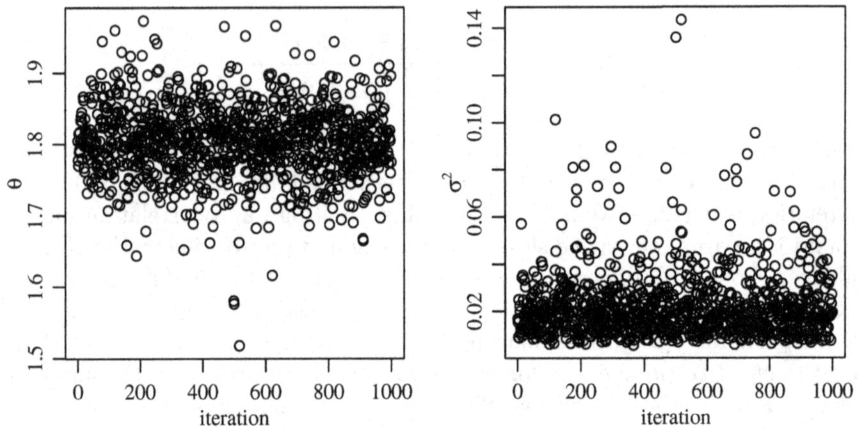

Fig. 6.7. Traceplots for θ and σ^2.

6.7 Discussion and further references

The term "Gibbs sampling" was coined by Geman and Geman (1984) in their paper on image analysis, but the algorithm appears earlier in the context of spatial statistics, for example, Besag (1974) or Ripley (1979). However, the general utility of the Gibbs sampler for Bayesian data analysis was not fully realized until the late 1980s (Gelfand and Smith, 1990). See Robert and Casella (2008) for a historical review.

Assessing the convergence of the Gibbs sampler and the accuracy of the MCMC approximation is difficult. Several authors have come up with convergence diagnostics (Gelman and Rubin, 1992; Geweke, 1992; Raftery and Lewis, 1992), although these can only highlight problems and not guarantee a good approximation (Geyer, 1992).

7

The multivariate normal model

Up until now all of our statistical models have been *univariate* models, that is, models for a single measurement on each member of a sample of individuals or each run of a repeated experiment. However, datasets are frequently *multivariate*, having multiple measurements for each individual or experiment. This chapter covers what is perhaps the most useful model for multivariate data, the multivariate normal model, which allows us to jointly estimate population means, variances and correlations of a collection of variables. After first calculating posterior distributions under semiconjugate prior distributions, we show how the multivariate normal model can be used to impute data that are missing at random.

7.1 The multivariate normal density

Example: Reading comprehension

A sample of twenty-two children are given reading comprehension tests before and after receiving a particular instructional method. Each student i will then have two scores, $Y_{i,1}$ and $Y_{i,2}$ denoting the pre- and post-instructional scores respectively. We denote each student's pair of scores as a 2×1 vector $\boldsymbol{Y}_i$, so that

$$\boldsymbol{Y}_i = \begin{pmatrix} Y_{i,1} \\ Y_{i,2} \end{pmatrix} = \begin{pmatrix} \text{score on first test} \\ \text{score on second test} \end{pmatrix}.$$

Things we might be interested in include the population mean $\boldsymbol{\theta}$,

$$\mathrm{E}[\boldsymbol{Y}] = \begin{pmatrix} \mathrm{E}[Y_{i,1}] \\ \mathrm{E}[Y_{i,2}] \end{pmatrix} = \begin{pmatrix} \theta_1 \\ \theta_2 \end{pmatrix}$$

and the population covariance matrix Σ,

$$\Sigma = \mathrm{Cov}[\boldsymbol{Y}] = \begin{pmatrix} \mathrm{E}[Y_1^2] - \mathrm{E}[Y_1]^2 & \mathrm{E}[Y_1Y_2] - \mathrm{E}[Y_1]\mathrm{E}[Y_2] \\ \mathrm{E}[Y_1Y_2] - \mathrm{E}[Y_1]\mathrm{E}[Y_2] & \mathrm{E}[Y_2^2] - \mathrm{E}[Y_2]^2 \end{pmatrix} = \begin{pmatrix} \sigma_1^2 & \sigma_{1,2} \\ \sigma_{1,2} & \sigma_2^2 \end{pmatrix},$$

P.D. Hoff, *A First Course in Bayesian Statistical Methods*,
Springer Texts in Statistics, DOI 10.1007/978-0-387-92407-6_7,

where the expectations above represent the unknown population averages. Having information about $\boldsymbol{\theta}$ and Σ may help us in assessing the effectiveness of the teaching method, possibly evaluated with $\theta_2 - \theta_1$, or the consistency of the reading comprehension test, which could be evaluated with the correlation coefficient $\rho_{1,2} = \sigma_{1,2}/\sqrt{\sigma_1^2\sigma_2^2}$.

The multivariate normal density

Notice that $\boldsymbol{\theta}$ and Σ are both functions of population *moments*, or population averages of powers of Y_1 and Y_2. In particular, $\boldsymbol{\theta}$ and Σ are functions of first- and second-order moments:

$$\begin{array}{ll} \text{first-order moments:} & \mathrm{E}[Y_1], \mathrm{E}[Y_2] \\ \text{second-order moments:} & \mathrm{E}[Y_1^2], \mathrm{E}[Y_1Y_2], \mathrm{E}[Y_2^2] \end{array}$$

Recall from Chapter 5 that a univariate normal model describes a population in terms of its mean and variance (θ, σ^2), or equivalently its first two moments $(\mathrm{E}[Y] = \theta, \mathrm{E}[Y^2] = \sigma^2 + \theta^2)$. The analogous model for describing first- and second-order moments of multivariate data is the *multivariate normal* model. We say a p-dimensional data vector $\boldsymbol{Y}$ has a multivariate normal distribution if its sampling density is given by

$$p(\boldsymbol{y}|\boldsymbol{\theta}, \Sigma) = (2\pi)^{-p/2}|\Sigma|^{-1/2} \exp\{-(\boldsymbol{y} - \boldsymbol{\theta})^T \Sigma^{-1}(\boldsymbol{y} - \boldsymbol{\theta})/2\}$$

where

$$\boldsymbol{y} = \begin{pmatrix} y_1 \\ y_2 \\ \vdots \\ y_p \end{pmatrix} \quad \boldsymbol{\theta} = \begin{pmatrix} \theta_1 \\ \theta_2 \\ \vdots \\ \theta_p \end{pmatrix} \quad \Sigma = \begin{pmatrix} \sigma_1^2 & \sigma_{1,2} & \cdots & \sigma_{1,p} \\ \sigma_{1,2} & \sigma_2^2 & \cdots & \sigma_{2,p} \\ \vdots & \vdots & & \vdots \\ \sigma_{1,p} & \cdots & \cdots & \sigma_p^2 \end{pmatrix}.$$

Calculating this density requires a few operations involving matrix algebra. For a matrix $\mathbf{A}$, the value of $|\mathbf{A}|$ is called the *determinant* of $\mathbf{A}$, and measures how "big" $\mathbf{A}$ is. The *inverse* of $\mathbf{A}$ is the matrix $\mathbf{A}^{-1}$ such that $\mathbf{A}\mathbf{A}^{-1}$ is equal to the identity matrix $\mathbf{I}_p$, the $p \times p$ matrix that has ones for its diagonal entries but is otherwise zero. For a $p \times 1$ vector $\boldsymbol{b}$, $\boldsymbol{b}^T$ is its *transpose*, and is simply the $1 \times p$ vector of the same values. Finally, the vector-matrix product $\boldsymbol{b}^T\mathbf{A}$ is equal to the $1 \times p$ vector $(\sum_{j=1}^p b_j a_{j,1}, \ldots, \sum_{j=1}^p b_j a_{j,p})$, and the value of $\boldsymbol{b}^T\mathbf{A}\boldsymbol{b}$ is the single number $\sum_{j=1}^p \sum_{k=1}^p b_j b_k a_{j,k}$. Fortunately, R can compute all of these quantities for us, as we shall see in the forthcoming example code.

Figure 7.1 gives contour plots and 30 samples from each of three different two-dimensional multivariate normal densities. In each one $\boldsymbol{\theta} = (50, 50)^T$, $\sigma_1^2 = 64$, $\sigma_2^2 = 144$, but the value of $\sigma_{1,2}$ varies from plot to plot, with $\sigma_{1,2} = -48$ for the left density, 0 for the middle and +48 for the density on the right (giving correlations of -.5, 0 and +.5 respectively). An interesting feature of the multivariate normal distribution is that the marginal distribution of each variable Y_j is a univariate normal distribution, with mean θ_j and variance σ_j^2.

This means that the marginal distributions for Y_1 from the three populations in Figure 7.1 are identical (the same holds for Y_2). The only thing that differs across the three populations is the relationship between Y_1 and Y_2, which is controlled by the covariance parameter $\sigma_{1,2}$.

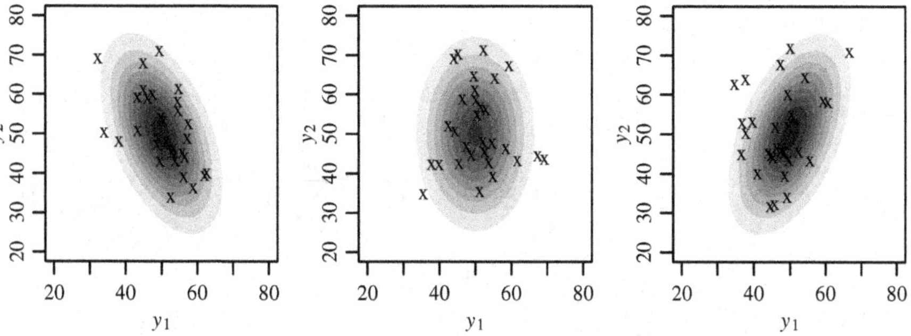

Fig. 7.1. Multivariate normal samples and densities.

7.2 A semiconjugate prior distribution for the mean

Recall from Chapters 5 and 6 that if $Y_1, \dots, Y_n$ are independent samples from a univariate normal population, then a convenient conjugate prior distribution for the population mean is also univariate normal. Similarly, a convenient prior distribution for the multivariate mean $\boldsymbol{\theta}$ is a multivariate normal distribution, which we will parameterize as

$$p(\boldsymbol{\theta}) = \text{multivariate normal}(\boldsymbol{\mu}_0, \Lambda_0),$$

where $\boldsymbol{\mu}_0$ and Λ_0 are the prior mean and variance of $\boldsymbol{\theta}$, respectively. What is the full conditional distribution of $\boldsymbol{\theta}$, given $\boldsymbol{y}_1, \dots, \boldsymbol{y}_n$ and Σ? In the univariate case, having normal prior and sampling distributions resulted in a normal full conditional distribution for the population mean. Let's see if this result holds for the multivariate case. We begin by examining the prior distribution as a function of $\boldsymbol{\theta}$:

$$\begin{aligned}
p(\boldsymbol{\theta}) &= (2\pi)^{-p/2}|\Lambda_0|^{-1/2}\exp\{-\frac{1}{2}(\boldsymbol{\theta}-\boldsymbol{\mu}_0)^T\Lambda_0^{-1}(\boldsymbol{\theta}-\boldsymbol{\mu}_0)\} \\
&= (2\pi)^{-p/2}|\Lambda_0|^{-1/2}\exp\{-\frac{1}{2}\boldsymbol{\theta}^T\Lambda_0^{-1}\boldsymbol{\theta}+\boldsymbol{\theta}^T\Lambda_0^{-1}\boldsymbol{\mu}_0-\frac{1}{2}\boldsymbol{\mu}_0^T\Lambda_0^{-1}\boldsymbol{\mu}_0\} \\
&\propto \exp\{-\frac{1}{2}\boldsymbol{\theta}^T\Lambda_0^{-1}\boldsymbol{\theta}+\boldsymbol{\theta}^T\Lambda_0^{-1}\boldsymbol{\mu}_0\} \\
&= \exp\{-\frac{1}{2}\boldsymbol{\theta}^T\mathbf{A}_0\boldsymbol{\theta}+\boldsymbol{\theta}^T\boldsymbol{b}_0\}, \qquad (7.1)
\end{aligned}$$

where $\mathbf{A}_0 = \Lambda_0^{-1}$ and $\boldsymbol{b}_0 = \Lambda_0^{-1}\boldsymbol{\mu}_0$. Conversely, Equation 7.1 says that if a random vector $\boldsymbol{\theta}$ has a density on $\mathbb{R}^p$ that is proportional to $\exp\{-\boldsymbol{\theta}^T\mathbf{A}\boldsymbol{\theta}/2 + \boldsymbol{\theta}^T\boldsymbol{b}\}$ for some matrix $\mathbf{A}$ and vector $\boldsymbol{b}$, then $\boldsymbol{\theta}$ must have a multivariate normal distribution with covariance $\mathbf{A}^{-1}$ and mean $\mathbf{A}^{-1}\boldsymbol{b}$.

If our sampling model is that $\{\boldsymbol{Y}_1, \ldots, \boldsymbol{Y}_n|\boldsymbol{\theta}, \Sigma\}$ are i.i.d. multivariate normal$(\boldsymbol{\theta}, \Sigma)$, then similar calculations show that the joint sampling density of the observed vectors $\boldsymbol{y}_1, \ldots, \boldsymbol{y}_n$ is

$$
\begin{aligned}
p(\boldsymbol{y}_1, \ldots, \boldsymbol{y}_n|\boldsymbol{\theta}, \Sigma) &= \prod_{i=1}^{n} (2\pi)^{-p/2}|\Sigma|^{-1/2} \exp\{-(\boldsymbol{y}_i - \boldsymbol{\theta})^T \Sigma^{-1}(\boldsymbol{y}_i - \boldsymbol{\theta})/2\} \\
&= (2\pi)^{-np/2}|\Sigma|^{-n/2} \exp\{-\frac{1}{2}\sum_{i=1}^{n}(\boldsymbol{y}_i - \boldsymbol{\theta})^T \Sigma^{-1}(\boldsymbol{y}_i - \boldsymbol{\theta})\} \\
&\propto \exp\{-\frac{1}{2}\boldsymbol{\theta}^T\mathbf{A}_1\boldsymbol{\theta} + \boldsymbol{\theta}^T\boldsymbol{b}_1\}, \qquad (7.2)
\end{aligned}
$$

where $\mathbf{A}_1 = n\Sigma^{-1}$, $\boldsymbol{b}_1 = n\Sigma^{-1}\bar{\boldsymbol{y}}$ and $\bar{\boldsymbol{y}}$ is the vector of variable-specific averages $\bar{\boldsymbol{y}} = (\frac{1}{n}\sum_{i=1}^{n} y_{i,1}, \ldots, \frac{1}{n}\sum_{i=1}^{n} y_{i,p})^T$. Combining Equations 7.1 and 7.2 gives

$$
\begin{aligned}
p(\boldsymbol{\theta}|\boldsymbol{y}_1, \ldots, \boldsymbol{y}_n, \Sigma) &\propto \exp\{-\frac{1}{2}\boldsymbol{\theta}^T\mathbf{A}_0\boldsymbol{\theta} + \boldsymbol{\theta}^T\boldsymbol{b}_0\} \times \exp\{-\frac{1}{2}\boldsymbol{\theta}^T\mathbf{A}_1\boldsymbol{\theta} + \boldsymbol{\theta}^T\boldsymbol{b}_1\} \\
&= \exp\{-\frac{1}{2}\boldsymbol{\theta}^T\mathbf{A}_n\boldsymbol{\theta} + \boldsymbol{\theta}^T\boldsymbol{b}_n\}, \text{ where} \qquad (7.3) \\
\mathbf{A}_n &= \mathbf{A}_0 + \mathbf{A}_1 = \Lambda_0^{-1} + n\Sigma^{-1} \text{ and} \\
\boldsymbol{b}_n &= \boldsymbol{b}_0 + \boldsymbol{b}_1 = \Lambda_0^{-1}\boldsymbol{\mu}_0 + n\Sigma^{-1}\bar{\boldsymbol{y}}.
\end{aligned}
$$

From the comments in the previous paragraph, Equation 7.3 implies that the conditional distribution of $\boldsymbol{\theta}$ therefore must be a multivariate normal distribution with covariance $\mathbf{A}_n^{-1}$ and mean $\mathbf{A}_n^{-1}\boldsymbol{b}_n$, so

$$
\begin{aligned}
\text{Cov}[\boldsymbol{\theta}|\boldsymbol{y}_1, \ldots, \boldsymbol{y}_n, \Sigma] &= \Lambda_n = (\Lambda_0^{-1} + n\Sigma^{-1})^{-1} \qquad (7.4) \\
\text{E}[\boldsymbol{\theta}|\boldsymbol{y}_1, \ldots, \boldsymbol{y}_n, \Sigma] &= \boldsymbol{\mu}_n = (\Lambda_0^{-1} + n\Sigma^{-1})^{-1}(\Lambda_0^{-1}\boldsymbol{\mu}_0 + n\Sigma^{-1}\bar{\boldsymbol{y}}) \qquad (7.5) \\
p(\boldsymbol{\theta}|\boldsymbol{y}_1, \ldots, \boldsymbol{y}_n, \Sigma) &= \text{multivariate normal}(\boldsymbol{\mu}_n, \Lambda_n). \qquad (7.6)
\end{aligned}
$$

It looks a bit complicated, but can be made more understandable by analogy with the univariate normal case: Equation 7.4 says that posterior precision, or inverse variance, is the sum of the prior precision and the data precision, just as in the univariate normal case. Similarly, Equation 7.5 says that the posterior expectation is a weighted average of the prior expectation and the sample mean. Notice that, since the sample mean is consistent for the population mean, the posterior mean also will be consistent for the population mean even if the true distribution of the data is not multivariate normal.

7.3 The inverse-Wishart distribution

Just as a variance σ^2 must be positive, a variance-covariance matrix Σ must be *positive definite*, meaning that

$$\boldsymbol{x}'\Sigma\boldsymbol{x} > 0 \text{ for all vectors } \boldsymbol{x}.$$

Positive definiteness guarantees that $\sigma_j^2 > 0$ for all j and that all correlations are between -1 and 1. Another requirement of our covariance matrix is that it is symmetric, which means that $\sigma_{j,k} = \sigma_{k,j}$. Any valid prior distribution for Σ must put all of its probability mass on this complicated set of symmetric, positive definite matrices. How can we formulate such a prior distribution?

Empirical covariance matrices

The *sum of squares* matrix of a collection of multivariate vectors $\boldsymbol{z}_1, \ldots, \boldsymbol{z}_n$ is given by

$$\sum_{i=1}^{n} \boldsymbol{z}_i \boldsymbol{z}_i^T = \mathbf{Z}^T\mathbf{Z},$$

where $\mathbf{Z}$ is the $n \times p$ matrix whose ith row is $\boldsymbol{z}_i^T$. Recall from matrix algebra that since $\boldsymbol{z}_i$ can be thought of as a $p \times 1$ matrix, $\boldsymbol{z}_i\boldsymbol{z}_i^T$ is the following $p \times p$ matrix:

$$\boldsymbol{z}_i\boldsymbol{z}_i^T = \begin{pmatrix} z_{i,1}^2 & z_{i,1}z_{i,2} & \cdots & z_{i,1}z_{i,p} \\ z_{i,2}z_{i,1} & z_{i,2}^2 & \cdots & z_{i,2}z_{i,p} \\ \vdots & & & \vdots \\ z_{i,p}z_{i,1} & z_{i,p}z_{i,2} & \cdots & z_{i,p}^2 \end{pmatrix}.$$

If the $\boldsymbol{z}_i$'s are samples from a population with zero mean, we can think of the matrix $\boldsymbol{z}_i\boldsymbol{z}_i^T/n$ as the contribution of vector $\boldsymbol{z}_i$ to the estimate of the covariance matrix of all of the observations. In this mean-zero case, if we divide $\mathbf{Z}^T\mathbf{Z}$ by n, we get a sample covariance matrix, an unbiased estimator of the population covariance matrix:

$$\tfrac{1}{n}[\mathbf{Z}^T\mathbf{Z}]_{j,j} = \tfrac{1}{n}\textstyle\sum_{i=1}^n z_{i,j}^2 = s_{j,j} = s_j^2$$
$$\tfrac{1}{n}[\mathbf{Z}^T\mathbf{Z}]_{j,k} = \tfrac{1}{n}\textstyle\sum_{i=1}^n z_{i,j}z_{i,k} = s_{j,k}\ .$$

If $n > p$ and the $\boldsymbol{z}_i$'s are linearly independent, then $\mathbf{Z}^T\mathbf{Z}$ will be positive definite and symmetric. This suggests the following construction of a "random" covariance matrix: For a given positive integer ν_0 and a $p \times p$ covariance matrix Φ_0,

1. sample $\boldsymbol{z}_1, \ldots, \boldsymbol{z}_{\nu_0} \sim$ i.i.d. multivariate normal$(\mathbf{0}, \Phi_0)$;
2. calculate $\mathbf{Z}^T\mathbf{Z} = \sum_{i=1}^{\nu_0} \boldsymbol{z}_i\boldsymbol{z}_i^T$.

We can repeat this procedure over and over again, generating matrices $\mathbf{Z}_1^T\mathbf{Z}_1, \ldots, \mathbf{Z}_S^T\mathbf{Z}_S$. The population distribution of these sum of squares matrices is called a *Wishart distribution* with parameters (ν_0, Φ_0), which has the following properties:

- If $\nu_0 > p$, then $\mathbf{Z}^T\mathbf{Z}$ is positive definite with probability 1.
- $\mathbf{Z}^T\mathbf{Z}$ is symmetric with probability 1.
- $\mathrm{E}[\mathbf{Z}^T\mathbf{Z}] = \nu_0\Phi_0$.

The Wishart distribution is a multivariate analogue of the gamma distribution (recall that if z is a mean-zero univariate normal random variable, then z^2 is a gamma random variable). In the univariate normal model, our prior distribution for the *precision* $1/\sigma^2$ is a gamma distribution, and our full conditional distribution for the *variance* is an inverse-gamma distribution. Similarly, it turns out that the Wishart distribution is a semi-conjugate prior distribution for the *precision matrix* Σ^{-1}, and so the inverse-Wishart distribution is our semi-conjugate prior distribution for the *covariance matrix* Σ. With a slight reparameterization, to sample a covariance matrix Σ from an inverse-Wishart distribution we perform the following steps:

1. sample $\boldsymbol{z}_1, \ldots, \boldsymbol{z}_{\nu_0} \sim$ i.i.d. multivariate normal$(\mathbf{0}, \mathbf{S}_0^{-1})$;
2. calculate $\mathbf{Z}^T\mathbf{Z} = \sum_{i=1}^{\nu_0} \boldsymbol{z}_i\boldsymbol{z}_i^T$;
3. set $\Sigma = (\mathbf{Z}^T\mathbf{Z})^{-1}$.

Under this simulation scheme, the precision matrix Σ^{-1} has a Wishart$(\nu_0, \mathbf{S}_0^{-1})$ distribution and the covariance matrix Σ has an inverse-Wishart$(\nu_0, \mathbf{S}_0^{-1})$ distribution. The expectations of Σ^{-1} and Σ are

$$\mathrm{E}[\Sigma^{-1}] = \nu_0\mathbf{S}_0^{-1}$$
$$\mathrm{E}[\Sigma] = \frac{1}{\nu_0 - p - 1}(\mathbf{S}_0^{-1})^{-1} = \frac{1}{\nu_0 - p - 1}\mathbf{S}_0.$$

If we are confident that the true covariance matrix is near some covariance matrix Σ_0, then we might choose ν_0 to be large and set $\mathbf{S}_0 = (\nu_0 - p - 1)\Sigma_0$, making the distribution of Σ concentrated around Σ_0. On the other hand, choosing $\nu_0 = p + 2$ and $\mathbf{S}_0 = \Sigma_0$ makes Σ only loosely centered around Σ_0.

Full conditional distribution of the covariance matrix

The inverse-Wishart$(\nu_0, \mathbf{S}_0^{-1})$ density is given by

$$p(\Sigma) = \left[2^{\nu_0 p/2}\pi^{\binom{p}{2}/2}|\mathbf{S}_0|^{-\nu_0/2}\prod_{j=1}^{p}\Gamma([\nu_0 + 1 - j]/2)\right]^{-1} \times$$
$$|\Sigma|^{-(\nu_0+p+1)/2} \times \exp\{-\mathrm{tr}(\mathbf{S}_0\Sigma^{-1})/2\}. \tag{7.7}$$

The normalizing constant is quite intimidating. Fortunately we will only have to work with the second line of the equation. The expression "tr" stands for *trace* and for a square $p \times p$ matrix $\mathbf{A}$, $\mathrm{tr}(\mathbf{A}) = \sum_{j=1}^{p} a_{j,j}$, the sum of the diagonal elements.

We now need to combine the above prior distribution with the sampling distribution for $\boldsymbol{Y}_1, \ldots, \boldsymbol{Y}_n$:

$$p(\boldsymbol{y}_1,\ldots,\boldsymbol{y}_n|\boldsymbol{\theta},\Sigma) = (2\pi)^{-np/2}|\Sigma|^{-n/2}\exp\{-\sum_{i=1}^{n}(\boldsymbol{y}_i-\boldsymbol{\theta})^T\Sigma^{-1}(\boldsymbol{y}_i-\boldsymbol{\theta})/2\}. \tag{7.8}$$

An interesting result from matrix algebra is that the sum $\sum_{k=1}^{K}\boldsymbol{b}_k^T\mathbf{A}\boldsymbol{b}_k = \text{tr}(\mathbf{B}^T\mathbf{B}\mathbf{A})$, where $\mathbf{B}$ is the matrix whose kth row is $\boldsymbol{b}_k^T$. This means that the term in the exponent of Equation 7.8 can be expressed as

$$\sum_{i=1}^{n}(\boldsymbol{y}_i-\boldsymbol{\theta})^T\Sigma^{-1}(\boldsymbol{y}_i-\boldsymbol{\theta}) = \text{tr}(\mathbf{S}_\theta\Sigma^{-1}), \text{ where}$$

$$\mathbf{S}_\theta = \sum_{i=1}^{n}(\boldsymbol{y}_i-\boldsymbol{\theta})(\boldsymbol{y}_i-\boldsymbol{\theta})^T.$$

The matrix $\mathbf{S}_\theta$ is the *residual sum of squares matrix* for the vectors $\boldsymbol{y}_1,\ldots,\boldsymbol{y}_n$ if the population mean is presumed to be $\boldsymbol{\theta}$. Conditional on $\boldsymbol{\theta}$, $\frac{1}{n}\mathbf{S}_\theta$ provides an unbiased estimate of the true covariance matrix $\text{Cov}[\boldsymbol{Y}]$ (more generally, when $\boldsymbol{\theta}$ is not conditioned on the sample covariance matrix is $\sum(\boldsymbol{y}_i-\bar{\boldsymbol{y}})(\boldsymbol{y}_i-\bar{\boldsymbol{y}})^T/(n-1)$ and is an unbiased estimate of Σ). Using the above result to combine Equations 7.7 and 7.8 gives the conditional distribution of Σ:

$$\begin{aligned}
&p(\Sigma|\boldsymbol{y}_1,\ldots,\boldsymbol{y}_n,\boldsymbol{\theta})\\
&\propto p(\Sigma)\times p(\boldsymbol{y}_1,\ldots\boldsymbol{y}_n|\boldsymbol{\theta},\Sigma)\\
&\propto \left(|\Sigma|^{-(\nu_0+p+1)/2}\exp\{-\text{tr}(\mathbf{S}_0\Sigma^{-1})/2\}\right)\times\left(|\Sigma|^{-n/2}\exp\{-\text{tr}(\mathbf{S}_\theta\Sigma^{-1})/2\}\right)\\
&= |\Sigma|^{-(\nu_0+n+p+1)/2}\exp\{-\text{tr}([\mathbf{S}_0+\mathbf{S}_\theta]\Sigma^{-1})/2\}.
\end{aligned}$$

Thus we have

$$\{\Sigma|\boldsymbol{y}_1,\ldots,\boldsymbol{y}_n,\boldsymbol{\theta}\}\sim\text{inverse-Wishart}(\nu_0+n,[\mathbf{S}_0+\mathbf{S}_\theta]^{-1}). \tag{7.9}$$

Hopefully this result seems somewhat intuitive: We can think of ν_0+n as the "posterior sample size," being the sum of the "prior sample size" ν_0 and the data sample size. Similarly, $\mathbf{S}_0+\mathbf{S}_\theta$ can be thought of as the "prior" residual sum of squares plus the residual sum of squares from the data. Additionally, the conditional expectation of the population covariance matrix is

$$\begin{aligned}
\text{E}[\Sigma|\boldsymbol{y}_1,\ldots,\boldsymbol{y}_n,\boldsymbol{\theta}] &= \frac{1}{\nu_0+n-p-1}(\mathbf{S}_0+\mathbf{S}_\theta)\\
&= \frac{\nu_0-p-1}{\nu_0+n-p-1}\frac{1}{\nu_0-p-1}\mathbf{S}_0+\frac{n}{\nu_0+n-p-1}\frac{1}{n}\mathbf{S}_\theta
\end{aligned}$$

and so the conditional expectation can be seen as a weighted average of the prior expectation and the unbiased estimator. Because it can be shown that $\mathbf{S}_\theta$ converges to the true population covariance matrix, the posterior expectation of Σ is a consistent estimator of the population covariance, even if the true population distribution is not multivariate normal.

7.4 Gibbs sampling of the mean and covariance

In the last two sections we showed that

$$\{\boldsymbol{\theta}|\boldsymbol{y}_1,\ldots,\boldsymbol{y}_n,\Sigma\} \sim \text{multivariate normal}(\boldsymbol{\mu}_n,\Lambda_n)$$
$$\{\Sigma|\boldsymbol{y}_1,\ldots,\boldsymbol{y}_n,\boldsymbol{\theta}\} \sim \text{inverse-Wishart}(\nu_n,\mathbf{S}_n^{-1}),$$

where $\{\Lambda_n,\boldsymbol{\mu}_n\}$ are defined in Equations 7.4 and 7.5, $\nu_n = \nu_0 + n$ and $\mathbf{S}_n = \mathbf{S}_0 + \mathbf{S}_\theta$. These full conditional distributions can be used to construct a Gibbs sampler, providing us with an MCMC approximation to the joint posterior distribution $p(\boldsymbol{\theta},\Sigma|\boldsymbol{y}_1,\ldots,\boldsymbol{y}_n)$. Given a starting value $\Sigma^{(0)}$, the Gibbs sampler generates $\{\boldsymbol{\theta}^{(s+1)},\Sigma^{(s+1)}\}$ from $\{\boldsymbol{\theta}^{(s)},\Sigma^{(s)}\}$ via the following two steps:

1. Sample $\boldsymbol{\theta}^{(s+1)}$ from its full conditional distribution:
   a) compute $\boldsymbol{\mu}_n$ and Λ_n from $\boldsymbol{y}_1,\ldots,\boldsymbol{y}_n$ and $\Sigma^{(s)}$;
   b) sample $\boldsymbol{\theta}^{(s+1)} \sim$ multivariate normal$(\boldsymbol{\mu}_n,\Lambda_n)$.
2. Sample $\Sigma^{(s+1)}$ from its full conditional distribution:
   a) compute $\mathbf{S}_n$ from $\boldsymbol{y}_1,\ldots,\boldsymbol{y}_n$ and $\boldsymbol{\theta}^{(s+1)}$;
   b) sample $\Sigma^{(s+1)} \sim$ inverse-Wishart$(\nu_0+n,\mathbf{S}_n^{-1})$.

Steps 1.a and 2.a highlight the fact that $\{\boldsymbol{\mu}_n,\Lambda_n\}$ depend on the value of Σ, and that $\mathbf{S}_n$ depends on the value of $\boldsymbol{\theta}$, and so these quantities need to be recalculated at every iteration of the sampler.

Example: Reading comprehension

Let's return to the example from the beginning of the chapter in which each of 22 children were given two reading comprehension exams, one before a certain type of instruction and one after. We'll model these 22 pairs of scores as i.i.d. samples from a multivariate normal distribution. The exam was designed to give average scores of around 50 out of 100, so $\boldsymbol{\mu}_0 = (50,50)^T$ would be a good choice for our prior expectation. Since the true mean cannot be below 0 or above 100, it is desirable to use a prior variance for $\boldsymbol{\theta}$ that puts little probability outside of this range. We'll take the prior variances on θ_1 and θ_2 to be $\lambda_{0,1}^2 = \lambda_{0,2}^2 = (50/2)^2 = 625$, so that the prior probability $\Pr(\theta_j \notin [0,100])$ is only 0.05. Finally, since the two exams are measuring similar things, whatever the true values of θ_1 and θ_2 are it is probable that they are close. We can reflect this with a prior correlation of 0.5, so that $\lambda_{1,2} = 312.5$. As for the prior distribution on Σ, some of the same logic about the range of exam scores applies. We'll take $\mathbf{S}_0$ to be the same as Λ_0, but only loosely center Σ around this value by taking $\nu_0 = p + 2 = 4$.

```
mu0<-c(50,50)
L0<-matrix(c(625,312.5,312.5,625),nrow=2,ncol=2)

nu0<-4
S0<-matrix(c(625,312.5,312.5,625),nrow=2,ncol=2)
```

The observed values $\boldsymbol{y}_1, \ldots, \boldsymbol{y}_{22}$ are plotted as dots in the second panel of Figure 7.2. The sample mean is $\bar{\boldsymbol{y}} = (47.18, 53.86)^T$, the sample variances are $s_1^2 = 182.16$ and $s_2^2 = 243.65$, and the sample correlation is $s_{1,2}/(s_1 s_2) = 0.70$. Let's use the Gibbs sampler described above to combine this sample information with our prior distributions to obtain estimates and confidence intervals for the population parameters. We begin by setting $\Sigma^{(0)}$ equal to the sample covariance matrix, and iterating from there. In the R-code below, `Y` is the 22×2 data matrix of the observed values.

```
data(chapter7) ; Y<-Y.reading
n<-dim(Y)[1] ; ybar<-apply(Y,2,mean)
Sigma<-cov(Y) ; THETA<-SIGMA<-NULL

set.seed(1)
for(s in 1:5000)
{

  ###update theta
  Ln<-solve( solve(L0) + n*solve(Sigma) )
  mun<-Ln%*%( solve(L0)%*%mu0 + n*solve(Sigma)%*%ybar )
  theta<-rmvnorm(1,mun,Ln)
  ###

  ###update Sigma
  Sn<- S0 + ( t(Y)-c(theta) )%*%t( t(Y)-c(theta) )
  Sigma<-solve( rwish(1, nu0+n, solve(Sn)) )
  ###

  ### save results
  THETA<-rbind(THETA, theta) ; SIGMA<-rbind(SIGMA, c(Sigma))
  ###

}
```

The above code generates 5,000 values $(\{\boldsymbol{\theta}^{(1)}, \Sigma^{(1)}\}), \ldots, \{\boldsymbol{\theta}^{(5000)}, \Sigma^{(5000)}\})$ whose empirical distribution approximates $p(\boldsymbol{\theta}, \Sigma | \boldsymbol{y}_1, \ldots, \boldsymbol{y}_n)$. It is left as an exercise to assess the convergence and autocorrelation of this Markov chain. From these samples we can approximate posterior probabilities and confidence regions of interest.

```
> quantile(  THETA[,2]-THETA[,1], prob=c(.025,.5,.975) )
     2.5%        50%       97.5%
 1.513573  6.668097 11.794824


> mean( THETA[,2]>THETA[,1])
[1] 0.9942
```

The posterior probability $\Pr(\theta_2 > \theta_1 | \boldsymbol{y}_1, \ldots, \boldsymbol{y}_n) = 0.99$ indicates strong evidence that, if we were to give exams and instruction to a large population of children, then the average score on the second exam would be higher than that on the first. This evidence is displayed graphically in the first panel of Figure 7.2, which shows 97.5%, 75%, 50%, 25% and 2.5% highest posterior density contours for the joint posterior distribution of $\boldsymbol{\theta} = (\theta_1, \theta_2)^T$. A highest posterior density contour is a two-dimensional analogue of a confidence interval. The contours for the posterior distribution of $\boldsymbol{\theta}$ are all mostly above the 45-degree line $\theta_1 = \theta_2$.

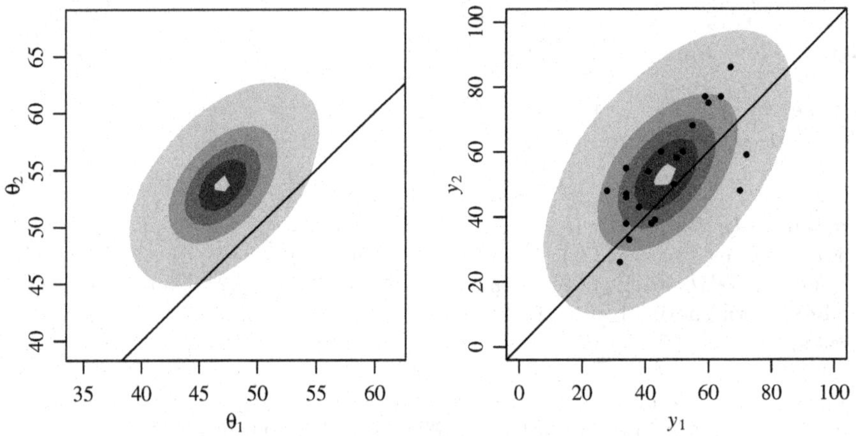

Fig. 7.2. Reading comprehension data and posterior distributions

Now let's ask a slightly different question - what is the probability that a randomly selected child will score higher on the second exam than on the first? The answer to this question is a function of the posterior predictive distribution of a new sample $(Y_1, Y_2)^T$, given the observed values. The second panel of Figure 7.2 shows highest posterior density contours of the posterior predictive distribution, which, while mostly being above the line $y_2 = y_1$, still has substantial overlap with the region below this line, and in fact $\Pr(Y_2 > Y_1 | \boldsymbol{y}_1, \ldots, \boldsymbol{y}_n) = 0.71$. How should we evaluate the effectiveness of the between-exam instruction? On one hand, the fact that $\Pr(\theta_2 > \theta_1 | \boldsymbol{y}_1, \ldots, \boldsymbol{y}_n) = 0.99$ seems to suggest that there is a "highly significant difference" in exam scores before and after the instruction, yet $\Pr(Y_2 > Y_1 | \boldsymbol{y}_1, \ldots, \boldsymbol{y}_n) = 0.71$ says that almost a third of the students will get a lower score on the second exam. The difference between these two probabilities is that the first is measuring the evidence that θ_2 is larger than θ_1 without regard to whether or not the magnitude of the difference $\theta_2 - \theta_1$ is large compared to the sampling variability of the data. Confusion over these two different ways of comparing populations is common in the reporting of results from experiments or surveys: studies with very large values of n often result in values of $\Pr(\theta_2 > \theta_1 | \boldsymbol{y}_1, \ldots, \boldsymbol{y}_n)$ that are very close to 1 (or p-values that are very close to zero), suggesting a "significant effect," even though such results say nothing about how large of an effect we expect to see for a randomly sampled individual.

7.5 Missing data and imputation

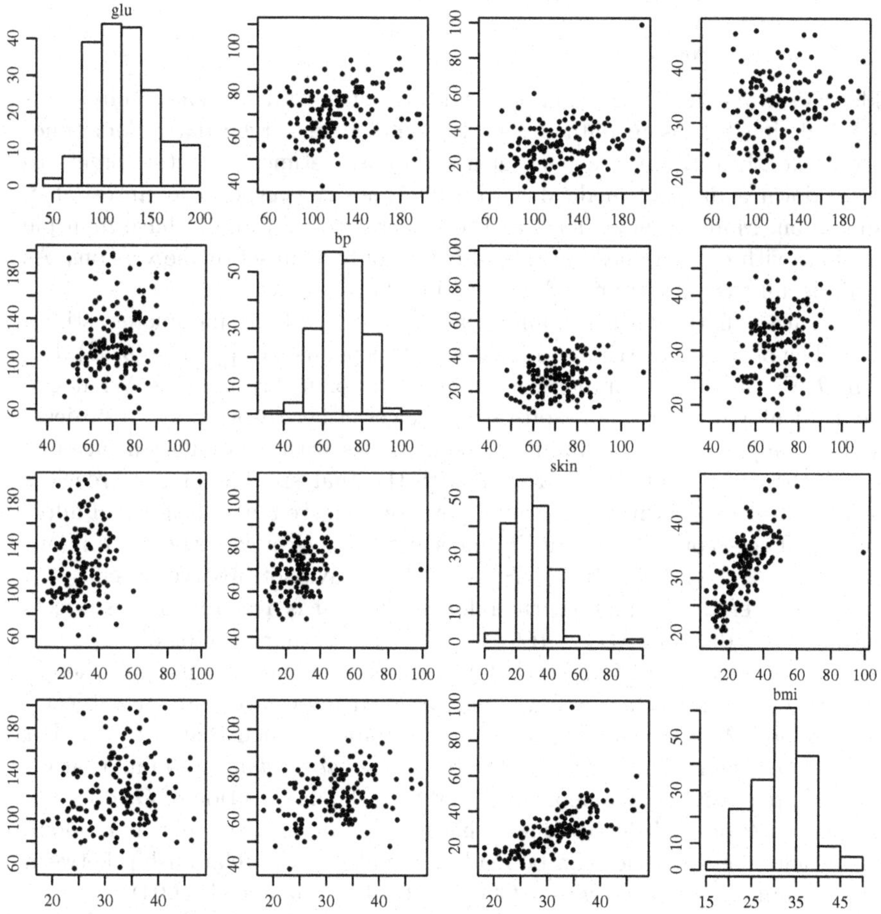

Fig. 7.3. Physiological data on 200 women.

Figure 7.3 displays univariate histograms and bivariate scatterplots for four variables taken from a dataset involving health-related measurements on 200 women of Pima Indian heritage living near Phoenix, Arizona (Smith et al, 1988). The four variables are `glu` (blood plasma glucose concentration), `bp` (diastolic blood pressure), `skin` (skin fold thickness) and `bmi` (body mass index). The first ten subjects in this dataset have the following entries:

	glu	bp	skin	bmi
1	86	68	28	30.2
2	195	70	33	NA
3	77	82	NA	35.8
4	NA	76	43	47.9
5	107	60	NA	NA
6	97	76	27	NA
7	NA	58	31	34.3
8	193	50	16	25.9
9	142	80	15	NA
10	128	78	NA	43.3

The NA's stand for "not available," and so some data for some individuals are "missing." Missing data are fairly common in survey data: Sometimes people accidentally miss a page of a survey, sometimes a doctor forgets to write down a piece of medical data, sometimes the response is unreadable, and so on. Many surveys (such as the General Social Survey) have multiple versions with certain questions appearing in only a subset of the versions. As a result, all the subjects may have missing data.

In such situations it is not immediately clear how to do parameter estimation. The posterior distribution for $\boldsymbol{\theta}$ and Σ depends on $\prod_{i=1}^{n} p(\boldsymbol{y}_i|\boldsymbol{\theta}, \Sigma)$, but $p(\boldsymbol{y}_i|\boldsymbol{\theta}, \Sigma)$ cannot be computed if components of $\boldsymbol{y}_i$ are missing. What can we do? Unfortunately, many software packages either throw away all subjects with incomplete data, or impute missing values with a population mean or some other fixed value, then proceed with the analysis. The first approach is bad because we are throwing away a potentially large amount of useful information. The second is statistically incorrect, as it says we are certain about the values of the missing data when in fact we have not observed them.

Let's carefully think about the information that is available from subjects with missing data. Let $\boldsymbol{O}_i = (O_1, \ldots, O_p)^T$ be a binary vector of zeros and ones such that $O_{i,j} = 1$ implies that $Y_{i,j}$ is observed and not missing, whereas $O_{i,j} = 0$ implies $Y_{i,j}$ is missing. Our observed information about subject i is therefore $\boldsymbol{O}_i = \boldsymbol{o}_i$ and $Y_{i,j} = y_{i,j}$ for variables j such that $o_{i,j} = 1$. For now, we'll assume that missing data are *missing at random*, meaning that $\boldsymbol{O}_i$ and $\boldsymbol{Y}_i$ are statistically independent and that the distribution of $\boldsymbol{O}_i$ does not depend on $\boldsymbol{\theta}$ or Σ. In cases where the data are missing but not at random, then sometimes inference can be made by modeling the relationship between $\boldsymbol{O}_i$, $\boldsymbol{Y}_i$ and the parameters (see Chapter 21 of Gelman et al (2004)).

In the case where data are missing at random, the sampling probability for the data from subject i is

$$
\begin{aligned}
p(\boldsymbol{o}_i, \{y_{i,j} : o_{i,j} = 1\}|\boldsymbol{\theta}, \Sigma) &= p(\boldsymbol{o}_i) \times p(\{y_{i,j} : o_{i,j} = 1\}|\boldsymbol{\theta}, \Sigma) \\
&= p(\boldsymbol{o}_i) \times \int \left\{ p(y_{i,1}, \ldots, y_{i,p}|\boldsymbol{\theta}, \Sigma) \prod_{y_{i,j}: o_{i,j}=0} dy_{i,j} \right\}.
\end{aligned}
$$

In words, our sampling probability for data from subject i is $p(\boldsymbol{o}_i)$ multiplied by the marginal probability of the observed variables, after integrating out the missing variables. To make this more concrete, suppose $\boldsymbol{y}_i = (y_{i,1}, \texttt{NA}, y_{i,3}, \texttt{NA})^T$, so $\boldsymbol{o}_i = (1,0,1,0)^T$. Then

$$\begin{aligned} p(\boldsymbol{o}_i, y_{i,1}, y_{i,3}|\boldsymbol{\theta}, \Sigma) &= p(\boldsymbol{o}_i) \times p(y_{i,1}, y_{i,3}|\boldsymbol{\theta}, \Sigma) \\ &= p(\boldsymbol{o}_i) \times \int p(\boldsymbol{y}_i|\boldsymbol{\theta}, \Sigma)\, dy_2\, dy_4 . \end{aligned}$$

So the correct thing to do when data are missing at random is to integrate over the missing data to obtain the marginal probability of the observed data. In this particular case of the multivariate normal model, this marginal probability is easily obtained: $p(y_{i,1}, y_{i,3}|\boldsymbol{\theta}, \Sigma)$ is simply a bivariate normal density with mean $(\theta_1, \theta_3)^T$ and a covariance matrix made up of $(\sigma_1^2, \sigma_{1,3}, \sigma_3^2)$. But combining marginal densities from subjects having different amounts of information can be notationally awkward. Fortunately, our integration can alternatively be done quite easily using Gibbs sampling.

Gibbs sampling with missing data

In Bayesian inference we use probability distributions to describe our information about unknown quantities. What are the unknown quantities for our multivariate normal model with missing data? The parameters $\boldsymbol{\theta}$ and Σ are unknown as usual, but the missing data are also an unknown but key component of our model. Treating it as such allows us to use Gibbs sampling to make inference on $\boldsymbol{\theta}, \Sigma$, as well as to make predictions for the missing values.

Let $\mathbf{Y}$ be the $n \times p$ matrix of all the potential data, observed and unobserved, and let $\mathbf{O}$ be the $n \times p$ matrix in which $o_{i,j} = 1$ if $Y_{i,j}$ is observed and $o_{i,j} = 0$ if $Y_{i,j}$ is missing. The matrix $\mathbf{Y}$ can then be thought of as consisting of two parts:

- $\mathbf{Y}_{\text{obs}} = \{y_{i,j} : o_{i,j} = 1\}$, the data that we do observe, and
- $\mathbf{Y}_{\text{miss}} = \{y_{i,j} : o_{i,j} = 0\}$, the data that we do not observe.

From our observed data we want to obtain $p(\boldsymbol{\theta}, \Sigma, \mathbf{Y}_{\text{miss}}|\mathbf{Y}_{\text{obs}})$, the posterior distribution of unknown and unobserved quantities. A Gibbs sampling scheme for approximating this posterior distribution can be constructed by simply adding one step to the Gibbs sampler presented in the previous section: Given starting values $\{\Sigma^{(0)}, \mathbf{Y}_{\text{miss}}^{(0)}\}$, we generate $\{\boldsymbol{\theta}^{(s+1)}, \Sigma^{(s+1)}, \mathbf{Y}_{\text{miss}}^{(s+1)}\}$ from $\{\boldsymbol{\theta}^{(s)}, \Sigma^{(s)}, \mathbf{Y}_{\text{miss}}^{(s)}\}$ by

1. sampling $\boldsymbol{\theta}^{(s+1)}$ from $p(\boldsymbol{\theta}|\mathbf{Y}_{\text{obs}}, \mathbf{Y}_{\text{miss}}^{(s)}, \Sigma^{(s)})$;
2. sampling $\Sigma^{(s+1)}$ from $p(\Sigma|\mathbf{Y}_{\text{obs}}, \mathbf{Y}_{\text{miss}}^{(s)}, \boldsymbol{\theta}^{(s+1)})$;
3. sampling $\mathbf{Y}_{\text{miss}}^{(s+1)}$ from $p(\mathbf{Y}_{\text{miss}}|\mathbf{Y}_{\text{obs}}, \boldsymbol{\theta}^{(s+1)}, \Sigma^{(s+1)})$.

Note that in steps 1 and 2, the fixed value of $\mathbf{Y}_{\text{obs}}$ combines with the current value of $\mathbf{Y}_{\text{miss}}^{(s)}$ to form a current version of a complete data matrix $\mathbf{Y}^{(s)}$ having no missing values. The n rows of the matrix of $\mathbf{Y}^{(s)}$ can then be plugged into formulae 7.6 and 7.9 to obtain the full conditional distributions of $\boldsymbol{\theta}$ and Σ. Step 3 is a bit more complicated:

$$\begin{aligned} p(\mathbf{Y}_{\text{miss}}|\mathbf{Y}_{\text{obs}}, \boldsymbol{\theta}, \Sigma) &\propto p(\mathbf{Y}_{\text{miss}}, \mathbf{Y}_{\text{obs}}|\boldsymbol{\theta}, \Sigma) \\ &= \prod_{i=1}^{n} p(\boldsymbol{y}_{i,\text{miss}}, \boldsymbol{y}_{i,\text{obs}}|\boldsymbol{\theta}, \Sigma) \\ &\propto \prod_{i=1}^{n} p(\boldsymbol{y}_{i,\text{miss}}|\boldsymbol{y}_{i,\text{obs}}, \boldsymbol{\theta}, \Sigma), \end{aligned}$$

so for each i we need to sample the missing elements of the data vector conditional on the observed elements. This is made possible via the following result about multivariate normal distributions: Let $\boldsymbol{y} \sim$ multivariate normal$(\boldsymbol{\theta}, \Sigma)$, let $\boldsymbol{a}$ be a subset of variable indices $\{1, \ldots, p\}$ and let $\boldsymbol{b}$ be the complement of $\boldsymbol{a}$. For example, if $p = 4$ then perhaps $\boldsymbol{a} = \{1, 2\}$ and $\boldsymbol{b} = \{3, 4\}$. If you know about inverses of partitioned matrices you can show that

$$\begin{aligned} \{\boldsymbol{y}_{[b]}|\boldsymbol{y}_{[a]}, \boldsymbol{\theta}, \Sigma\} &\sim \text{multivariate normal}(\boldsymbol{\theta}_{b|a}, \Sigma_{b|a}), \text{ where} \\ \boldsymbol{\theta}_{b|a} &= \boldsymbol{\theta}_{[b]} + \Sigma_{[b,a]}(\Sigma_{[a,a]})^{-1}(\boldsymbol{y}_{[a]} - \boldsymbol{\theta}_{[a]}) \qquad (7.10) \\ \Sigma_{b|a} &= \Sigma_{[b,b]} - \Sigma_{[b,a]}(\Sigma_{[a,a]})^{-1}\Sigma_{[a,b]}. \qquad (7.11) \end{aligned}$$

In the above formulae, $\boldsymbol{\theta}_{[b]}$ refers to the elements of $\boldsymbol{\theta}$ corresponding to the indices in $\boldsymbol{b}$, and $\Sigma_{[a,b]}$ refers to the matrix made up of the elements that are in rows $\boldsymbol{a}$ and columns $\boldsymbol{b}$ of Σ.

Let's try to gain a little bit of intuition about what is going on in Equations 7.10 and 7.11. Suppose $\boldsymbol{y}$ is a sample from our population of four variables `glu`, `bp`, `skin` and `bmi`. If we have `glu` and `bp` data for someone ($\boldsymbol{a} = \{1, 2\}$) but are missing `skin` and `bmi` measurements ($\boldsymbol{b} = \{3, 4\}$), then we would be interested in the conditional distribution of these missing measurements $\boldsymbol{y}_{[b]}$ given the observed information $\boldsymbol{y}_{[a]}$. Equation 7.10 says that the conditional mean of `skin` and `bmi` start off at their unconditional mean $\boldsymbol{\theta}_{[b]}$, but then are modified by $(\boldsymbol{y}_{[a]} - \boldsymbol{\theta}_{[a]})$. For example, if a person had higher than average values of `glu` and `bp`, then $(\boldsymbol{y}_{[a]} - \boldsymbol{\theta}_{[a]})$ would be a 2×1 vector of positive numbers. For our data the 2×2 matrix $\Sigma_{[b,a]}(\Sigma_{[a,a]})^{-1}$ has all positive entries, and so $\boldsymbol{\theta}_{b|a} > \boldsymbol{\theta}_{[b]}$. This makes sense: If all four variables are positively correlated, then if we observe higher than average values of `glu` and `bp`, we should also expect higher than average values of `skin` and `bmi`. Also note that $\Sigma_{b|a}$ is equal to the unconditional variance $\Sigma_{[b,b]}$ but with something subtracted off, suggesting that the conditional variance is less than the unconditional variance. Again, this makes sense: having information about some variables should decrease, or at least not increase, our uncertainty about the others.

The R code below implements the Gibbs sampling scheme for missing data described in steps 1, 2 and 3 above:

```
data(chapter7) ; Y<-Y.pima.miss
### prior parameters
n<-dim(Y)[1] ; p<-dim(Y)[2]
mu0<-c(120,64,26,26)
sd0<-(mu0/2)
L0<-matrix(.1,p,p) ; diag(L0)<-1 ; L0<-L0*outer(sd0,sd0)
nu0<-p+2 ; S0<-L0
###

### starting values
Sigma<-S0
Y.full<-Y
O<-1*(!is.na(Y))
for(j in 1:p)
{
  Y.full[is.na(Y.full[,j]),j]<-mean(Y.full[,j],na.rm=TRUE)
}
###

### Gibbs sampler
THETA<-SIGMA<-Y.MISS<-NULL
set.seed(1)
for(s in 1:1000)
{

  ###update theta
  ybar<-apply(Y.full,2,mean)
  Ln<-solve( solve(L0) + n*solve(Sigma) )
  mun<-Ln%*%( solve(L0)%*%mu0 + n*solve(Sigma)%*%ybar )
  theta<-rmvnorm(1,mun,Ln)
  ###

  ###update Sigma
  Sn<- S0 + ( t(Y.full)-c(theta) )%*%t( t(Y.full)-c(theta) )
  Sigma<-solve( rwish(1, nu0+n, solve(Sn)) )
  ###

  ###update missing data
  for(i in 1:n)
  {
    b <- ( O[i,]==0 )
    a <- ( O[i,]==1 )
    iSa<- solve(Sigma[a,a])
    beta.j <- Sigma[b,a]%*%iSa
    Sigma.j   <- Sigma[b,b] - Sigma[b,a]%*%iSa%*%Sigma[a,b]
    theta.j<- theta[b] + beta.j%*%(t(Y.full[i,a])-theta[a])
    Y.full[i,b] <- rmvnorm(1,theta.j,Sigma.j )
  }
```

```
  ### save results
  THETA<-rbind(THETA,theta) ; SIGMA<-rbind(SIGMA,c(Sigma))
  Y.MISS<-rbind(Y.MISS, Y.full[O==0] )
  ###
}
###
```

The prior mean of $\boldsymbol{\mu}_0 = (120, 64, 26, 26)^T$ was obtained from national averages, and the prior variances were based primarily on keeping most of the prior mass on values that are above zero. These prior distributions are likely much more diffuse than more informed prior distributions that could be provided by someone who is familiar with this population or these variables.

The Monte Carlo approximation of $\mathrm{E}[\boldsymbol{\theta}|\boldsymbol{y}_1,\ldots,\boldsymbol{y}_n]$ is (123.46, 71.03, 29.35, 32.18), obtained by averaging the 1,000 $\boldsymbol{\theta}$-values generated by the Gibbs sampler. Posterior confidence intervals and other quantities can additionally be obtained in the usual way from the Gibbs samples. We can also average the 1,000 values of Σ to obtain $\mathrm{E}[\Sigma|\boldsymbol{y}_1,\ldots,\boldsymbol{y}_n]$, the posterior expectation of Σ. However, when looking at associations among a set of variables, it is often the correlations that are of interest and not the covariances. To each covariance matrix Σ there corresponds a correlation matrix $\mathbf{C}$, given by

$$\mathbf{C} = \left\{ c_{j,k} : c_{j,k} = \Sigma_{[j,k]} / \sqrt{\Sigma_{[j,j]} \Sigma_{[k,k]}} \right\}.$$

We can convert our 1,000 posterior samples of Σ into 1,000 posterior samples of $\mathbf{C}$ using the following R-code:

```
COR <- array( dim=c(p,p,1000) )
for(s in 1:1000)
{
  Sig<-matrix( SIGMA[s,] ,nrow=p,ncol=p)
  COR[,,s] <- Sig/sqrt( outer( diag(Sig),diag(Sig) ) )
}
```

This code generates a $4\times4\times1000$ array, where each "slice" is a 4×4 correlation matrix generated from the posterior distribution. The posterior expectation of $\mathbf{C}$ is

$$\mathrm{E}[\mathbf{C}|\boldsymbol{y}_1,\ldots,\boldsymbol{y}_n] = \begin{pmatrix} 1.00 & 0.23 & 0.25 & 0.19 \\ 0.23 & 1.00 & 0.25 & 0.24 \\ 0.25 & 0.25 & 1.00 & 0.65 \\ 0.19 & 0.24 & 0.65 & 1.00 \end{pmatrix}$$

and marginal posterior 95% quantile-based confidence intervals can be obtained with the command `apply(COR, c(1,2), quantile,prob=c(.025,.975) )`. These are displayed graphically in the left panel of Figure 7.4.

Prediction and regression

Multivariate models are often used to predict one or more variables given the others. Consider, for example, a predictive model of `glu` based on measurements of `bp`, `skin` and `bmi`. Using $\boldsymbol{a} = \{2, 3, 4\}$ and $\boldsymbol{b} = \{1\}$ in Equation 7.10,

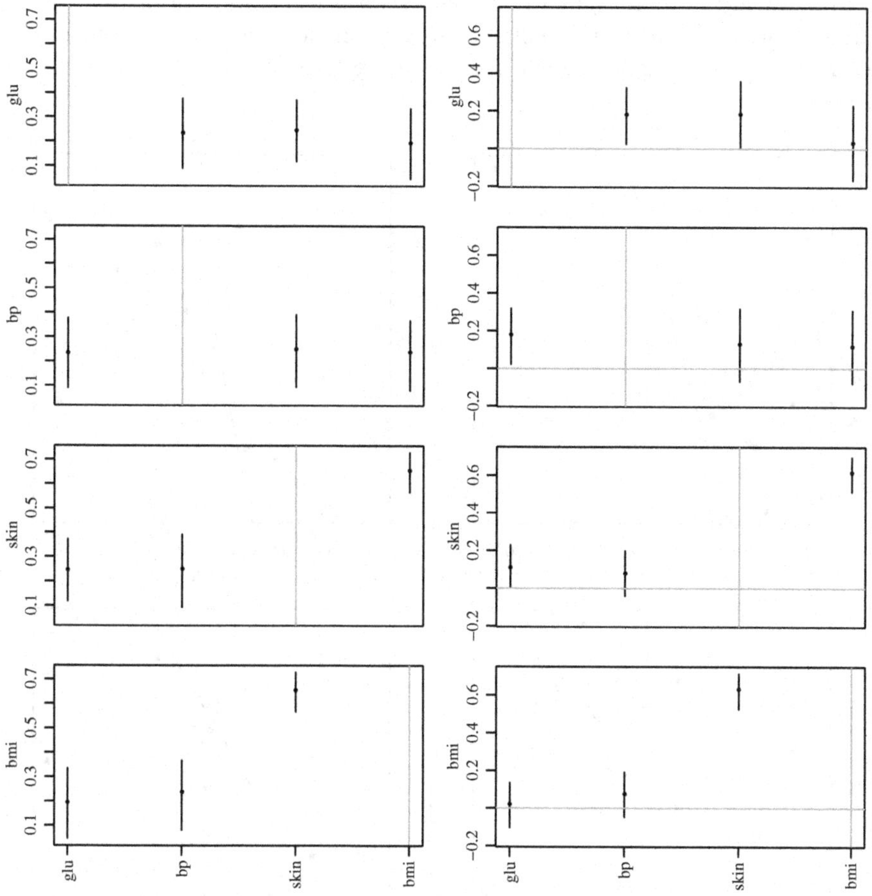

Fig. 7.4. Ninety-five percent posterior confidence intervals for correlations (left) and regression coefficients derived from the correlation matrix (right).

the conditional mean of $\boldsymbol{y}_{[b]}$ =`glu`, given numerical values of $\boldsymbol{y}_{[a]} = \{$`bp`, `skin`, `bmi` $\}$, is given by

$$\mathrm{E}[\boldsymbol{y}_{[b]}|\boldsymbol{\theta}, \Sigma, \boldsymbol{y}_{[a]}] = \boldsymbol{\theta}_{[b]} + \boldsymbol{\beta}^T_{b|a}(\boldsymbol{y}_{[a]} - \boldsymbol{\theta}_{[a]})$$

where $\boldsymbol{\beta}^T_{b|a} = \Sigma_{[b,a]}(\Sigma_{[a,a]})^{-1}$. Since this takes the form of a linear regression model, we call the value of $\boldsymbol{\beta}_{b|a}$ the regression coefficient for $\boldsymbol{y}_{[b]}$ given $\boldsymbol{y}_{[a]}$ based on Σ. Values of $\boldsymbol{\beta}_{b|a}$ can be computed for each posterior sample of Σ, allowing us to obtain posterior expectations and confidence intervals for these regression coefficients. Quantile-based 95% confidence intervals for each of $\{\boldsymbol{\beta}_{1|234}, \boldsymbol{\beta}_{2|134}, \boldsymbol{\beta}_{3|124}, \boldsymbol{\beta}_{4|123}\}$ are shown graphically in the second column of Figure 7.4. The regression coefficients often tell a different story than the correlations: The bottom row of plots, for example, shows that while there is strong evidence that the correlations between `bmi` and each of the other variables are all positive, the plots on the right-hand side suggest that `bmi` is nearly conditionally independent of `glu` and `bp` given `skin`.

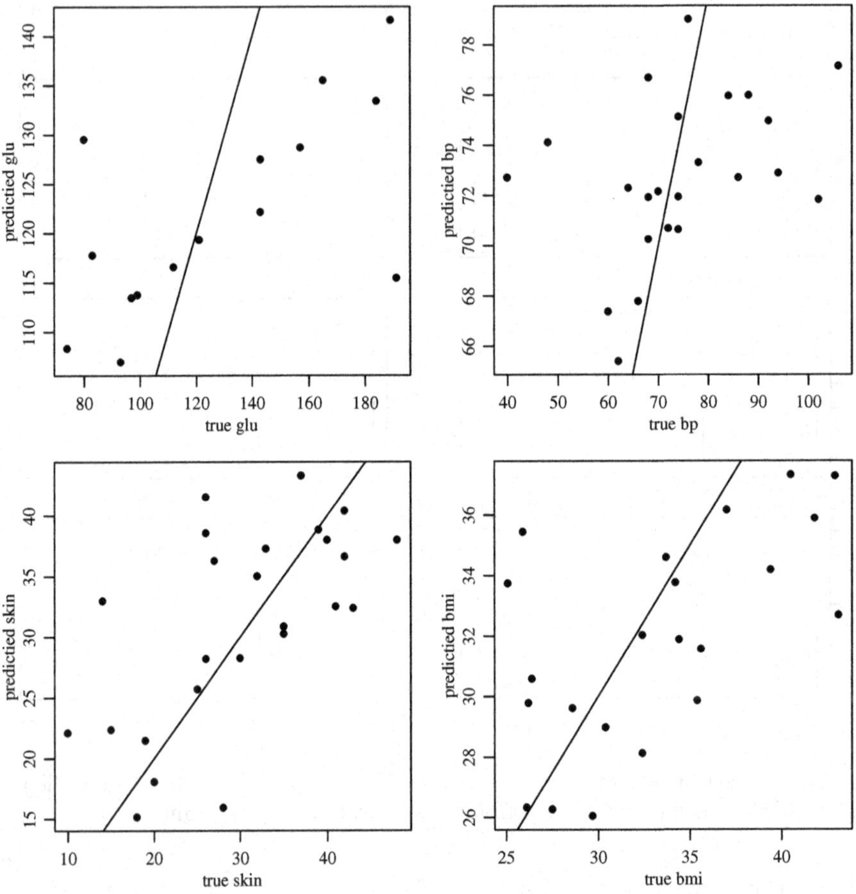

Fig. 7.5. True values of the missing data versus their posterior expectations.

Out-of-sample validation

Actually, the dataset we just analyzed was created by taking a complete data matrix with no missing values and randomly replacing 10% of the entries with `NA`'s. Since the original dataset is available, we can compare values predicted by the model to the actual sample values. This comparison is made graphically in Figure 7.5, which plots the true value of $y_{i,j}$ against its posterior mean for each $\{i,j\}$ such that $o_{i,j} = 0$. It looks like we are able to do a better job predicting missing values of `skin` and `bmi` than the other two variables. This makes sense, as these two variables have the highest correlation. If `skin` is missing, we can make a good prediction for it based on the observed value of `bmi`, and vice-versa. Such a procedure, where we evaluate how well a model does at predicting data that were not used to estimate the parameters, is called *out-of-sample validation*, and is often used to quantify the predictive performance of a model.

7.6 Discussion and further references

The multivariate normal model can be justified as a sampling model for reasons analogous to those for the univariate normal model (see Section 5.7): It is characterized by independence between the sample mean and sample variance (Rao, 1958), it is a maximum entropy distribution and it provides consistent estimation of the population mean and variance, even if the population is not multivariate normal.

The multivariate normal and Wishart distributions form the foundation of multivariate data analysis. A classic text on the subject is Mardia et al (1979), and one with more coverage of Bayesian approaches is Press (1982). An area of much current Bayesian research involving the multivariate normal distribution is the study of *graphical models* (Lauritzen, 1996; Jordan, 1998). A graphical model allows elements of the precision matrix to be exactly equal to zero, implying some variables are conditionally independent of each other. A generalization of the Wishart distribution, known as the hyper-inverse-Wishart distribution, has been developed for such models (Dawid and Lauritzen, 1993; Letac and Massam, 2007).

8

Group comparisons and hierarchical modeling

In this chapter we discuss models for the comparison of means across groups. In the two-group case, we parameterize the two population means by their average and their difference. This type of parameterization is extended to the multigroup case, where the average group mean and the differences across group means are described by a normal sampling model. This model, together with a normal sampling model for variability among units within a group, make up a hierarchical normal model that describes both within-group and between-group variability. We also discuss an extension to this normal hierarchical model which allows for across-group heterogeneity in variances in addition to heterogeneity in means.

8.1 Comparing two groups

The first panel of Figure 8.1 shows math scores from a sample of 10th grade students from two public U.S. high schools. Thirty-one students from school 1 and 28 students from school 2 were randomly selected to participate in a math test. Both schools have a total enrollment of around 600 10th graders each, and both are in urban neighborhoods.

Suppose we are interested in estimating θ_1, the average score we would obtain if all 10th graders in school 1 were tested, and possibly comparing it to θ_2, the corresponding average from school 2. The results from the sample data are $\bar{y}_1 = 50.81$ and $\bar{y}_2 = 46.15$, suggesting that θ_1 is larger than θ_2. However, if different students had been sampled from each of the two schools, then perhaps $\bar{y}_2$ would have been larger than $\bar{y}_1$. To assess whether or not the observed mean difference of $\bar{y}_1 - \bar{y}_2 = 4.66$ is large compared to the sampling variability it is standard practice to compute the t-statistic, which is the ratio of the observed difference to an estimate of its standard deviation:

P.D. Hoff, *A First Course in Bayesian Statistical Methods*,
Springer Texts in Statistics, DOI 10.1007/978-0-387-92407-6_8,

$$t(\boldsymbol{y}_1, \boldsymbol{y}_2) = \frac{\bar{y}_1 - \bar{y}_2}{s_p\sqrt{1/n_1 + 1/n_2}} = \frac{50.81 - 46.15}{10.44\sqrt{1/31 + 1/28}} = 1.74,$$

where $s_p^2 = [(n_1 - 1)s_1^2 + (n_2 - 1)s_2^2]/(n_1 + n_2 - 2)$, the pooled estimate of the population variance of the two groups. Is this value of 1.74 large? From introductory statistics, we know that if the population of scores from the two schools are both normally distributed with the same mean and variance, then the sampling distribution of the t-statistic $t(\boldsymbol{Y}_1, \boldsymbol{Y}_2)$ is a t-distribution with $n_1 + n_2 - 2 = 57$ degrees of freedom. The density of this distribution is plotted in the second panel of Figure 8.1, along with the observed value of the t-statistic. If the two populations indeed follow the same normal population, then the *pre-experimental* probability of sampling a dataset that would generate a value of $t(\boldsymbol{Y}_1, \boldsymbol{Y}_2)$ greater in absolute value than 1.74 is $p = 0.087$. You may recall that this latter number is called the (two-sided) *p-value*. While a small p-value is generally considered as indicating evidence that θ_1 and θ_2 are different, the p-value should not be confused with the probability that $\theta_1 = \theta_2$. Although not completely justified by statistical theory for this purpose, p-values are often used in parameter estimation and model selection. For example, the following is a commonly taught data analysis procedure for comparing the population means of two groups:

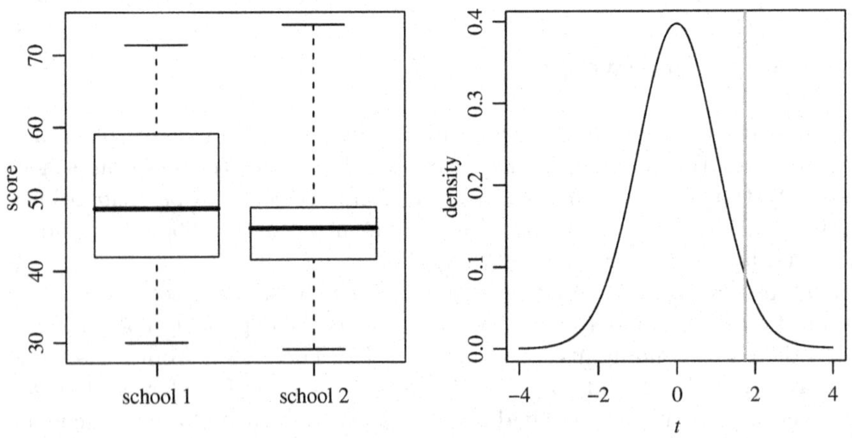

Fig. 8.1. Boxplots of samples of 10th grade math scores from two schools, and the null distribution for testing equality of the population means. The gray line indicates the observed value of the t-statistic.

Model selection based on p-values:

If $p < 0.05$,
- reject the model that the two groups have the same distribution;
- conclude that $\theta_1 \neq \theta_2$;
- use the estimates $\hat{\theta}_1 = \bar{y}_1$, $\hat{\theta}_2 = \bar{y}_2$.

If $p > 0.05$
- accept the model that the two groups have the same distribution;
- conclude that $\theta_1 = \theta_2$;
- use the estimates $\hat{\theta}_1 = \hat{\theta}_2 = (\sum y_{i,1} + \sum y_{i,2})/(n_1 + n_2)$.

This data analysis procedure results in either treating the two populations as completely distinct or treating them as exactly identical. Do these rather extreme alternatives make sense? For our math score data, the above procedure would take the p-value of 0.087 and tell us to treat the population means of the two groups as being numerically equivalent, although there seems to be some evidence of a difference. Conversely, it is not too hard to imagine a scenario where the sample from school 1 might have included a few more high-performing students, the sample from school 2 a few more low-performing students, in which case we could have observed a p-value of 0.04 or 0.05. In this latter case we would have treated each population separately, using only data from school 1 to estimate θ_1 and similarly for school 2. This latter approach seems somewhat inefficient: Since the two samples are both measuring the same thing on similar populations of students, it might make sense to use some of the information from one group to help estimate the mean in the other.

The p-value-based procedure described above can be re-expressed as estimating θ_1 as $\hat{\theta}_1 = w\bar{y}_1 + (1 - w)\bar{y}_2$, where $w = 1$ if $p < 0.05$ and $w = n_1/(n_1 + n_2)$ otherwise. Instead of using such an extreme procedure, it might make more sense to allow w to vary continuously and have a value that depends on such things as the relative sample sizes n_1 and n_2, the sampling variability σ^2 and our prior information about the similarities of the two populations. An estimator similar to this is produced by a Bayesian analysis that allows for information to be shared across the groups. Consider the following sampling model for data from the two groups:

$$\begin{aligned} Y_{i,1} &= \mu + \delta + \epsilon_{i,1} \\ Y_{i,2} &= \mu - \delta + \epsilon_{i,2} \\ \{\epsilon_{i,j}\} &\sim \text{i.i.d. normal}(0, \sigma^2) \ . \end{aligned}$$

Using this parameterization where $\theta_1 = \mu + \delta$ and $\theta_2 = \mu - \delta$, we see that δ represents half the population difference in means, as $(\theta_1 - \theta_2)/2 = \delta$, and μ represents the pooled average, as $(\theta_1 + \theta_2)/2 = \mu$. Convenient conjugate prior distributions for the unknown parameters are

$$\begin{aligned} p(\mu, \delta, \sigma^2) &= p(\mu) \times p(\delta) \times p(\sigma^2) \\ \mu &\sim \text{normal}(\mu_0, \gamma_0^2) \\ \delta &\sim \text{normal}(\delta_0, \tau_0^2) \\ \sigma^2 &\sim \text{inverse-gamma}(\nu_0/2, \nu_0\sigma_0^2/2). \end{aligned}$$

It is left as an exercise to show that the full conditional distributions of these parameters are as follows:

$$\begin{aligned} &\{\mu|\boldsymbol{y}_1, \boldsymbol{y}_2, \delta, \sigma^2\} \sim \text{normal}(\mu_n, \gamma_n^2), \text{ where} \\ &\quad \mu_n = \gamma_n^2 \times [\mu_0/\gamma_0^2 + \textstyle\sum_{i=1}^{n_1}(y_{i,1} - \delta)/\sigma^2 + \sum_{i=1}^{n_2}(y_{i,2} + \delta)/\sigma^2] \\ &\quad \gamma_n^2 = [1/\gamma_0^2 + (n_1 + n_2)/\sigma^2]^{-1} \end{aligned}$$

$$\begin{aligned} &\{\delta|\boldsymbol{y}_1, \boldsymbol{y}_2, \mu, \sigma^2\} \sim \text{normal}(\delta_n, \tau_n^2), \text{ where} \\ &\quad \delta_n = \tau_n^2 \times [\delta_0/\tau_0^2 + \textstyle\sum(y_{i,1} - \mu)/\sigma^2 - \sum(y_{i,2} - \mu)/\sigma^2] \\ &\quad \tau_n^2 = [1/\tau_0^2 + (n_1 + n_2)/\sigma^2]^{-1} \end{aligned}$$

$$\begin{aligned} &\{\sigma^2|\boldsymbol{y}_1, \boldsymbol{y}_2, \mu, \delta\} \sim \text{inverse-gamma}(\nu_n/2, \nu_n\sigma_n^2/2), \text{ where} \\ &\quad \nu_n = \nu_0 + n_1 + n_2 \\ &\quad \nu_n\sigma_n^2 = \nu_0\sigma_0^2 + \textstyle\sum(y_{i,1} - [\mu + \delta])^2 + \sum(y_{i,2} - [\mu - \delta])^2 \end{aligned}$$

Although these formulae seem quite involved, you should try to convince yourself that they make sense. One way to do this is to plug in extreme values for the prior parameters. For example, if $\nu_0 = 0$ then

$$\sigma_n^2 = \frac{\sum(y_{i,1} - [\mu + \delta])^2 + \sum(y_{i,2} - [\mu - \delta])^2}{n_1 + n_2}$$

which is a pooled-sample estimate of the variance if the values of μ and δ were known. Similarly, if $\mu_0 = \delta_0 = 0$ and $\gamma_0^2 = \tau_0^2 = \infty$, then (defining $0/\infty = 0$)

$$\mu_n = \frac{\sum(y_{i,1} - \delta) + \sum(y_{i,2} + \delta)}{n_1 + n_2}, \quad \delta_n = \frac{\sum(y_{i,1} - \mu) - \sum(y_{i,2} - \mu)}{n_1 + n_2}$$

and if you plug in μ_n for μ and δ_n for δ, you get $\bar{y}_1 = \mu_n + \delta_n$ and $\bar{y}_2 = \mu_n - \delta_n$.

Analysis of the math score data

The math scores were based on results of a national exam in the United States, standardized to produce a nationwide mean of 50 and a standard deviation of 10. Unless these two schools were known in advance to be extremely exceptional, reasonable prior parameters can be based on this information. For the prior distributions of μ and σ^2, we'll take $\mu_0 = 50$ and $\sigma_0^2 = 10^2 = 100$, although this latter value is likely to be an overestimate of the within-school sampling variability. We'll make these prior distributions somewhat diffuse, with $\gamma_0^2 = 25^2 = 625$ and $\nu_0 = 1$. For the prior distribution on δ, choosing $\delta_0 = 0$ represents the prior opinion that $\theta_1 > \theta_2$ and $\theta_2 > \theta_1$ are equally probable. Finally, since the scores are bounded between 0 an 100, half the difference between θ_1 and θ_2 must be less than 50 in absolute value, so a value of $\tau_0^2 = 25^2 = 625$ seems reasonably diffuse.

Using the full-conditional distributions given above, we can construct a Gibbs sampler to approximate the posterior distribution $p(\mu, \delta, \sigma^2 | \boldsymbol{y}_1, \boldsymbol{y}_2)$. R-code to do the approximation appears below:

```
data(chapter8)
y1<-y.school1 ; n1<-length(y1)
y2<-y.school2 ; n2<-length(y2)
##### prior parameters
mu0<-50 ; g02<-625
del0<-0 ; t02<-625
s20<-100; nu0<-1
#####

##### starting values
mu<- ( mean(y1) + mean(y2) )/2
del<- ( mean(y1) - mean(y2) )/2
#####

##### Gibbs sampler
MU<-DEL<-S2<-NULL

set.seed(1)
for(s in 1:5000)
{

  ##update s2
  s2<-1/rgamma(1,(nu0+n1+n2)/2,
        (nu0*s20+sum((y1-mu-del)^2)+sum((y2-mu+del)^2) )/2)
  ##

  ##update mu
  var.mu<-  1/(1/g02+ (n1+n2)/s2 )
  mean.mu<-var.mu*(mu0/g02+sum(y1-del)/s2+sum(y2+del)/s2)
  mu<-rnorm(1,mean.mu,sqrt(var.mu))
  ##

  ##update del
  var.del<-  1/(1/t02+ (n1+n2)/s2 )
  mean.del<-var.del*(del0/t02+sum(y1-mu)/s2-sum(y2-mu)/s2)
  del<-rnorm(1,mean.del,sqrt(var.del))
  ##

  ##save parameter values
  MU<-c(MU,mu) ; DEL<-c(DEL,del) ; S2<-c(S2,s2)
}
#####
```

Figure 8.2 shows the marginal posterior distributions of μ and δ, and how they are much more concentrated than their corresponding prior distributions. In particular, a 95% quantile-based posterior confidence interval for $2 \times \delta$, the difference in average scores between the two schools, is $(-.61, 9.98)$. Although this interval contains zero, the differences between the prior and posterior distributions indicate that we have gathered substantial evidence that the population mean for school 1 is higher than that of school 2. Additionally, the posterior probability $\Pr(\theta_1 > \theta_2 | \boldsymbol{y}_1, \boldsymbol{y}_2) = \Pr(\delta > 0 | \boldsymbol{y}_1, \boldsymbol{y}_2) \approx 0.96$, whereas the corresponding prior probability was $\Pr(\delta > 0) = 0.50$. However, we should be careful not to confuse this probability with the probability that a randomly selected student from school 1 has a higher score than one sampled from school 2. This latter probability can be obtained from the joint posterior predictive distribution, which gives $\Pr(Y_1 > Y_2 | \boldsymbol{y}_1, \boldsymbol{y}_2) \approx 0.62$.

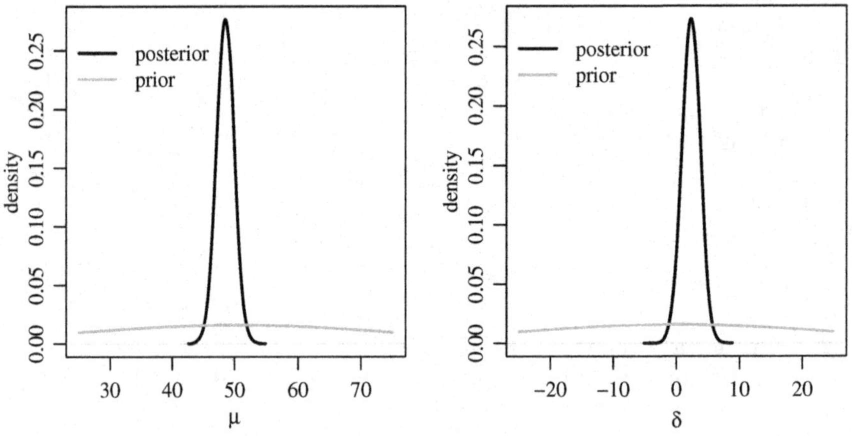

Fig. 8.2. Prior and posterior distributions for μ and δ.

8.2 Comparing multiple groups

The data in the previous section was part of the 2002 Educational Longitudinal Study (ELS), a survey of students from a large sample of schools across the United States. This dataset includes a population of schools as well as a population of students within each school. Datasets like this, where there is a hierarchy of nested populations, are often called *hierarchical* or *multilevel*. Other situations having the same sort of data structure include data on

- patients within several hospitals,
- genes within a group of animals, or

- people within counties within regions within countries.

The simplest type of multilevel data has two levels, in which one level consists of *groups* and the other consists of *units within groups*. In this case we denote $y_{i,j}$ as the data on the ith unit in group j.

8.2.1 Exchangeability and hierarchical models

Recall from Chapter 2 that a sequence of random variables $Y_1, \ldots, Y_n$ is exchangeable if the probability density representing our information about the sequence satisfies $p(y_1, \ldots, y_n) = p(y_{\pi_1}, \ldots, y_{\pi_n})$ for any permutation π. Exchangeability is a reasonable property of $p(y_1, \ldots, y_n)$ if we lack information distinguishing the random variables. For example, if $Y_1, \ldots, Y_n$ were math scores from n randomly selected students from a particular school, then in the absence of other information about the students we might treat their math scores as exchangeable. If exchangeability holds for all values of n, then de Finetti's theorem says that an equivalent formulation of our information is that

$$\phi \sim p(\phi)$$
$$\{Y_1, \ldots, Y_n|\phi\} \sim \text{i.i.d. } p(y|\phi).$$

In other words, the random variables can be thought of as independent samples from a population described by some fixed but unknown population feature ϕ. In the normal model, for example, we take $\phi = \{\theta, \sigma^2\}$ and model the data as conditionally i.i.d. normal(θ, σ^2).

Now let's consider a model describing our information about a hierarchical data $\{\boldsymbol{Y}_1, \ldots, \boldsymbol{Y}_m\}$, where $\boldsymbol{Y}_j = \{Y_{1,j}, \ldots, Y_{n_j,j}\}$. What properties should a model $p(\boldsymbol{y}_1, \ldots, \boldsymbol{y}_m)$ have? Let's consider first $p(\boldsymbol{y}_j) = p(y_{1,j}, \ldots, y_{n_j,j})$, the marginal probability density of data from a single group j. The discussion in the preceding paragraph suggests that we should not treat $Y_{1,j}, \ldots, Y_{n_j,j}$ as being independent, as doing so would imply, for example, that $p(y_{n_j,j}| y_{1,j}, \ldots, y_{n_j-1,j}) = p(y_{n_j,j})$, and that the values of $Y_{1,j}, \ldots, Y_{n_j-1,j}$ would give us no information about $Y_{n_j,j}$. However, if all that is known about $Y_{1,j}, \ldots, Y_{n_j,j}$ is that they are random samples from group j, then treating $Y_{1,j}, \ldots, Y_{n_j,j}$ as exchangeable makes sense. If group j is large compared to the sample size n_j, then de Finetti's theorem and results of Diaconis and Freedman (1980) say that we can model the data within group j as conditionally i.i.d. given some group-specific parameter ϕ_j:

$$\{Y_{1,j}, \ldots, Y_{n_j,j}|\phi_j\} \sim \text{ i.i.d. } p(y|\phi_j)\,.$$

But how should we represent our information about $\phi_1, \ldots, \phi_m$? As before, we might not want to treat these parameters as independent, because doing so would imply that knowing the values of $\phi_1, \ldots, \phi_{m-1}$ does not change our information about the value of ϕ_m. However, if the groups themselves are samples from some larger population of groups, then exchangeability of the group-specific parameters might be appropriate. Applying de Finetti's theorem a second time gives

$$\{\phi_1, \ldots, \phi_m | \psi\} \sim \text{ i.i.d. } p(\phi|\psi)$$

for some sampling model $p(\phi|\psi)$ and an unknown parameter ψ. This double application of de Finetti's theorem has led us to three probability distributions:

$$\begin{aligned} \{y_{1,j}, \ldots, y_{n_j,j} | \phi_j\} &\sim \text{ i.i.d. } p(y|\phi_j) && \text{(within-group sampling variability)} \\ \{\phi_1, \ldots, \phi_m | \psi\} &\sim \text{ i.i.d. } p(\phi|\psi) && \text{(between-group sampling variability)} \\ \psi &\sim p(\psi) && \text{(prior distribution)} \end{aligned}$$

It is important to recognize that the distributions $p(y|\phi)$ and $p(\phi|\psi)$ both represent sampling variability among populations of objects: $p(y|\phi)$ represents variability among measurements within a group and $p(\phi|\psi)$ represents variability across groups. In contrast, $p(\psi)$ represents information about a single fixed but unknown quantity. For this reason, we refer to $p(y|\phi)$ and $p(\phi|\psi)$ as *sampling distributions*, and are conceptually distinct from $p(\psi)$, which is a *prior distribution*. In particular, the data will be used to estimate the within- and between-group sampling distributions $p(y|\phi)$ and $p(\phi|\psi)$, whereas the prior distribution $p(\psi)$ is not estimated from the data.

8.3 The hierarchical normal model

A popular model for describing the heterogeneity of means across several populations is the hierarchical normal model, in which the within- and between-group sampling models are both normal:

$$\phi_j = \{\theta_j, \sigma^2\}, \; p(y|\phi_j) = \text{normal}(\theta_j, \sigma^2) \qquad \text{(within-group model)} \quad (8.1)$$

$$\psi = \{\mu, \tau^2\}, \; p(\theta_j|\psi) = \text{normal}(\mu, \tau^2) \qquad \text{(between-group model)} \quad (8.2)$$

It might help to visualize this setup as in Figure 8.3. Note that $p(\phi|\psi)$ only describes the heterogeneity across group means, and not any heterogeneity in group-specific variances. In fact, the within-group sampling variability σ^2 is assumed to be constant across groups. At the end of this chapter we will eliminate this assumption by adding a component to the model that allows for group-specific variances.

The fixed but unknown parameters in this model are μ, τ^2 and σ^2. For convenience we will use standard semiconjugate normal and inverse-gamma prior distributions for these parameters:

$$\begin{aligned} 1/\sigma^2 &\sim \text{ gamma } (\nu_0/2, \nu_0\sigma_0^2/2) \\ 1/\tau^2 &\sim \text{ gamma } (\eta_0/2, \eta_0\tau_0^2/2) \\ \mu &\sim \text{ normal } (\mu_0, \gamma_0^2) \end{aligned}$$

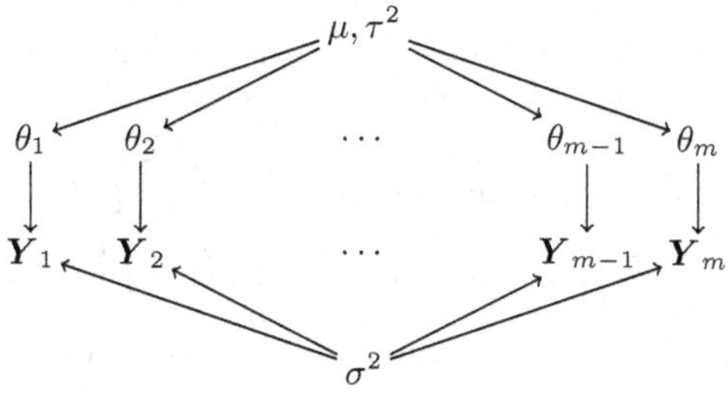

Fig. 8.3. A graphical representation of the basic hierarchical normal model.

8.3.1 Posterior inference

The unknown quantities in our system include the group-specific means $\{\theta_1, \ldots, \theta_m\}$, the within-group sampling variability σ^2 and the mean and variance (μ, τ^2) of the population of group-specific means. Joint posterior inference for these parameters can be made by constructing a Gibbs sampler which approximates the posterior distribution $p(\theta_1, \ldots, \theta_m, \mu, \tau^2, \sigma^2 | \boldsymbol{y}_1, \ldots, \boldsymbol{y}_m)$.

The Gibbs sampler proceeds by iteratively sampling each parameter from its full conditional distribution. Deriving the full conditional distributions in this highly parameterized system may seem like a daunting task, but it turns out that all of the necessary technical details have been covered in Chapters 5 and 6. All that is required of us at this point is that we recognize certain analogies between the current model and the univariate normal model. Useful for this will be the following factorization:

$$
\begin{aligned}
& p(\theta_1, \ldots, \theta_m, \mu, \tau^2, \sigma^2 | \boldsymbol{y}_1, \ldots, \boldsymbol{y}_m) \\
& \propto p(\mu, \tau^2, \sigma^2) \times p(\theta_1, \ldots, \theta_m | \mu, \tau^2, \sigma^2) \times p(\boldsymbol{y}_1, \ldots, \boldsymbol{y}_m | \theta_1, \ldots, \theta_m, \mu, \tau^2, \sigma^2) \\
& = p(\mu) p(\tau^2) p(\sigma^2) \left\{ \prod_{j=1}^{m} p(\theta_j | \mu, \tau^2) \right\} \left\{ \prod_{j=1}^{m} \prod_{i=1}^{n_j} p(y_{i,j} | \theta_j, \sigma^2) \right\}. \qquad (8.3)
\end{aligned}
$$

The term in the second pair of brackets is the result of an important conditional independence feature of our model. Conditionally on $\{\theta_1, \ldots, \theta_m, \mu, \tau^2, \sigma^2\}$, the random variables $Y_{1,j}, \ldots, Y_{n_j,j}$ are independent with a distribution that depends only on θ_j and σ^2 and not on μ or τ^2. It is helpful to think about this fact in terms of the diagram in Figure 8.3: The existence of a path from (μ, τ^2) to each $\boldsymbol{Y}_j$ indicates that while (μ, τ^2) provides information about $\boldsymbol{Y}_j$, it only does so indirectly through θ_j, which separates the two quantities in the graph.

Full conditional distributions of μ and τ^2

As a function of μ and τ^2, the term in Equation 8.3 is proportional to

$$
p(\mu) p(\tau^2) \prod_{j=1}^{m} p(\theta_j | \mu, \tau^2),
$$

and so the full conditional distributions of μ and τ^2 are also proportional to this quantity. In particular, this must mean that

$$p(\mu|\theta_1,\ldots,\theta_m,\tau^2,\sigma^2,\boldsymbol{y}_1,\ldots,\boldsymbol{y}_m) \propto p(\mu)\prod p(\theta_j|\mu,\tau^2)$$
$$p(\tau^2|\theta_1,\ldots,\theta_m,\mu,\sigma^2,\boldsymbol{y}_1,\ldots,\boldsymbol{y}_m) \propto p(\tau^2)\prod p(\theta_j|\mu,\tau^2).$$

These distributions are exactly the full conditional distributions from the one-sample normal problem in Chapter 6. In Chapter 6 we derived the full conditionals of the population mean and variance of a normal population, assuming independent normal and inverse-gamma prior distributions. In our current situation, $\theta_1,\ldots,\theta_m$ are the i.i.d. samples from a normal population, and μ and τ^2 are the unknown population mean and variance. In Chapter 6, we saw that if $y_1,\ldots,y_n$ were i.i.d. normal(θ,σ^2) and θ had a normal prior distribution, then the conditional distribution of θ was also normal. Since our current situation is exactly analogous, the fact that $\theta_1,\ldots,\theta_m$ are i.i.d. normal(μ,τ^2) and μ has a normal prior distribution implies that the conditional distribution of μ must be normal as well. Similarly, just as σ^2 had an inverse-gamma conditional distribution in Chapter 6, τ^2 has an inverse-gamma distribution in the current situation. Applying the results of Chapter 6 with the appropriate symbolic replacements, we have

$$\{\mu|\theta_1,\ldots,\theta_m,\tau^2\} \sim \text{normal}\left(\frac{m\bar{\theta}/\tau^2+\mu_0/\gamma_0^2}{m/\tau^2+1/\gamma_0^2}, [m/\tau^2+1/\gamma_0^2]^{-1}\right)$$

$$\{1/\tau^2|\theta_1,\ldots,\theta_m,\mu\} \sim \text{gamma}\left(\frac{\eta_0+m}{2}, \frac{\eta_0\tau_0^2+\sum(\theta_j-\mu)^2}{2}\right).$$

Full conditional of θ_j

Collecting the terms in Equation 8.3 that depend on θ_j shows that the full conditional distribution of θ_j must be proportional to

$$p(\theta_j|\mu,\tau^2,\sigma^2,\boldsymbol{y}_1,\ldots,\boldsymbol{y}_m) \propto p(\theta_j|\mu,\tau^2)\prod_{i=1}^{n_j} p(y_{i,j}|\theta_j,\sigma^2).$$

This says that, conditional on $\{\mu,\tau^2,\sigma^2,\boldsymbol{y}_j\}$, θ_j must be conditionally independent of the other θ's as well as independent of the data from groups other than j. Again, it is helpful to refer to Figure 8.3: While there is a path from each θ_j to every other θ_k, the paths go through (μ,τ^2) or σ^2. We can think of this as meaning that the θ's contribute no information about each other beyond that contained in μ,τ^2 and σ^2.

The terms in the above equation include a normal density for θ_j multiplied by a product of normal densities where θ_j is the mean. Mathematically, this is exactly the same setup as the one-sample normal model, in which $p(\theta_j|\mu,\tau^2)$ is the prior distribution instead of the sampling model for the θ's. The full conditional distribution is therefore

$$\{\theta_j|y_{1,j},\ldots,y_{n_j,j},\sigma^2\} \sim \text{normal}(\frac{n_j\bar{y}_j/\sigma^2+\mu/\tau^2}{n_j/\sigma^2+1/\tau^2},[n_j/\sigma^2+1/\tau^2]^{-1}).$$

Full conditional of σ^2

Using Figure 8.3 and the arguments in the previous two paragraphs, you should convince yourself that σ^2 is conditionally independent of $\{\mu,\tau^2\}$ given $\{\boldsymbol{y}_1,\ldots,\boldsymbol{y}_m,\theta_1,\ldots,\theta_m\}$. The derivation of the full conditional of σ^2 is similar to that in the one-sample normal model, except now we have information about σ^2 from m separate groups:

$$p(\sigma^2|\theta_1,\ldots,\theta_m,\boldsymbol{y}_1,\ldots,\boldsymbol{y}_m) \propto p(\sigma^2)\prod_{j=1}^{m}\prod_{i=1}^{n_j}p(y_{i,j}|\theta_j,\sigma^2)$$

$$\propto (\sigma^2)^{-\nu_0/2+1}e^{-\frac{\nu_0\sigma_0^2}{2\sigma^2}}(\sigma^2)^{-\sum n_j/2}e^{-\frac{\sum\sum(y_{i,j}-\theta_j)^2}{2\sigma^2}}.$$

Adding the powers of σ^2 and collecting the terms in the exponent, we recognize this as proportional to an inverse-gamma density, giving

$$\{1/\sigma^2|\boldsymbol{\theta},\boldsymbol{y}_1,\ldots,\boldsymbol{y}_n\} \sim \text{gamma}(\frac{1}{2}[\nu_0+\sum_{j=1}^{m}n_j],\frac{1}{2}[\nu_0\sigma_0^2+\sum_{j=1}^{m}\sum_{i=1}^{n_j}(y_{i,j}-\theta_j)^2]).$$

Note that $\sum\sum(y_{i,j}-\theta_j)^2$ is the sum of squared residuals across all groups, conditional on the within-group means, and so the conditional distribution concentrates probability around a pooled-sample estimate of the variance.

8.4 Example: Math scores in U.S. public schools

Let's return to the 2002 ELS data described previously. This survey included 10th grade children from 100 different large urban public high schools, all having a 10th grade enrollment of 400 or greater. Data from these schools are shown in Figure 8.4, with scores from students within the same school plotted along a common vertical bar.

A histogram of the sample averages is shown in the first panel of Figure 8.5. The range of average scores is quite large, with the lowest average being 36.6 and the highest 65.0. The second panel of the figure shows the relationship between the sample average and the sample size. This plot seems to indicate that very extreme sample averages tend to be associated with schools with small sample sizes. For example, the school with the highest sample average has the lowest sample size, and many schools with low sample averages also have low sample sizes. This relationship between sample averages and sample size is fairly common in hierarchical datasets. To understand this phenomenon, consider a situation in which all θ_j's were equal to some common value, say θ_0, but the sample sizes were different. The expected value of each sample

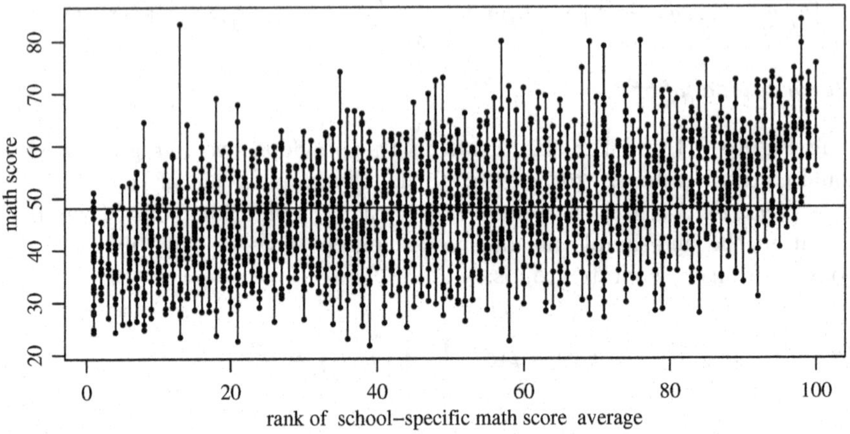

Fig. 8.4. A graphical representation of the ELS data.

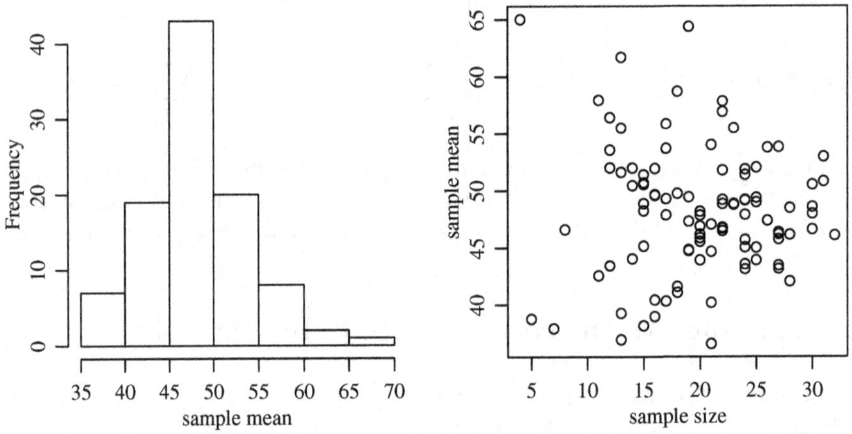

Fig. 8.5. Empirical distribution of sample means, and the relationship between sample mean and sample size.

average would then be $\mathrm{E}[\bar{Y}_j|\theta_j, \sigma^2] = \theta_j = \theta_0$, but the variances would depend on the sample size, since $\mathrm{Var}[\bar{Y}_j|\sigma_j^2] = \sigma^2/n_j$. As a result, sample averages for groups with large sample sizes would be very close to θ_0, whereas the sample averages for groups with small sample sizes would be farther away, both less than and greater than θ_0. For this reason, it is not uncommon that groups with very high or very low sample averages are also those groups with low sample sizes.

8.4.1 Prior distributions and posterior approximation

The prior parameters we need to specify are

(ν_0, σ_0^2) for $p(\sigma^2)$,
(η_0, τ_0^2) for $p(\tau_0^2)$ and
(μ_0, γ_0^2) for $p(\mu)$.

As described above, the math exam was designed to give a nationwide variance of 100. Since this variance includes both within-school and between-school variance, the within-school variance should be at most 100, which we take as σ_0^2. This is likely to be an overestimate, and so we only weakly concentrate the prior distribution around this value by taking $\nu_0 = 1$. Similarly, the between-school variance should not be more than 100, and so we take $\tau_0^2 = 100$ and $\eta_0 = 1$. Finally, the nationwide mean over all schools is 50. Although the mean for large urban public schools may be different than the nationwide average, it should not differ by too much. We take $\mu_0 = 50$ and $\gamma_0^2 = 25$, so that the prior probability that μ is in the interval $(40, 60)$ is about 95%.

Posterior approximation proceeds by iterative sampling of each unknown quantity from its full conditional distribution. Given a current state of the unknowns $\{\theta_1^{(s)}, \ldots, \theta_m^{(s)}, \mu^{(s)}, \tau^{2(s)}, \sigma^{2(s)}\}$, a new state is generated as follows:

1. sample $\mu^{(s+1)} \sim p(\mu|\theta_1^{(s)}, \ldots, \theta_m^{(s)}, \tau^{2(s)})$;
2. sample $\tau^{2(s+1)} \sim p(\tau^2|\theta_1^{(s)}, \ldots, \theta_m^{(s)}, \mu^{(s+1)})$;
3. sample $\sigma^{2(s+1)} \sim p(\sigma^2|\theta_1^{(s)}, \ldots, \theta_m^{(s)}, \boldsymbol{y}_1, \ldots, \boldsymbol{y}_m)$;
4. for each $j \in \{1, \ldots, m\}$, sample $\theta_j^{(s+1)} \sim p(\theta_j|\mu^{(s+1)}, \tau^{2(s+1)}, \sigma^{2(s+1)}, \boldsymbol{y}_j)$.

The order in which the new parameters are generated does not matter. What does matter is that each parameter is updated conditional upon *the most current* value of the other parameters. This Gibbs sampling procedure can be implemented in R with the code below:

```
data(chapter8) ; Y<-Y.school.mathscore

### weakly informative priors
nu0<-1 ; s20<-100
eta0<-1 ; t20<-100
mu0<-50 ; g20<-25
###

### starting values
m<-length(unique(Y[,1]))
n<-sv<-ybar<-rep(NA,m)
for(j in 1:m)
{
  ybar[j]<-mean(Y[Y[,1]==j,2])
  sv[j]<-var(Y[Y[,1]==j,2])
  n[j]<-sum(Y[,1]==j)
```

```
}
theta<-ybar ; sigma2<-mean(sv)
mu<-mean(theta) ; tau2<-var(theta)
###

### setup MCMC
set.seed(1)
S<-5000
THETA<-matrix( nrow=S,ncol=m)
SMT<-matrix( nrow=S,ncol=3)
###

### MCMC algorithm
for(s in 1:S)
{

  # sample new values of the thetas
  for(j in 1:m)
  {
    vtheta<-1/(n[j]/sigma2+1/tau2)
    etheta<-vtheta*(ybar[j]*n[j]/sigma2+mu/tau2)
    theta[j]<-rnorm(1,etheta,sqrt(vtheta))
  }

  #sample new value of sigma2
  nun<-nu0+sum(n)
  ss<-nu0*s20
  for(j in 1:m){ss<-ss+sum((Y[Y[,1]==j,2]-theta[j])^2)}
  sigma2<-1/rgamma(1,nun/2,ss/2)

  #sample a new value of mu
  vmu<- 1/(m/tau2+1/g20)
  emu<- vmu*(m*mean(theta)/tau2 + mu0/g20)
  mu<-rnorm(1,emu,sqrt(vmu))

  # sample a new value of tau2
  etam<-eta0+m
  ss<- eta0*t20 + sum( (theta-mu)^2 )
  tau2<-1/rgamma(1,etam/2,ss/2)

  #store results
  THETA[s,]<-theta
  SMT[s,]<-c(sigma2,mu,tau2)

}
###
```

Running this algorithm produces an $S \times m$ matrix `THETA`, containing a value of the within-school mean for each school at each iteration of the Markov chain. Additionally, the $S \times 3$ matrix `SMT` stores values of σ^2, μ and τ^2, representing approximate, correlated draws from the posterior distribution of these parameters.

MCMC diagnostics

Before we make inference using these MCMC samples we should determine if there might be any problems with the Gibbs sampler. The first thing we want to do is to see if there are any indications that the chain is not stationary, i.e. if the simulated parameter values are moving in a consistent direction. One way to do this is with traceplots, or plots of the parameter values versus iteration number. However, when the number of samples is large, such plots can be difficult to read because of the high density of the plotted points (see, for example, Figure 6.6). Standard practice is to plot only a subsequence of MCMC samples, such as every 100th sample. Another approach is to produce boxplots of sequential groups of samples, as is done in Figure 8.6. The first boxplot in the first plot, for example, represents the empirical distribution of $\{\mu^{(1)}, \ldots, \mu^{(500)}\}$, the second boxplot represents the distribution of $\{\mu^{(501)}, \ldots, \mu^{(1000)}\}$, and so on. Each of the 10 boxplots represents 1/10th of the MCMC samples. If stationarity has been achieved, then the distribution of samples in any one boxplot should be the same as that in any other. If we were to see that the medians or interquartile ranges of the boxplots were moving in a consistent direction with iteration number, then we would suspect that stationarity had not been achieved and we would have to run the chain longer.

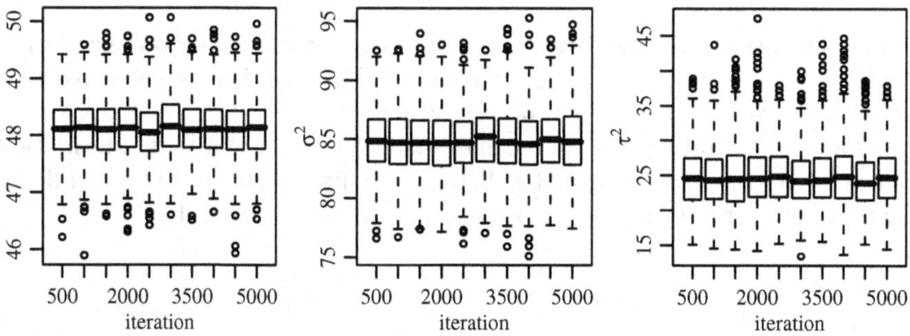

Fig. 8.6. Stationarity plots of the MCMC samples of μ, σ^2 and τ^2.

There does not seem to be any evidence that the chain has not achieved stationarity, so we move on to see how quickly the Gibbs sampler is moving around the parameter space. Lag-1 autocorrelations for the sequences of μ, σ^2 and τ^2 are 0.15, 0.053 and 0.312, and the effective sample sizes are 3706, 4499 and 2503, respectively. Approximate Monte Carlo standard errors can be obtained by dividing the approximated posterior standard deviations by the square root of the effective sample sizes, giving values of (0.009, 0.04, 0.09) for μ, σ^2 and τ^2 respectively. These are quite small compared to the scale of the approximated posterior expectations of these parameters, (48.12, 84.85, 24.84). Diagnostics should also be performed for the θ-parameters: The effective sample sizes for the 100 sequences of θ-values ranged between 3,627 and 5,927, with Monte Carlo standard errors ranging between 0.02 and 0.05.

8.4.2 Posterior summaries and shrinkage

Figure 8.7 shows Monte Carlo approximations to the posterior densities of $\{\mu, \sigma^2, \tau^2\}$. The posterior means of μ, σ and τ are 48.12, 9.21 and 4.97 respectively, indicating that roughly 95% of the scores within a school are within $4 \times 9.21 \approx 37$ points of each other, whereas 95% of the school-specific averages are within $4 \times 4.97 \approx 20$ points of each other.

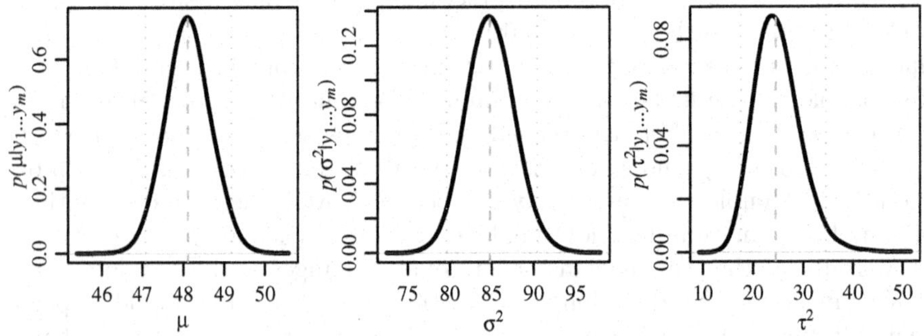

Fig. 8.7. Marginal posterior distributions, with 2.5%, 50% and 97.5% quantiles given by vertical lines.

One of the motivations behind hierarchical modeling is that information can be shared across groups. Recall that, conditional on μ, τ^2, σ^2 and the data, the expected value of θ_j is a weighted average of $\bar{y}_j$ and μ:

$$\mathrm{E}[\theta_j | \boldsymbol{y}_j, \mu, \tau, \sigma] = \frac{\bar{y}_j n_j/\sigma^2 + \mu/\tau^2}{n_j/\sigma^2 + 1/\tau^2} .$$

As a result, the expected value of θ_j is pulled a bit from $\bar{y}_j$ towards μ by an amount depending on n_j. This effect is called *shrinkage*. The first panel of Figure 8.8 plots $\bar{y}_j$ versus $\hat{\theta}_j = \mathrm{E}[\theta_j | \boldsymbol{y}_1, \ldots, \boldsymbol{y}_m]$ for each group. Notice that the relationship roughly follows a line with a slope that is less than one, indicating that high values of $\bar{y}_j$ correspond to slightly less high values of $\hat{\theta}_j$, and low values of $\bar{y}_j$ correspond to slightly less low values of $\hat{\theta}_j$. The second panel of the plot shows the amount of shrinkage as a function of the group-specific sample size. Groups with low sample sizes get shrunk the most, whereas groups with large sample sizes hardly get shrunk at all. This makes sense: The larger the sample size for a group, the more information we have for that group and the less information we need to "borrow" from the rest of the population.

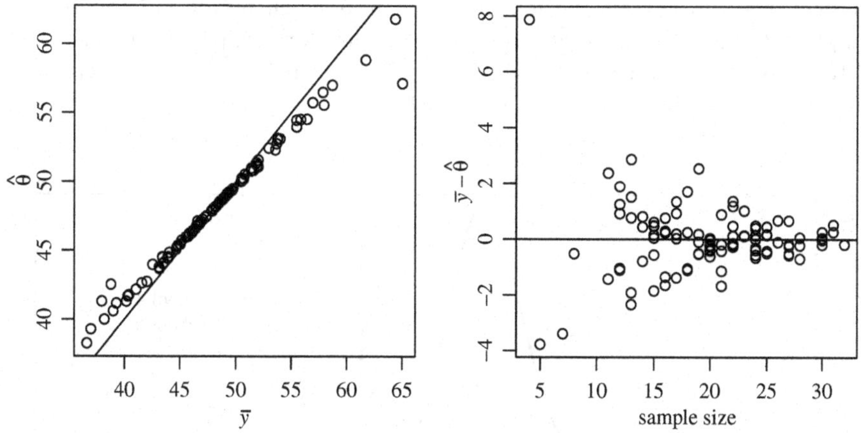

Fig. 8.8. Shrinkage as a function of sample size.

Suppose our task is to rank the schools according to what we think their performances would be if every student in each school took the math exam. In this case, it makes sense to rank the schools according to the school-specific posterior expectations $\{\mathrm{E}[\theta_1|\boldsymbol{y}_1,\ldots,\boldsymbol{y}_m],\ldots,\mathrm{E}[\theta_m|\boldsymbol{y}_1,\ldots,\boldsymbol{y}_m]\}$. Alternatively, we could ignore the results of the hierarchical model and just use the school-specific sample averages $\{\bar{y}_1,\ldots,\bar{y}_m\}$. The two methods will give similar but not exactly the same rankings. Consider the posterior distributions of θ_{46} and θ_{82}, as shown in Figure 8.9. Both of these schools have exceptionally low sample means, in the bottom 10% of all schools. The first thing to note is that the posterior density for school 46 is more peaked than that of school 82. This is because the sample size for school 46 is 21 students, whereas that of school 82 is only 5 students. Therefore, our degree of certainty for θ_{46} is much higher than that for θ_{82}.

The raw data for the two schools are shown in dotplots below the posterior densities, with the large dots representing the sample means $\bar{y}_{46}$ and $\bar{y}_{82}$. Note that while the posterior expectation for school 82 is higher than that of 46 (42.53 compared to 41.31), the sample mean for school 82 is lower than that of 46 (38.76 compared to 40.18). Does this make sense? Suppose on the day of the exam the student who got the lowest exam score in school 82 did not come to class. Then the sample mean for school 82 would have been 41.99 as opposed to 38.76, a change of more than three points. In contrast, if the lowest performing student in school 46 had not shown up, $\bar{y}_{46}$ would have been 40.9 as opposed to 40.18, a change of only three quarters of a point. In other words, the low value of the sample mean for school 82 can be explained by either θ_{82} being very low, or just the possibility that a few of the five sampled students were among the poorer performing students in the school. In contrast, for school 46 this latter possibility cannot explain the low value of the sample mean: Even if a few of the sampled students were unrepresentative of their school-specific average, it would not affect the sample mean as much because of the larger sample size. For this reason, it makes sense to shrink the expectation of school 82 towards the population expectation $\mathrm{E}[\mu|\boldsymbol{y}_1,\ldots,\boldsymbol{y}_n] = 48.11$ by a greater amount than for the expectation of school 46.

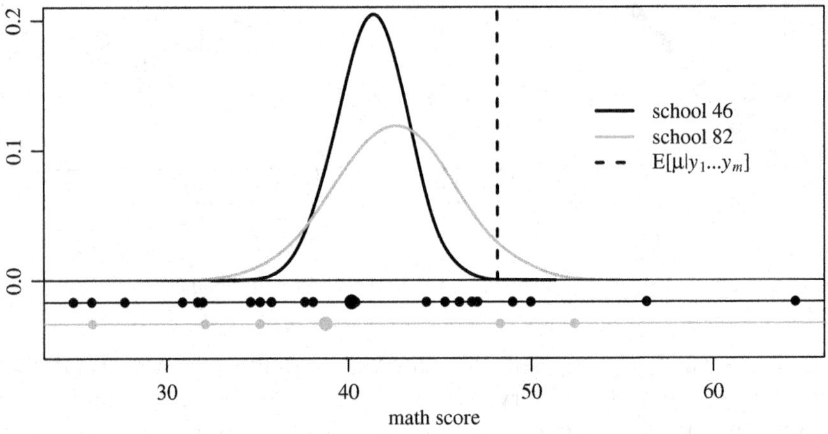

Fig. 8.9. Data and posterior distributions for two schools.

To some people this reversal of rankings may seem strange or "unfair:" The performance by the sampled students in school 46 was better on average than those sampled in school 82, so why should they be ranked lower? While "fairness" may be debated, the hierarchical model reflects the objective fact that there is more evidence that θ_{46} is exceptionally low than there is evidence that θ_{82} is exceptionally low. There are many other real-life situations where differing amounts of evidence results in a switch of a ranking. For example, on any given basketball team there are "bench" players who play very few minutes during any given game. As such, many bench players have taken only a few free throws in their entire career, and many have an observed free throw shooting percentage of 100%. Under certain circumstances during a basketball game (a "technical foul") the coach has the opportunity to choose from among any of his or her players to take a free throw and hopefully score a point. In practice, coaches always choose an experienced veteran player with a percentage of around 87% over a bench player who has made, for example three of three free throws in his career. While it may seem "unfair," it is the right decision to make: The coaches recognize that it is very unlikely that the bench player's true free throw shooting percentage is anywhere near 100%.

8.5 Hierarchical modeling of means and variances

If the population means vary across groups, shouldn't we allow for the possibility that the population variances also vary across groups? Letting σ_j^2 be the variance for group j, our sampling model would then become $Y_{1,j}, \ldots, Y_{n_j,j} \sim$ i.i.d. normal(θ_j, σ_j^2), and our full conditional distribution for each θ_j would be

$$\{\theta_j | y_{1,j}, \ldots, y_{n_j,j}, \sigma_j^2\} \sim \text{normal}\left(\frac{n_j \bar{y}_j / \sigma_j^2 + \mu/\tau^2}{n_j/\sigma_j^2 + 1/\tau^2}, [n_j/\sigma_j^2 + 1/\tau^2]^{-1} \right).$$

How does σ_j^2 get estimated? If we were to specify that

$$1/\sigma_1^2, \ldots, 1/\sigma_m^2 \sim \text{i.i.d. gamma}(\nu_0/2, \nu_0\sigma_0^2/2), \tag{8.4}$$

then as is shown in Chapter 6 the full conditional distribution of σ_j^2 is

$$\{1/\sigma_j^2 | y_{1,j}, \ldots, y_{n_j,j}, \theta_j\} \sim \text{gamma}\left([\nu_0 + n_j]/2, [\nu_0\sigma_0^2 + \sum (y_{i,j} - \theta_j)^2]/2 \right),$$

and estimation for $\sigma_1^2, \ldots, \sigma_m^2$ can proceed by iteratively sampling their values along with the other parameters in a Gibbs sampler.

If ν_0 and σ_0^2 are fixed in advance at some particular values, then the distribution in (8.4) represents a prior distribution on variances such that, for example, $p(\sigma_m^2 | \sigma_1^2, \ldots, \sigma_{m-1}^2) = p(\sigma_m^2)$, and so the information we may have about $\sigma_1^2, \ldots, \sigma_{m-1}^2$ is not used to help us estimate σ_m^2. This seems inefficient: If the sample size in group m were small and we saw that $\sigma_1^2, \ldots, \sigma_{m-1}^2$ were tightly concentrated around a particular value, then we would want to use this fact to improve our estimation of σ_m^2. In other words, we want to be able to learn about the sampling distribution of the σ_j^2's and use this information to improve our estimation for groups that may have low sample sizes. This can be done by treating ν_0 and σ_0^2 as parameters to be estimated, in which case (8.4) is properly thought of as a sampling model for across-group heterogeneity in population variances, and not as a prior distribution. Putting this together with our model for heterogeneity in population means gives a hierarchical model for both means and variances, which is depicted graphically in Figure 8.10.

The unknown parameters to be estimated include $\{(\theta_1, \sigma_1^2), \ldots, (\theta_m, \sigma_m^2)\}$ representing the within-group sampling distributions, $\{\mu, \tau^2\}$ representing across-group heterogeneity in means and $\{\nu_0, \sigma_0^2\}$ representing across-group heterogeneity in variances. As before, the joint posterior distribution for all of

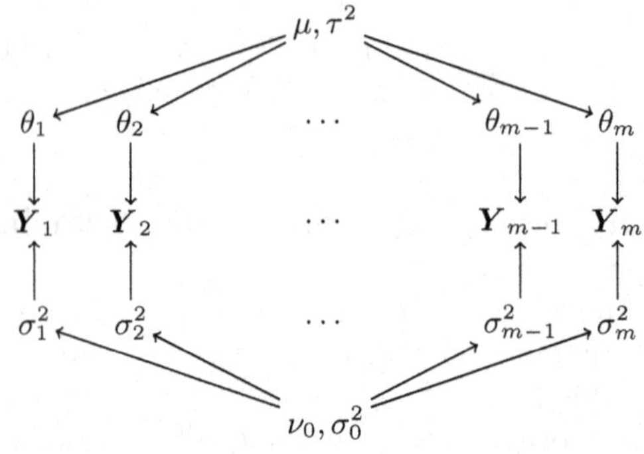

Fig. 8.10. A graphical representation of the hierarchical normal model with heterogeneous means and variances.

these parameters can be approximated by iteratively sampling each parameter from its full conditional distribution given the others. The full conditional distributions for μ and τ^2 are unchanged from the previous section and the full conditional distributions of θ_j and σ_j^2 are given above. What remains to do is to specify the prior distributions for ν_0 and σ_0^2 and obtain their full conditional distributions. A conjugate class of prior densities for σ_0^2 are the gamma densities: If $p(\sigma_0^2) \sim \text{gamma}(a, b)$, then it is straightforward to show that

$$p(\sigma_0^2|\sigma_1^2, \ldots, \sigma_m^2, \nu_0) = \text{dgamma}(a + \frac{1}{2}m\nu_0, b + \frac{1}{2}\nu_0 \sum_{j=1}^{m}(1/\sigma_j^2)).$$

Notice that for small values of a and b the conditional mean of σ_0^2 is approximately the harmonic mean of $\sigma_1^2, \ldots, \sigma_m^2$.

A simple conjugate prior for ν_0 does not exist, but if we restrict ν_0 to be a whole number, then it is easy to sample from its full conditional distribution. For example, if we let the prior on ν_0 be the geometric distribution on $\{1, 2, \ldots\}$ so that $p(\nu_0) \propto e^{-\alpha\nu_0}$, then the full conditional distribution of ν_0 is proportional to

$$\begin{aligned}
&p(\nu_0|\sigma_0^2, \sigma_1^2, \ldots, \sigma_m^2) \\
&\propto p(\nu_0) \times p(\sigma_1^2, \ldots, \sigma_m^2|\nu_0, \sigma_0^2) \\
&\propto \left(\frac{(\nu_0\sigma_0^2/2)^{\nu_0/2}}{\Gamma(\nu_0/2)}\right)^m \left(\prod_{j=1}^{m}\frac{1}{\sigma_j^2}\right)^{\nu_0/2-1} \times \exp\{-\nu_0(\alpha + \frac{1}{2}\sigma_0^2\sum(1/\sigma_j^2))\}.
\end{aligned}$$

While not pretty, this unnormalized probability distribution can be computed for a large range of ν_0-values and then sampled from. For example, the R-code to sample from this distribution is as follows:

```
# NUMAX, alpha must be specified ,
# sigma2.schools is the vector of current values
# of the school specific population variances.

x<-1:NUMAX

lpnu0<- m*( .5*x*log(s20*x/2)-lgamma(x/2)  ) +
          (x/2-1)*sum(log(1/sigma2.schools))        +
          -x*(alpha + .5*s20*sum(1/sigma2.schools) )

nu0<-sample(x,1,prob=exp(lpnu0-max(lpnu0)))
```

8.5.1 Analysis of math score data

Let's re-analyze the math score data with our hierarchical model for school-specific means and variances. We'll take the parameters in our prior distributions to be the same as in the previous section, with $\alpha = 1$ and $\{a = 1, b = 1/100\}$ for the prior distributions on ν_0 and σ_0^2. After running a Gibbs sampler for 5,000 iterations, the posterior distributions of $\{\mu, \tau^2, \nu_0, \sigma_0^2\}$ are approximated and plotted in Figure 8.11. Additionally, the posterior distributions of μ and τ^2 under the hierarchical model with constant group variance is shown in gray lines for comparison.

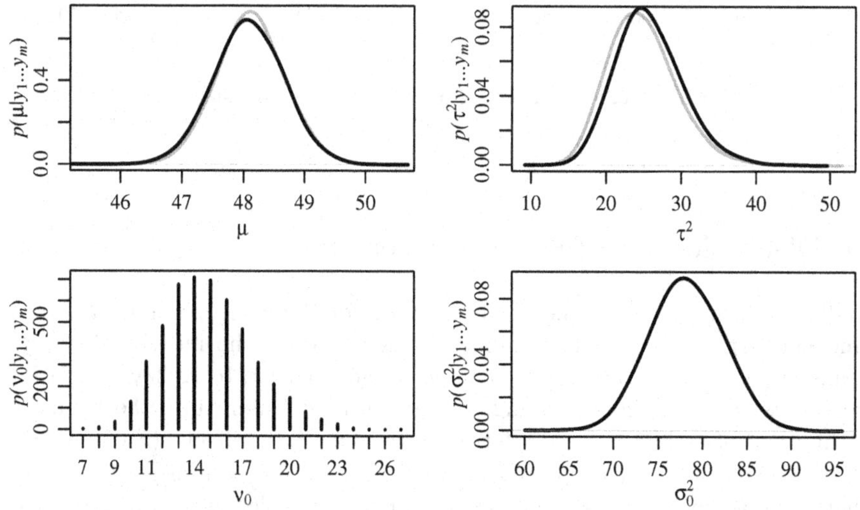

Fig. 8.11. Posterior distributions of between-group heterogeneity parameters.

The hierarchical model of Section 8.3, in which all within-group variances are forced to be equal, is equivalent to a value of $\nu_0 = \infty$ in this hierarchical model. In contrast, a low value of $\nu_0 = 1$ would indicate that the variances are quite unequal, and little information about variances should be shared across groups. Our hierarchical analysis indicates that neither of these extremes is appropriate, as the posterior distribution of ν_0 is concentrated around a moderate value of 14 or 15. This estimated distribution of $\sigma_1^2, \ldots, \sigma_n^2$ is used to shrink extreme sample variances towards an across-group center, as is shown in Figure 8.12. The relationship between sample size and the amount of variance shrinkage is shown in the second panel of the plot. As with estimation for the group means, the larger amounts of shrinkage generally occur for groups with smaller sample sizes.

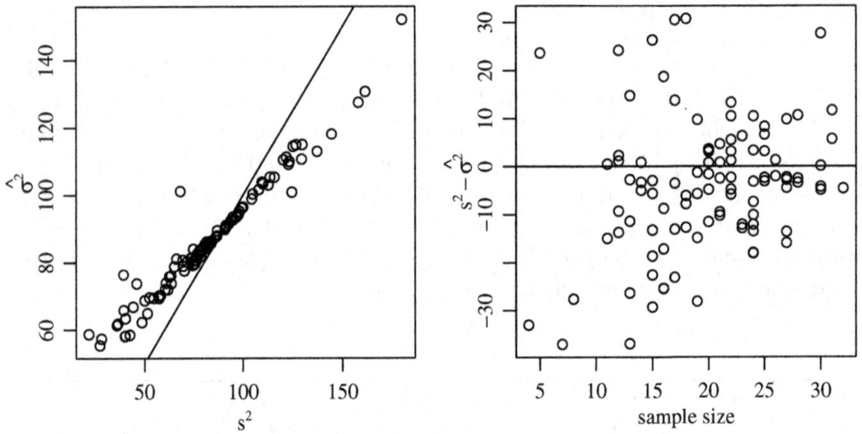

Fig. 8.12. Shrinkage as a function of sample size.

8.6 Discussion and further references

Lindley and Smith (1972) laid the foundation for Bayesian hierarchical modeling, although the idea of shrinking the estimates of the individual group means towards an across-group mean goes back at least to Kelley (1927) in the context of educational testing. In the statistical literature, the benefits of this type of estimation are referred to as the "Stein effect" (Stein, 1956, 1981). Estimators of this type generally take the form $\hat{\theta}_j = w_j \bar{y}_j + (1 - w_j)\bar{y}$, where $\bar{y}$ is an average over all groups and the w_j's depend on n_j, σ^2 and τ^2. So-called *empirical Bayes* procedures obtain estimates of σ^2 and τ^2 from the data, then plug these values into the formula for $\hat{\theta}_j$ (Efron and Morris, 1973; Casella, 1985). Such procedures often yield estimates of the θ_j's that are nearly equivalent to those from Bayesian procedures, but ignore uncertainty in the values of σ^2 and τ^2. For a detailed treatment of empirical Bayes methods, see Carlin and Louis (1996).

Terminology for hierarchical models is inconsistent in the literature. For the simple hierarchical model $y_{i,j} = \theta_j + \epsilon_{i,j}$, $\theta_j = \mu + \gamma_j$, the θ_j's (or γ_j's) may be referred to as either "fixed effects" or "random effects," usually depending on how they are estimated. The distribution of the θ_j's is unfortunately often referred to as a prior distribution, which mischaracterizes Bayesian inference and renders the distinction between prior information and population distribution somewhat meaningless. For a discussion of some of this confusion, see Gelman and Hill (2007, pp. 245-246).

Hierarchical modeling of variances is not common, perhaps due to the mean parameters being of greater interest. However, erroneously assuming a common within-group variance could lead to improper pooling of information, or to the shrinkage of group-specific parameters by inappropriate amounts.

9

Linear regression

Linear regression modeling is an extremely powerful data analysis tool, useful for a variety of inferential tasks such as prediction, parameter estimation and data description. In this section we give a very brief introduction to the linear regression model and the corresponding Bayesian approach to estimation. Additionally, we discuss the relationship between Bayesian and ordinary least squares regression estimates.

One difficult aspect of regression modeling is deciding which explanatory variables to include in a model. This variable selection problem has a natural Bayesian solution: Any collection of models having different sets of regressors can be compared via their Bayes factors. When the number of possible regressors is small, this allows us to assign a posterior probability to each regression model. When the number of regressors is large, the space of models can be explored with a Gibbs sampling algorithm.

9.1 The linear regression model

Regression modeling is concerned with describing how the sampling distribution of one random variable Y varies with another variable or set of variables $\boldsymbol{x} = (x_1, \ldots, x_p)$. Specifically, a regression model postulates a form for $p(y|\boldsymbol{x})$, the conditional distribution of Y given $\boldsymbol{x}$. Estimation of $p(y|\boldsymbol{x})$ is made using data $y_1, \ldots, y_n$ that are gathered under a variety of conditions $\boldsymbol{x}_1, \ldots, \boldsymbol{x}_n$.

Example: Oxygen uptake (from Kuehl (2000))

Twelve healthy men who did not exercise regularly were recruited to take part in a study of the effects of two different exercise regimen on oxygen uptake. Six of the twelve men were randomly assigned to a 12-week flat-terrain running program, and the remaining six were assigned to a 12-week step aerobics program. The maximum oxygen uptake of each subject was measured (in liters per minute) while running on an inclined treadmill, both before and

P.D. Hoff, *A First Course in Bayesian Statistical Methods*,
Springer Texts in Statistics, DOI 10.1007/978-0-387-92407-6_9,

after the 12-week program. Of interest is how a subject's change in maximal oxygen uptake may depend on which program they were assigned to. However, other factors, such as age, are expected to affect the change in maximal uptake as well.

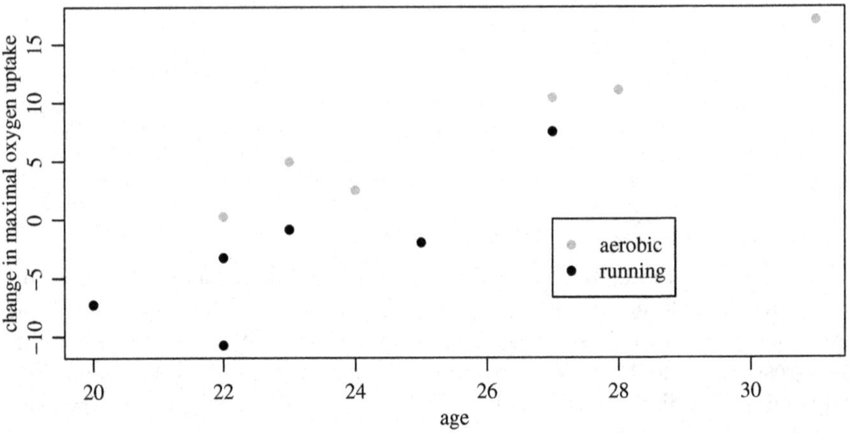

Fig. 9.1. Change in maximal oxygen uptake as a function of age and exercise program.

How might we estimate the conditional distribution of oxygen uptake for a given exercise program and age? One possibility would be to estimate a population mean and variance for each age and program combination. For example, we could estimate a mean and variance from the 22-year-olds in the study who were assigned the running program, and a separate mean and variance for the 22-year-olds assigned to the aerobics program. The data from this study, shown in Figure 9.1, indicate that such an approach is problematic. For example, there is only one 22-year-old assigned to the aerobics program, which is not enough data to provide information about a population variance. Furthermore, there are many age/program combinations for which there are no data.

One solution to this problem is to assume that the conditional distribution $p(y|\boldsymbol{x})$ changes smoothly as a function of $\boldsymbol{x}$, so that data we have at one value of $\boldsymbol{x}$ can inform us about what might be going on at a different value. A *linear regression model* is a particular type of smoothly changing model for $p(y|\boldsymbol{x})$ that specifies that the conditional expectation $\mathrm{E}[Y|\boldsymbol{x}]$ has a form that is linear in a set of parameters:

$$\int yp(y|\boldsymbol{x})\,dy = \mathrm{E}[Y|\boldsymbol{x}] = \beta_1 x_1 + \cdots + \beta_p x_p = \boldsymbol{\beta}^T \boldsymbol{x}\,.$$

It is important to note that such a model allows a great deal of freedom for $x_1, \ldots, x_p$. For example, in the oxygen uptake example we could let $x_1 = \text{age}$ and $x_2 = \text{age}^2$ if we thought there might be a quadratic relationship between maximal uptake and age. However, Figure 9.1 does not indicate any quadratic relationships, and so a reasonable model for $p(y|\boldsymbol{x})$ could include two different linear relationships between age and uptake, one for each group:

$$\begin{aligned} Y_i &= \beta_1 x_{i,1} + \beta_2 x_{i,2} + \beta_3 x_{i,3} + \beta_4 x_{i,4} + \epsilon_i \text{ , where} \quad (9.1) \\ x_{i,1} &= 1 \text{ for each subject } i \\ x_{i,2} &= 0 \text{ if subject } i \text{ is on the running program, 1 if on aerobic} \\ x_{i,3} &= \text{age of subject } i \\ x_{i,4} &= x_{i,2} \times x_{i,3} \end{aligned}$$

Under this model the conditional expectations of Y for the two different levels of $x_{i,2}$ are

$$\begin{aligned} \mathrm{E}[Y|\boldsymbol{x}] &= \beta_1 + \beta_3 \times \text{age} \text{ if } x_2 = 0, \text{ and} \\ \mathrm{E}[Y|\boldsymbol{x}] &= (\beta_1 + \beta_2) + (\beta_3 + \beta_4) \times \text{age} \text{ if } x_2 = 1. \end{aligned}$$

In other words, the model assumes that the relationship is linear in age for both exercise groups, with the difference in intercepts given by β_2 and the difference in slopes given by β_4. If we assumed that $\beta_2 = \beta_4 = 0$, then we would have an identical line for both groups. If we assumed $\beta_4 = 0$ then we would have a different line for each group but they would be parallel. Allowing all coefficients to be non-zero gives us two unrelated lines. Some different possibilities are depicted graphically in Figure 9.2.

We still have not specified anything about $p(y|\boldsymbol{x})$ beyond $\mathrm{E}[Y|\boldsymbol{x}]$. The *normal linear regression model* specifies that, in addition to $\mathrm{E}[Y|\boldsymbol{x}]$ being linear, the sampling variability around the mean is i.i.d. from a normal distribution:

$$\begin{aligned} \epsilon_1, \ldots, \epsilon_n &\sim \text{i.i.d. normal}(0, \sigma^2) \\ Y_i &= \boldsymbol{\beta}^T \boldsymbol{x}_i + \epsilon_i \,. \end{aligned}$$

This model provides a complete specification of the joint probability density of observed data $y_1, \ldots, y_n$ conditional upon $\boldsymbol{x}_1, \ldots, \boldsymbol{x}_n$ and values of $\boldsymbol{\beta}$ and σ^2:

$$\begin{aligned} & p(y_1, \ldots, y_n | \boldsymbol{x}_1, \ldots \boldsymbol{x}_n, \boldsymbol{\beta}, \sigma^2) \quad (9.2) \\ &= \prod_{i=1}^n p(y_i | \boldsymbol{x}_i, \boldsymbol{\beta}, \sigma^2) \\ &= (2\pi\sigma^2)^{-n/2} \exp\{-\frac{1}{2\sigma^2} \sum_{i=1}^n (y_i - \boldsymbol{\beta}^T \boldsymbol{x}_i)^2\}. \quad (9.3) \end{aligned}$$

Another way to write this joint probability density is in terms of the multivariate normal distribution: Let $\boldsymbol{y}$ be the n-dimensional column vector

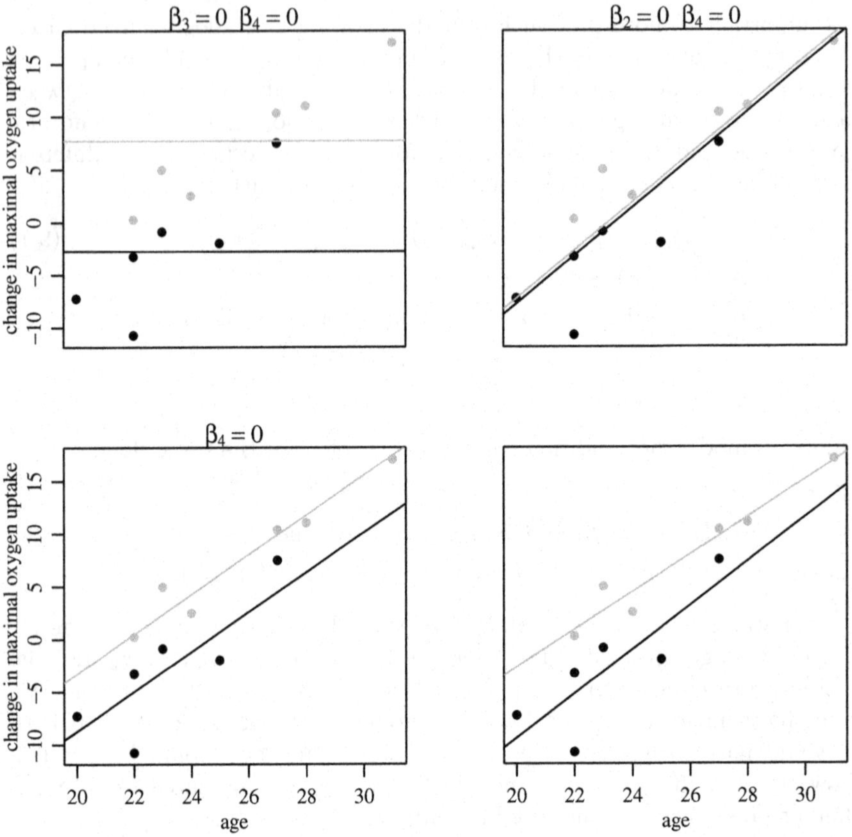

Fig. 9.2. Least squares regression lines for the oxygen uptake data, under four different models.

$(y_1, \ldots, y_n)^T$, and let $\mathbf{X}$ be the $n \times p$ matrix whose ith row is $\boldsymbol{x}_i$. Then the normal regression model is that

$$\{\boldsymbol{y}|\mathbf{X}, \boldsymbol{\beta}, \sigma^2\} \sim \text{ multivariate normal } (\mathbf{X}\boldsymbol{\beta}, \sigma^2\mathbf{I}),$$

where $\mathbf{I}$ is the $p \times p$ identity matrix and

$$\mathbf{X}\boldsymbol{\beta} = \begin{pmatrix} \boldsymbol{x}_1 \rightarrow \\ \boldsymbol{x}_2 \rightarrow \\ \vdots \\ \boldsymbol{x}_n \rightarrow \end{pmatrix} \begin{pmatrix} \beta_1 \\ \vdots \\ \beta_p \end{pmatrix} = \begin{pmatrix} \beta_1 x_{1,1} + \cdots + \beta_p x_{1,p} \\ \vdots \\ \beta_1 x_{n,1} + \cdots + \beta_p x_{n,p} \end{pmatrix} = \begin{pmatrix} \mathrm{E}[Y_1|\boldsymbol{\beta}, \boldsymbol{x}_1] \\ \vdots \\ \mathrm{E}[Y_n|\boldsymbol{\beta}, \boldsymbol{x}_n] \end{pmatrix}.$$

The density (9.3) depends on $\boldsymbol{\beta}$ through the residuals $(y_i - \boldsymbol{\beta}^T \boldsymbol{x}_i)$. Given the observed data, the term in the exponent is maximized when the sum of squared residuals, $\mathrm{SSR}(\boldsymbol{\beta}) = \sum_{i=1}^n (y_i - \boldsymbol{\beta}^T \boldsymbol{x}_i)^2$ is minimized. To find the value of $\boldsymbol{\beta}$ at which this minimum occurs it is helpful to rewrite SSR($\boldsymbol{\beta}$) in matrix notation:

$$\begin{aligned}\text{SSR}(\boldsymbol{\beta}) &= \sum_{i=1}^{n}(y_i - \boldsymbol{\beta}^T\boldsymbol{x}_i)^2 = (\boldsymbol{y} - \mathbf{X}\boldsymbol{\beta})^T(\boldsymbol{y} - \mathbf{X}\boldsymbol{\beta}) \\ &= \boldsymbol{y}^T\boldsymbol{y} - 2\boldsymbol{\beta}^T\mathbf{X}^T\boldsymbol{y} + \boldsymbol{\beta}^T\mathbf{X}^T\mathbf{X}\boldsymbol{\beta}\,.\end{aligned}$$

Recall from calculus that

1. a minimum of a function $g(z)$ occurs at a value z such that $\frac{d}{dz}g(z) = 0$;
2. the derivative of $g(z) = az$ is a and the derivative of $g(z) = bz^2$ is $2bz$.

These facts translate over to the multivariate case, and can be used to obtain the minimizer of SSR($\boldsymbol{\beta}$):

$$\begin{aligned}\frac{d}{d\boldsymbol{\beta}}\text{SSR}(\boldsymbol{\beta}) &= \frac{d}{d\boldsymbol{\beta}}\left(\boldsymbol{y}^T\boldsymbol{y} - 2\boldsymbol{\beta}^T\mathbf{X}^T\boldsymbol{y} + \boldsymbol{\beta}^T\mathbf{X}^T\mathbf{X}\boldsymbol{\beta}\right) \\ &= -2\mathbf{X}^T\boldsymbol{y} + 2\mathbf{X}^T\mathbf{X}\boldsymbol{\beta}\ \text{, therefore} \\ \frac{d}{d\boldsymbol{\beta}}\text{SSR}(\boldsymbol{\beta}) = 0 &\Leftrightarrow -2\mathbf{X}^T\boldsymbol{y} + 2\mathbf{X}^T\mathbf{X}\boldsymbol{\beta} = 0 \\ &\Leftrightarrow \mathbf{X}^T\mathbf{X}\boldsymbol{\beta} = \mathbf{X}^T\boldsymbol{y} \\ &\Leftrightarrow \boldsymbol{\beta} = (\mathbf{X}^T\mathbf{X})^{-1}\mathbf{X}^T\boldsymbol{y}\,.\end{aligned}$$

The value $\hat{\boldsymbol{\beta}}_{\text{ols}} = (\mathbf{X}^T\mathbf{X})^{-1}\mathbf{X}^T\boldsymbol{y}$ is called the "ordinary least squares" (OLS) estimate of $\boldsymbol{\beta}$, as it provides the value of $\boldsymbol{\beta}$ that minimizes the sum of squared residuals. This value is unique as long as the inverse $(\mathbf{X}^T\mathbf{X})^{-1}$ exists. The value $\hat{\boldsymbol{\beta}}_{\text{ols}}$ also plays a role in Bayesian estimation, as we shall see in the next section.

9.1.1 Least squares estimation for the oxygen uptake data

Let's find the least squares regression estimates for the model in Equation 9.1, and use the results to evaluate differences between the two exercise groups. The ages of the 12 subjects, along with their observed changes in maximal oxygen uptake, are

$$\begin{aligned}\boldsymbol{x}_3 &= (23, 22, 22, 25, 27, 20, 31, 23, 27, 28, 22, 24) \\ \boldsymbol{y} &= (-0.87, -10.74, -3.27, -1.97, 7.50, -7.25, 17.05, 4.96, 10.40, \\ &\qquad 11.05, 0.26, 2.51)\,,\end{aligned}$$

with the first six elements of each vector corresponding to the subjects in the running group and the latter six corresponding to subjects in the aerobics group. After constructing the 12×4 matrix $\mathbf{X}$ out of the vectors $\boldsymbol{x}_1, \boldsymbol{x}_2, \boldsymbol{x}_3, \boldsymbol{x}_4$ defined as in (9.1), the matrices $\mathbf{X}^T\mathbf{X}$ and $\mathbf{X}^T\boldsymbol{y}$ can be computed:

$$\mathbf{X}^T\mathbf{X} = \begin{pmatrix} 12 & 6 & 294 & 155 \\ 6 & 6 & 155 & 155 \\ 294 & 155 & 7314 & 4063 \\ 155 & 155 & 4063 & 4063 \end{pmatrix} \quad \mathbf{X}^T\boldsymbol{y} = \begin{pmatrix} 29.63 \\ 46.23 \\ 978.81 \\ 1298.79 \end{pmatrix}.$$

Inverting the $\mathbf{X}^T\mathbf{X}$ matrix and multiplying the result by $\mathbf{X}^T\boldsymbol{y}$ give the vector $\hat{\boldsymbol{\beta}}_{\text{ols}} = (-51.29, 13.11, 2.09, -.32)^T$. This means that the estimated linear relationship between uptake and age has an intercept and slope of -51.29 and 2.09 for the running group, and $-51.29 + 13.11 = -38.18$ and $2.09 - 0.32 = 1.77$ for the aerobics group. These two lines are plotted in the fourth panel of Figure 9.2. An unbiased estimate of σ^2 can be obtained from $\text{SSR}(\hat{\boldsymbol{\beta}}_{\text{ols}})/(n-p)$, which for these data gives $\hat{\sigma}^2_{\text{ols}} = 8.54$. The sampling variance of the vector $\hat{\boldsymbol{\beta}}_{\text{ols}}$ can be shown to be equal to $(\mathbf{X}^T\mathbf{X})^{-1}\sigma^2$. We do not know the true value of σ^2, but the value of $\hat{\sigma}^2_{\text{ols}}$ can be plugged in to give standard errors for the components of $\hat{\boldsymbol{\beta}}_{\text{ols}}$. These are 12.25, 15.76, 0.53 and 0.65 for the four regression coefficients in order. Comparing the values of $\hat{\boldsymbol{\beta}}_{\text{ols}}$ to their standard errors suggests that the evidence for differences between the two exercise regimen is not very strong. We will explore this further in the next few sections.

9.2 Bayesian estimation for a regression model

We begin with a simple semiconjugate prior distribution for $\boldsymbol{\beta}$ and σ^2 to be used when there is information available about the parameters. In situations where prior information is unavailable or difficult to quantify, an alternative "default" class of prior distributions is given.

9.2.1 A semiconjugate prior distribution

The sampling density of the data (Equation 9.3), as a function of $\boldsymbol{\beta}$, is

$$\begin{aligned} p(\boldsymbol{y}|\mathbf{X}, \boldsymbol{\beta}, \sigma^2) &\propto \exp\{-\frac{1}{2\sigma^2}\text{SSR}(\boldsymbol{\beta})\} \\ &= \exp\{-\frac{1}{2\sigma^2}[\boldsymbol{y}^T\boldsymbol{y} - 2\boldsymbol{\beta}^T\mathbf{X}^T\boldsymbol{y} + \boldsymbol{\beta}^T\mathbf{X}^T\mathbf{X}\boldsymbol{\beta}]\}. \end{aligned}$$

The role that $\boldsymbol{\beta}$ plays in the exponent looks very similar to that played by $\boldsymbol{y}$, and the distribution of $\boldsymbol{y}$ is multivariate normal. This suggests that a multivariate normal prior distribution for $\boldsymbol{\beta}$ is conjugate. Let's see if this is correct: If $\boldsymbol{\beta} \sim$ multivariate normal$(\boldsymbol{\beta}_0, \Sigma_0)$, then

$$\begin{aligned} &p(\boldsymbol{\beta}|\boldsymbol{y}, \mathbf{X}, \sigma^2) \\ &\propto p(\boldsymbol{y}|\mathbf{X}, \boldsymbol{\beta}, \sigma^2) \times p(\boldsymbol{\beta}) \\ &\propto \exp\{-\frac{1}{2}(-2\boldsymbol{\beta}^T\mathbf{X}^T\boldsymbol{y}/\sigma^2 + \boldsymbol{\beta}^T\mathbf{X}^T\mathbf{X}\boldsymbol{\beta}/\sigma^2) - \frac{1}{2}(-2\boldsymbol{\beta}^T\Sigma_0^{-1}\boldsymbol{\beta}_0 + \boldsymbol{\beta}^T\Sigma_0^{-1}\boldsymbol{\beta})\} \\ &= \exp\{\boldsymbol{\beta}^T(\Sigma_0^{-1}\boldsymbol{\beta}_0 + \mathbf{X}^T\boldsymbol{y}/\sigma^2) - \frac{1}{2}\boldsymbol{\beta}^T(\Sigma_0^{-1} + \mathbf{X}^T\mathbf{X}/\sigma^2)\boldsymbol{\beta}\}. \end{aligned}$$

Referring back to Chapter 7, we recognize this as being proportional to a multivariate normal density, with

$$\mathrm{Var}[\boldsymbol{\beta}|\boldsymbol{y},\mathbf{X},\sigma^2] = (\Sigma_0^{-1} + \mathbf{X}^T\mathbf{X}/\sigma^2)^{-1} \tag{9.4}$$
$$\mathrm{E}[\boldsymbol{\beta}|\boldsymbol{y},\mathbf{X},\sigma^2] = (\Sigma_0^{-1} + \mathbf{X}^T\mathbf{X}/\sigma^2)^{-1}(\Sigma_0^{-1}\boldsymbol{\beta}_0 + \mathbf{X}^T\boldsymbol{y}/\sigma^2). \tag{9.5}$$

As usual, we can gain some understanding of these formulae by considering some limiting cases. If the elements of the prior precision matrix Σ_0^{-1} are small in magnitude, then the conditional expectation $\mathrm{E}[\boldsymbol{\beta}|\boldsymbol{y},\mathbf{X},\sigma^2]$ is approximately equal to $(\mathbf{X}^T\mathbf{X})^{-1}\mathbf{X}^T\boldsymbol{y}$, the least squares estimate. On the other hand, if the measurement precision is very small (σ^2 is very large), then the expectation is approximately $\boldsymbol{\beta}_0$, the prior expectation.

As in most normal sampling problems, the semiconjugate prior distribution for σ^2 is an inverse-gamma distribution. Letting $\gamma = 1/\sigma^2$ be the measurement precision, if $\gamma \sim \mathrm{gamma}(\nu_0/2, \nu_0\sigma_0^2/2)$, then

$$\begin{aligned} p(\gamma|\boldsymbol{y},\mathbf{X},\boldsymbol{\beta}) &\propto p(\gamma)p(\boldsymbol{y}|\mathbf{X},\boldsymbol{\beta},\gamma) \\ &\propto \left[\gamma^{\nu_0/2-1}\exp(-\gamma\times\nu_0\sigma_0^2/2)\right] \times \left[\gamma^{n/2}\exp(-\gamma\times \mathrm{SSR}(\boldsymbol{\beta})/2)\right] \\ &= \gamma^{(\nu_0+n)/2-1}\exp(-\gamma[\nu_0\sigma_0^2 + \mathrm{SSR}(\boldsymbol{\beta})]/2), \end{aligned}$$

which we recognize as a gamma density, so that

$$\{\sigma^2|\boldsymbol{y},\mathbf{X},\boldsymbol{\beta}\} \sim \text{inverse-gamma}([\nu_0+n]/2, [\nu_0\sigma_0^2 + \mathrm{SSR}(\boldsymbol{\beta})]/2).$$

Constructing a Gibbs sampler to approximate the joint posterior distribution $p(\boldsymbol{\beta},\sigma^2|\boldsymbol{y},\mathbf{X})$ is then straightforward: Given current values $\{\boldsymbol{\beta}^{(s)},\sigma^{2(s)}\}$, new values can be generated by

1. updating $\boldsymbol{\beta}$:
   a) compute $\mathbf{V} = \mathrm{Var}[\boldsymbol{\beta}|\boldsymbol{y},\mathbf{X},\sigma^{2(s)}]$ and $\mathbf{m} = \mathrm{E}[\boldsymbol{\beta}|\boldsymbol{y},\mathbf{X},\sigma^{2(s)}]$
   b) sample $\boldsymbol{\beta}^{(s+1)} \sim$ multivariate normal$(\mathbf{m},\mathbf{V})$
2. updating σ^2:
   a) compute $\mathrm{SSR}(\boldsymbol{\beta}^{(s+1)})$
   b) sample $\sigma^{2(s+1)} \sim$ inverse-gamma$([\nu_0+n]/2, [\nu_0\sigma_0^2 + \mathrm{SSR}(\boldsymbol{\beta}^{(s+1)})]/2)$.

9.2.2 Default and weakly informative prior distributions

A Bayesian analysis of a regression model requires specification of the prior parameters $(\boldsymbol{\beta}_0, \Sigma_0)$ and (ν_0, σ_0^2). Finding values of these parameters that represent actual prior information can be difficult. In the oxygen uptake experiment, for example, a quick scan of a few articles on exercise physiology indicates that males in their 20s have an oxygen uptake of around 150 liters per minute with a standard deviation of 15. If we take $150 \pm 2\times 15 = (120, 180)$ as a prior expected range of the oxygen uptake distribution, then the changes in oxygen uptake should lie within (-60,60) with high probability. Considering our subjects in the running group, this means that the line $\beta_1 + \beta_3 x$ should produce values between -60 and 60 for all values of x between 20 and 30. A little algebra then shows that we need a prior distribution on (β_1, β_3) such that $-300 < \beta_1 < 300$ and $-12 < \beta_3 < 12$ with high probability. This could be done by taking $\Sigma_{0,1,1} = 150^2$ and $\Sigma_{0,2,2} = 6^2$, for example. However, we would still need to specify the prior variances of the other parameters, as well as the six prior correlations between the parameters. The task of constructing an informative prior distribution only gets harder as the number of regressors increases, as the number of prior correlation parameters is $\binom{p}{2}$, which increases quadratically in p.

Sometimes an analysis must be done in the absence of precise prior information, or information that is easily converted into the parameters of a conjugate prior distribution. In these situations one could stick to least squares estimation, with the drawback that probability statements about $\boldsymbol{\beta}$ would be unavailable. Alternatively, it is sometimes possible to justify a prior distribution with other criteria. One idea is that, if the prior distribution is not going to represent real prior information about the parameters, then it should be as minimally informative as possible. The resulting posterior distribution would then represent the posterior information of someone who began with little knowledge of the population being studied. To some, such an analysis would give a "more objective" result than using an informative prior distribution, especially one that did not actually represent real prior information. One type of weakly informative prior is the *unit information prior* (Kass and Wasserman, 1995). A unit information prior is one that contains the same amount of information as that would be contained in only a single observation. For example, the precision of $\hat{\boldsymbol{\beta}}_{\text{ols}}$ is its inverse variance, or $(\mathbf{X}^T\mathbf{X})/\sigma^2$. Since this can be viewed as the amount of information in n observations, the amount of information in one observation should be "one nth" as much, i.e. $(\mathbf{X}^T\mathbf{X})/(n\sigma^2)$. The unit information prior thus sets $\Sigma_0^{-1} = (\mathbf{X}^T\mathbf{X})/(n\sigma^2)$. Kass and Wasserman (1995) further suggest setting $\boldsymbol{\beta}_0 = \hat{\boldsymbol{\beta}}_{\text{ols}}$, thus centering the prior distribution of $\boldsymbol{\beta}$ around the OLS estimate. Such a distribution cannot be strictly considered a real *prior* distribution, as it requires knowledge of $\boldsymbol{y}$ to be constructed. However, it only uses a small amount of the information in $\boldsymbol{y}$, and can be loosely thought of as the prior distribution of a person with unbiased but weak prior information. In a similar way, the prior distribution of σ^2 can be weakly centered around $\hat{\sigma}^2_{\text{ols}}$ by taking $\nu_0 = 1$ and $\sigma_0^2 = \hat{\sigma}^2_{\text{ols}}$.

Another principle for constructing a prior distribution for $\boldsymbol{\beta}$ is based on the idea that the parameter estimation should be invariant to changes in the scale of the regressors. For example, suppose someone were to analyze the oxygen uptake data using $\tilde{x}_{i,3}$ = age in months, instead of $x_{i,3}$ = age in years. It makes sense that our posterior distribution for $12 \times \tilde{\beta}_3$ in the model with $\tilde{x}_{i,3}$ should be the same as the posterior distribution for β_3 based on the model with $x_{i,3}$. This condition requires, for example, that the posterior expected change in y for a year change in age is the same, whether age is recorded in terms of months or years. More generally, suppose $\mathbf{X}$ is a given set of regressors and $\tilde{\mathbf{X}} = \mathbf{XH}$ for some $p \times p$ matrix $\mathbf{H}$. If we obtain the posterior distribution of $\boldsymbol{\beta}$ from $\boldsymbol{y}$ and $\mathbf{X}$, and the posterior distribution of $\tilde{\boldsymbol{\beta}}$ from $\boldsymbol{y}$ and $\tilde{\mathbf{X}}$, then, according to this principle of invariance, the posterior distributions of $\boldsymbol{\beta}$ and $\mathbf{H}\tilde{\boldsymbol{\beta}}$ should be the same. Some linear algebra shows that this condition will be met if $\boldsymbol{\beta}_0 = \mathbf{0}$ and $\Sigma_0 = k(\mathbf{X}^T\mathbf{X})^{-1}$ for any positive value k. A popular specification of k is to relate it to the error variance σ^2, so that $k = g\sigma^2$ for some positive value g. These choices of prior parameters result in a version of the so-called "g-prior" (Zellner, 1986), a widely studied and used prior distribution for regression parameters (Zellner's original g-prior allowed $\boldsymbol{\beta}_0$ to be non-zero). Under this invariant g-prior the conditional distribution of $\boldsymbol{\beta}$ given $(\boldsymbol{y}, \mathbf{X}, \sigma^2)$ is still multivariate normal, but Equations 9.4 and 9.5 reduce to the following simpler forms:

$$\begin{aligned}
\mathrm{Var}[\boldsymbol{\beta}|\boldsymbol{y}, \mathbf{X}, \sigma^2] &= [\mathbf{X}^T\mathbf{X}/(g\sigma^2) + \mathbf{X}^T\mathbf{X}/\sigma^2]^{-1} \\
&= \frac{g}{g+1}\sigma^2(\mathbf{X}^T\mathbf{X})^{-1} &(9.6)\\
\mathrm{E}[\boldsymbol{\beta}|\boldsymbol{y}, \mathbf{X}, \sigma^2] &= [\mathbf{X}^T\mathbf{X}/(g\sigma^2) + \mathbf{X}^T\mathbf{X}/\sigma^2]^{-1}\mathbf{X}^T\boldsymbol{y}/\sigma^2 \\
&= \frac{g}{g+1}(\mathbf{X}^T\mathbf{X})^{-1}\mathbf{X}^T\boldsymbol{y}. &(9.7)
\end{aligned}$$

Parameter estimation under the g-prior is simplified as well: It turns out that, under this prior distribution, $p(\sigma^2|\boldsymbol{y}, \mathbf{X})$ is an inverse-gamma distribution, which means that we can directly sample $(\sigma^2, \boldsymbol{\beta})$ from their posterior distribution by first sampling from $p(\sigma^2|\boldsymbol{y}, \mathbf{X})$ and then from $p(\boldsymbol{\beta}|\sigma^2, \boldsymbol{y}, \mathbf{X})$.

Derivation of $p(\sigma^2|\boldsymbol{y}, \mathbf{X})$

The marginal posterior density of σ^2 is proportional to $p(\sigma^2) \times p(\boldsymbol{y}|\mathbf{X}, \sigma^2)$. Using the rules of marginal probability, the latter term in this product can be expressed as the following integral:

$$p(\boldsymbol{y}|\mathbf{X}, \sigma^2) = \int p(\boldsymbol{y}|\mathbf{X}, \boldsymbol{\beta}, \sigma^2)p(\boldsymbol{\beta}|\mathbf{X}, \sigma^2)\, d\boldsymbol{\beta}.$$

Writing out the two densities inside the integral, we have

$$\begin{aligned}
p(\boldsymbol{y}|\mathbf{X}, \boldsymbol{\beta}, \sigma^2)p(\boldsymbol{\beta}|\mathbf{X}, \sigma^2) = (2\pi\sigma^2)^{-n/2}\exp[-\frac{1}{2\sigma^2}(\boldsymbol{y} - \mathbf{X}\boldsymbol{\beta})^T(\boldsymbol{y} - \mathbf{X}\boldsymbol{\beta})] \times \\
|2\pi g\sigma^2(\mathbf{X}^T\mathbf{X})^{-1}|^{-1}\exp[-\frac{1}{2g\sigma^2}\boldsymbol{\beta}^T\mathbf{X}^T\mathbf{X}\boldsymbol{\beta}].
\end{aligned}$$

Combining the terms in the exponents gives

$$\begin{aligned}
-\frac{1}{2\sigma^2}&\left[(\boldsymbol{y} - \mathbf{X}\boldsymbol{\beta})^T(\boldsymbol{y} - \mathbf{X}\boldsymbol{\beta}) + \boldsymbol{\beta}^T\mathbf{X}^T\mathbf{X}\boldsymbol{\beta}/g\right] \\
&= -\frac{1}{2\sigma^2}\left[\boldsymbol{y}^T\boldsymbol{y} - 2\boldsymbol{y}^T\mathbf{X}\boldsymbol{\beta} + \boldsymbol{\beta}^T\mathbf{X}^T\mathbf{X}\boldsymbol{\beta}(1 + 1/g)\right] \\
&= -\frac{1}{2\sigma^2}\boldsymbol{y}^T\boldsymbol{y} - \frac{1}{2}(\boldsymbol{\beta} - \mathrm{m})^T\mathbf{V}^{-1}(\boldsymbol{\beta} - \mathrm{m}) + \frac{1}{2}\mathrm{m}^T\mathbf{V}^{-1}\mathrm{m}\,,
\end{aligned}$$

where

$$\mathbf{V} = \frac{g}{g+1}\sigma^2(\mathbf{X}^T\mathbf{X})^{-1} \text{ and } \mathbf{m} = \frac{g}{g+1}(\mathbf{X}^T\mathbf{X})^{-1}\mathbf{X}^T\boldsymbol{y}.$$

This means that we can write $p(\boldsymbol{y}|\boldsymbol{\beta},\mathbf{X},\sigma^2)p(\boldsymbol{\beta}|\mathbf{X},\sigma^2)$ as

$$\left[(2\pi\sigma^2)^{-n/2}\exp(-\frac{1}{2\sigma^2}\boldsymbol{y}^T\boldsymbol{y})\right] \times \left[(1+g)^{-p/2}\exp(\frac{1}{2}\mathbf{m}^T\mathbf{V}^{-1}\mathbf{m})\right] \times$$
$$\left[|2\pi\mathbf{V}|^{-1/2}\exp[-\frac{1}{2}(\boldsymbol{\beta}-\mathbf{m})^T\mathbf{V}^{-1}(\boldsymbol{\beta}-\mathbf{m})]\right].$$

The third term in the product is the only term that depends on $\boldsymbol{\beta}$. This term is exactly the multivariate normal density with mean $\mathbf{m}$ and variance $\mathbf{V}$, which as a probability density must integrate to 1. This means that if we integrate the whole thing with respect to $\boldsymbol{\beta}$ we are left with only the first two terms:

$$\begin{aligned} p(\boldsymbol{y}|\mathbf{X},\sigma^2) &= \int p(\boldsymbol{y}|\boldsymbol{\beta},\mathbf{X},\sigma^2)p(\boldsymbol{\beta}|\mathbf{X},\sigma^2)\, d\boldsymbol{\beta} \\ &= \left[(2\pi\sigma^2)^{-n/2}\exp(-\frac{1}{2\sigma^2}\boldsymbol{y}^T\boldsymbol{y})\right] \times \left[(1+g)^{-p/2}\exp(\frac{1}{2}\mathbf{m}^T\mathbf{V}^{-1}\mathbf{m})\right], \end{aligned}$$

which, after combining the terms in the exponents, is

$$p(\boldsymbol{y}|\mathbf{X},\sigma^2) = (2\pi)^{-n/2}(1+g)^{-p/2}(\sigma^2)^{-n/2}\exp(-\frac{1}{2\sigma^2}\mathrm{SSR}_g),$$

where SSR_g is defined as

$$\mathrm{SSR}_g = \boldsymbol{y}^T\boldsymbol{y} - \mathbf{m}^T\mathbf{V}^{-1}\mathbf{m} = \boldsymbol{y}^T(\mathbf{I} - \frac{g}{g+1}\mathbf{X}(\mathbf{X}^T\mathbf{X})^{-1}\mathbf{X}^T)\boldsymbol{y}.$$

The term SSR_g decreases to $\mathrm{SSR}_{\mathrm{ols}} = \sum(y_i - \hat{\boldsymbol{\beta}}_{\mathrm{ols}}\boldsymbol{x}_i)^2$ as $g \to \infty$. The effect of g is that it shrinks down the magnitude of the regression coefficients and can prevent overfitting of the data.

The last step in identifying $p(\sigma^2|\boldsymbol{y},\mathbf{X})$ is to multiply $p(\boldsymbol{y}|\mathbf{X},\sigma^2)$ by the prior distribution. Letting $\gamma = 1/\sigma^2 \sim \text{gamma}(\nu_0/2, \nu_0\sigma_0^2/2)$, we have

$$\begin{aligned} p(\gamma|\boldsymbol{y},\mathbf{X}) &\propto p(\gamma)p(\boldsymbol{y}|\mathbf{X},\gamma) \\ &\propto \left[\gamma^{\nu_0/2-1}\exp(-\gamma\times\nu_0\sigma_0^2/2)\right] \times \left[\gamma^{n/2}\exp(-\gamma\times\mathrm{SSR}_g/2)\right] \\ &= \gamma^{(\nu_0+n)/2-1}\exp[-\gamma\times(\nu_0\sigma_0^2+\mathrm{SSR}_g)/2] \\ &\propto \text{dgamma}(\gamma, [\nu_0+n]/2, [\nu_0\sigma_0^2+\mathrm{SSR}_g]/2)\ , \end{aligned}$$

and so $\{\sigma^2|\boldsymbol{y},\mathbf{X}\} \sim \text{inverse-gamma}([\nu_0+n]/2, [\nu_0\sigma_0^2+\mathrm{SSR}_g]/2)$. These calculations, along with Equations 9.6 and 9.7, show that under this prior distribution, $p(\sigma^2|\boldsymbol{y},\mathbf{X})$ and $p(\boldsymbol{\beta}|\boldsymbol{y},\mathbf{X},\sigma^2)$ are inverse-gamma and multivariate normal distributions respectively. Since we can sample from both of these distributions, samples from the joint posterior distribution $p(\sigma^2,\boldsymbol{\beta}|\boldsymbol{y},\mathbf{X})$ can be made with Monte Carlo approximation, and Gibbs sampling is unnecessary. A sample value of $(\sigma^2,\boldsymbol{\beta})$ from $p(\sigma^2,\boldsymbol{\beta}|\boldsymbol{y},\mathbf{X})$ can be made as follows:

1. sample $1/\sigma^2 \sim \text{gamma}([\nu_0 + n]/2, [\nu_0\sigma_0^2 + \text{SSR}_g]/2)$;
2. sample $\boldsymbol{\beta} \sim$ multivariate normal$(\frac{g}{g+1}\hat{\boldsymbol{\beta}}_{\text{ols}}, \frac{g}{g+1}\sigma^2[\mathbf{X}^T\mathbf{X}]^{-1})$.

R-code to generate multiple independent Monte Carlo samples from the posterior distribution is below:

```
data(chapter9) ; y<-yX.o2uptake[,1] ; X<-yX.o2uptake[,-1]
g<-length(y) ; nu0<-1 ; s20<-8.54
S<-1000

## data: y, X
## prior parameters: g, nu0, s20
## number of independent samples to generate: S

n<-dim(X)[1] ; p<-dim(X)[2]
Hg<- (g/(g+1)) * X%*%solve(t(X)%*%X)%*%t(X)
SSRg<- t(y)%*%( diag(1,nrow=n) - Hg ) %*%y

s2<-1/rgamma(S, (nu0+n)/2, (nu0*s20+SSRg)/2 )

Vb<- g*solve(t(X)%*%X)/(g+1)
Eb<- Vb%*%t(X)%*%y

E<-matrix(rnorm(S*p,0,sqrt(s2)),S,p)
beta<-t( t(E%*%chol(Vb)) +c(Eb))
```

Bayesian analysis of the oxygen uptake data

We will use the invariant g-prior with $g = n = 12$, $\nu_0 = 1$ and $\sigma_0^2 = \hat{\sigma}_{\text{ols}}^2 = 8.54$. The posterior mean of $\boldsymbol{\beta}$ can be obtained directly from Equation 9.7: Since $\text{E}[\boldsymbol{\beta}|\boldsymbol{y}, \mathbf{X}, \sigma^2]$ does not depend on σ^2, we have $\text{E}[\boldsymbol{\beta}|\boldsymbol{y}, \mathbf{X}] = \text{E}[\boldsymbol{\beta}|\boldsymbol{y}, \mathbf{X}, \sigma^2] = \frac{g}{g+1}\hat{\boldsymbol{\beta}}_{\text{ols}}$, so the posterior means of the four regression parameters are $12 \times (-51.29, 13.11, 2.09, -0.32)/13 = (-47.35, 12.10, 1.93, -0.29)$. Posterior standard deviations of these parameters are (14.41, 18.62, 0.62, 0.77), based on 1,000 independent Monte Carlo samples generated using the R-code above. The marginal and joint posterior distributions for (β_2, β_4) are given in Figure 9.3, along with the (marginal) prior distributions for comparison. The posterior distributions seem to suggest only weak evidence of a difference between the two groups, as the 95% quantile-based posterior intervals for β_2 and β_4 both contain zero. However, these parameters taken by themselves do not quite tell the whole story. According to the model, the average difference in y between two people of the same age x but in different exercise programs is $\beta_2 + \beta_4 x$. Thus the posterior distribution for the effect of the aerobics program over the running program is obtained via the posterior distribution of $\beta_2 + \beta_4 x$ for each age x. Boxplots of these posterior distributions are shown in Figure 9.4, which indicates reasonably strong evidence of a difference at young ages, and less evidence at the older ones.

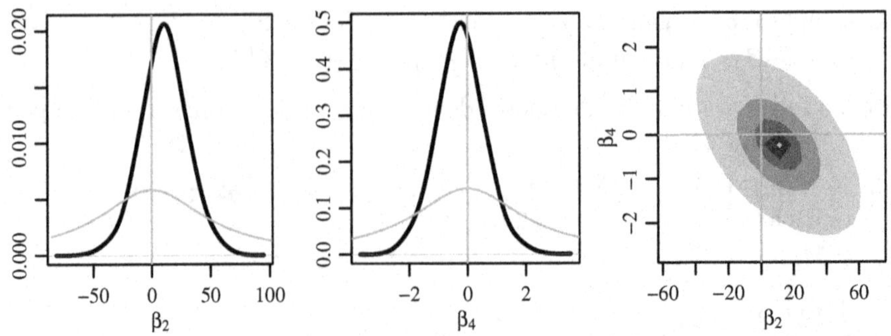

Fig. 9.3. Posterior distributions for β_2 and β_4, with marginal prior distributions in the first two plots for comparison.

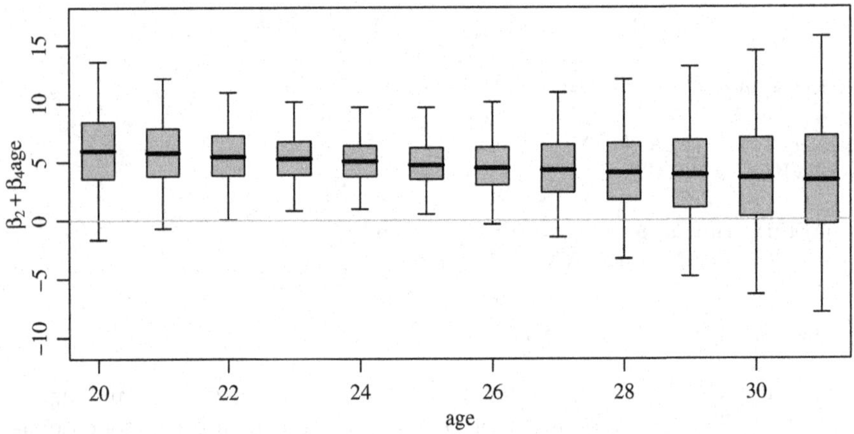

Fig. 9.4. Ninety-five percent confidence intervals for the difference in expected change scores between aerobics subjects and running subjects.

9.3 Model selection

Often in regression analysis we are faced with a large number of possible regressor variables, even though we suspect that a majority of the regressors have no true relationship to the variable Y. In these situations, including all of the possible variables in a regression model can lead to poor statistical performance. Standard statistical advice is that we should include in our regression model only those variables for which there is substantial evidence of an association with y. Doing so not only produces simpler, more aesthetically pleasing data analyses, but also generally provides models with better statistical properties in terms of prediction and estimation.

Example: Diabetes

Baseline data for ten variables $x_1, \ldots, x_{10}$ on a group of 442 diabetes patients were gathered, as well as a measure y of disease progression taken one year after the baseline measurements. From these data we hope to make a predictive model for y based on the baseline measurements. While a regression model with ten variables would not be overwhelmingly complex, it is suspected that the relationship between y and the x_j's may not be linear, and that including second-order terms like x_j^2 and $x_j x_k$ in the regression model might aid in prediction. The regressors therefore include ten *main effects* $x_1, \ldots, x_{10}$, $\binom{10}{2} = 45$ *interactions* of the form $x_j x_k$ and nine quadratic terms x_j^2 (one of the regressors, $x_2 =$ sex, is binary, so $x_2 = x_2^2$, making it unnecessary to include x_2^2). This gives a total of $p = 64$ regressors. To help with the interpretation of the parameters and to put the regressors on a common scale, all of the variables have been centered and scaled so that $\boldsymbol{y}$ and the columns of $\mathbf{X}$ all have mean zero and variance one.

In this section we will build predictive regression models for y based on the 64 regressor variables. To evaluate the models, we will randomly divide the 442 diabetes subjects into 342 *training samples* and 100 *test samples*, providing us with a training dataset $(\boldsymbol{y}, \mathbf{X})$ and a test dataset $(\boldsymbol{y}_{\text{test}}, \mathbf{X}_{\text{test}})$. We will fit the regression model using the training data and then use the estimated regression coefficients to generate $\hat{\boldsymbol{y}}_{\text{test}} = \mathbf{X}_{\text{test}}\hat{\boldsymbol{\beta}}$. The performance of the predictive model can then be evaluated by comparing $\boldsymbol{y}_{\text{test}}$ to $\hat{\boldsymbol{y}}_{\text{test}}$. Let's begin by building a predictive model with ordinary least squares regression with all 64 variables. The first panel of Figure 9.5 plots the true values of the 100 test samples $\boldsymbol{y}_{\text{test}}$ versus their predicted values $\hat{\boldsymbol{y}}_{\text{test}} = \mathbf{X}_{\text{test}}\hat{\boldsymbol{\beta}}$, where $\hat{\boldsymbol{\beta}}$ was estimated using the 342 training samples. While there is clearly a positive relationship between the true values and the predictions, there is quite a bit of error: The average squared predictive error is $\frac{1}{100}\sum(y_{\text{test},i} - \hat{y}_{\text{test},i})^2 = 0.67$, whereas if we just predicted each test case to be zero, our predictive error would be $\frac{1}{100}\sum y_{\text{test,i}}^2 = 0.97$.

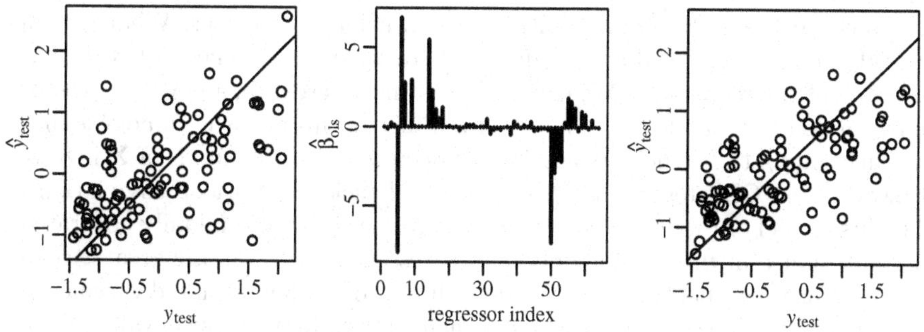

Fig. 9.5. Predicted values and regression coefficients for the diabetes data.

The second panel of Figure 9.5 shows the estimated values of each of the 64 regression coefficients. Of note is that the majority of coefficients are estimated to be quite small. Perhaps our predictions could be improved by removing from the regression model those variables that show little evidence of being non-zero. By doing so, we hope to remove from the predictive model any regressors that have spurious associations to Y (i.e. those associations specific only to the training data), leaving only those regressors that would have associations for any group of subjects (i.e. both the training and test data). One standard way to assess the evidence that the true value of a regression coefficient β_j is not zero is with a t-statistic, which is obtained by dividing the OLS estimate $\hat{\beta}_j$ by its standard error, so $t_j = \hat{\beta}_j / [\hat{\sigma}^2 (\mathbf{X}^T\mathbf{X})^{-1}_{j,j}]^{1/2}$. We might then consider removing from the model those regressor variables with small absolute values of t_j. For example, consider the following procedure:

1. Obtain the estimator $\hat{\boldsymbol{\beta}}_{\text{ols}} = (\mathbf{X}^T\mathbf{X})^{-1}\mathbf{X}^T\boldsymbol{y}$ and its t-statistics.
2. If there are any regressors j such that $|t_j| < t_{\text{cutoff}}$,
   a) find the regressor j_{min} having the smallest value of $|t_j|$ and remove column j_{min} from $\mathbf{X}$.
   b) return to step 1.
3. If $|t_j| > t_{\text{cutoff}}$ for all variables j remaining in the model, then stop.

Such procedures, in which a potentially large set of regressors is reduced to a smaller set, are called *model selection procedures*. The procedure defined in steps 1, 2 and 3 above describes a type of *backwards elimination* procedure, in which all regressors are initially included but then are iteratively removed until the remaining regressors satisfy some criterion. A standard choice for t_{cutoff} is an upper quantile of a t or standard normal distribution. If we apply the above procedure to the diabetes data with $t_{\text{cutoff}} = 1.65$ (corresponding roughly to a p-value of 0.10), then 44 of the 64 variables are eliminated, leaving 20 variables in the regression model. The third plot of Figure 9.5 shows $\boldsymbol{y}_{\text{test}}$ versus predicted values based on the reduced-model regression coefficients. The plot indicates that the predicted values from this model are more accurate than those from the full model, and indeed the average squared predictive error is $\frac{1}{100}\sum(y_{\text{test},i} - \hat{y}_{\text{test},i})^2 = 0.53$.

Backwards selection is not without its drawbacks, however. What sort of model would this procedure produce if there were no association between Y and any of the regressors? We can evaluate this by creating a new data vector $\tilde{\boldsymbol{y}}$ by randomly permuting the values of $\boldsymbol{y}$. Since in this case the value of $\boldsymbol{x}_i$ has no effect on $\tilde{y}_i$, the "true" association between $\tilde{\boldsymbol{y}}$ and the columns of $\mathbf{X}$ is zero. However, the OLS regression model will still pick up spurious associations: The first panel of Figure 9.6 shows the t-statistics for one randomly generated permutation $\tilde{\boldsymbol{y}}$ of $\boldsymbol{y}$. Initially, only one regressor has a t-statistic greater than 1.65, but as we sequentially remove the columns of $\mathbf{X}$ the estimated variance of the remaining regressors decreases and their t-statistics increase in value. With $t_{\text{cutoff}} = 1.65$, the procedure arrives at a regression model with 18 regressors, 17 of which have t-statistics greater than 2 in absolute value, and four of which have statistics greater than 3. Even though $\tilde{\boldsymbol{y}}$ was generated without regard to $\mathbf{X}$, the backwards selection procedure erroneously suggests that many of the regressors do have an association. Such misleading results are fairly common with backwards elimination and other sequential model selection procedures (Berk, 1978).

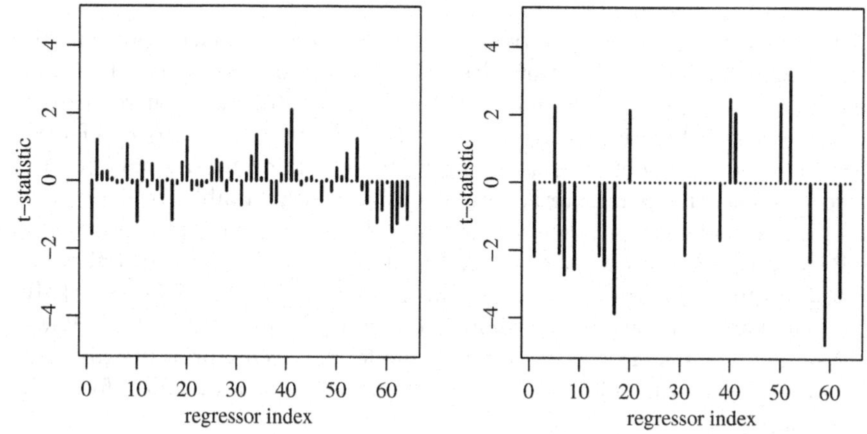

Fig. 9.6. t-statistics for the regression of $\tilde{y}$ on $\mathbf{X}$, before and after backwards elimination.

9.3.1 Bayesian model comparison

The Bayesian solution to the model selection problem is conceptually straightforward: If we believe that many of the regression coefficients are potentially equal to zero, then we simply come up with a prior distribution that reflects this possibility. This can be accomplished by specifying that each regression coefficient has some non-zero probability of being exactly zero. A convenient way to represent this is to write the regression coefficient for variable j as $\beta_j = z_j \times b_j$, where $z_j \in \{0, 1\}$ and b_j is some real number. With this parameterization, our regression equation becomes

$$y_i = z_1 b_1 x_{i,1} + \cdots + z_p b_p x_{i,p} + \epsilon_i.$$

The z_j's indicate which regression coefficients are non-zero. For example, in the oxygen uptake problem,

$$\begin{aligned}
\mathrm{E}[Y|\boldsymbol{x},\boldsymbol{b},\boldsymbol{z}=(1,0,1,0)] &= b_1x_1 + b_3x_3 \\
&= b_1 + b_3 \times \text{age} \\
\mathrm{E}[Y|\boldsymbol{x},\boldsymbol{b},\boldsymbol{z}=(1,1,0,0)] &= b_1x_1 + b_2x_2 \\
&= b_1 + b_2 \times \text{group} \\
\mathrm{E}[Y|\boldsymbol{x},\boldsymbol{b},\boldsymbol{z}=(1,1,1,0)] &= b_1x_1 + b_2x_2 + b_3x_3 \\
&= b_1 + b_2 \times \text{group} + b_3 \times \text{age}.
\end{aligned}$$

Each value of $\boldsymbol{z} = (z_1, \ldots, z_p)$ corresponds to a different model, or more specifically, a different collection of variables having non-zero regression coefficients. For example, we say that the model with $\boldsymbol{z} = (1,0,1,0)$ is a linear regression model for y as a function of age. The model with $\boldsymbol{z} = (1,1,1,0)$ is referred to as a regression model for y as a function of age, but with a group-specific intercept. With this parameterization, choosing which variables to include in a regression model is equivalent to choosing which z_j's are 0 and which are 1.

Bayesian model selection proceeds by obtaining a posterior distribution for $\boldsymbol{z}$. Of course, doing so requires a joint prior distribution on $\{\boldsymbol{z}, \boldsymbol{\beta}, \sigma^2\}$. It turns out that a version of the g-prior described in the previous section allows us to evaluate $p(\boldsymbol{y}|\mathbf{X}, \boldsymbol{z})$ for each possible model $\boldsymbol{z}$. Given a prior distribution $p(\boldsymbol{z})$ over models, this allows us to compute a posterior probability for each regression model:

$$p(\boldsymbol{z}|\boldsymbol{y}, \mathbf{X}) = \frac{p(\boldsymbol{z})p(\boldsymbol{y}|\mathbf{X}, \boldsymbol{z})}{\sum_{\tilde{\boldsymbol{z}}} p(\tilde{\boldsymbol{z}})p(\boldsymbol{y}|\mathbf{X}, \tilde{\boldsymbol{z}})}.$$

Alternatively, we can compare the evidence for any two models with the posterior odds:

$$\begin{aligned}
\text{odds}(\boldsymbol{z}_a, \boldsymbol{z}_b|\boldsymbol{y}, \mathbf{X}) = \frac{p(\boldsymbol{z}_a|\boldsymbol{y}, \mathbf{X})}{p(\boldsymbol{z}_b|\boldsymbol{y}, \mathbf{X})} &= \frac{p(\boldsymbol{z}_a)}{p(\boldsymbol{z}_b)} \times \frac{p(\boldsymbol{y}|\mathbf{X}, \boldsymbol{z}_a)}{p(\boldsymbol{y}|\mathbf{X}, \boldsymbol{z}_b)} \\
\text{posterior odds} &= \text{prior odds} \times \text{“Bayes factor”}
\end{aligned}$$

The Bayes factor can be interpreted as how much the data favor model $\boldsymbol{z}_a$ over model $\boldsymbol{z}_b$. In order to obtain a posterior distribution over models, we will have to compute $p(\boldsymbol{y}|\mathbf{X}, \boldsymbol{z})$ for each model $\boldsymbol{z}$ under consideration.

Computing the marginal probability

The marginal probability is obtained from the integral

$$\begin{aligned}
p(\boldsymbol{y}|\mathbf{X}, \boldsymbol{z}) &= \int\int p(\boldsymbol{y}, \boldsymbol{\beta}, \sigma^2|\mathbf{X}, \boldsymbol{z})\; d\boldsymbol{\beta} d\sigma^2 \\
&= \int\int p(\boldsymbol{y}|\boldsymbol{\beta}, \mathbf{X}, \boldsymbol{z}, \sigma^2) p(\boldsymbol{\beta}|\mathbf{X}, \boldsymbol{z}, \sigma^2) p(\sigma^2)\; d\boldsymbol{\beta}\; d\sigma^2. \qquad (9.8)
\end{aligned}$$

Using a version of the g-prior distribution for $\boldsymbol{\beta}$, we will be able to compute this integral without needing much calculus. For any given $\boldsymbol{z}$ with p_z non-zero entries, let $\mathbf{X}_z$ be the $n \times p_z$ matrix corresponding to the variables j for which

$z_j = 1$, and similarly let $\boldsymbol{\beta}_z$ be the $p_z \times 1$ vector consisting of the entries of $\boldsymbol{\beta}$ for which $z_j = 1$. Our modified g-prior distribution for $\boldsymbol{\beta}$ is that $\beta_j = 0$ for j's such that $z_j = 0$, and that

$$\{\boldsymbol{\beta}_z | \mathbf{X}_z, \sigma^2\} \sim \text{ multivariate normal } (\mathbf{0}, g\sigma^2[\mathbf{X}_z{}^T\mathbf{X}_z]^{-1}).$$

If we integrate (9.8) with respect to $\boldsymbol{\beta}$ first, we have

$$\begin{aligned} p(\boldsymbol{y}|\mathbf{X}, \boldsymbol{z}) &= \int \left(\int p(\boldsymbol{y}|\mathbf{X}, \boldsymbol{z}, \sigma^2, \boldsymbol{\beta}) p(\boldsymbol{\beta}|\mathbf{X}, \boldsymbol{z}, \sigma^2) \, d\boldsymbol{\beta} \right) p(\sigma^2) \, d\sigma^2 \\ &= \int p(\boldsymbol{y}|\mathbf{X}, \boldsymbol{z}, \sigma^2) p(\sigma^2) \, d\sigma^2 . \end{aligned}$$

The form for the marginal probability $p(\boldsymbol{y}|\mathbf{X}, \boldsymbol{z}, \sigma^2)$ was computed in the last section. Using those results, writing $\gamma = 1/\sigma^2$ and letting $p(\gamma)$ be the gamma density with parameters $(\nu_0/2, \nu_0\sigma_0^2/2)$, we can show that conditional density of $(\boldsymbol{y}, \gamma)$ given $(\mathbf{X}, \boldsymbol{z})$ is

$$\begin{aligned} p(\boldsymbol{y}|\mathbf{X}, \boldsymbol{z}, \gamma) \times p(\gamma) = (2\pi)^{-n/2}(1+g)^{-p_z/2} \times \left[\gamma^{n/2} e^{-\gamma \mathrm{SSR}_g^z/2}\right] \times \\ (\nu_0\sigma_0^2/2)^{\nu_0/2}\Gamma(\nu_0/2)^{-1}\left[\gamma^{\nu_0/2-1}e^{-\gamma\nu_0\sigma_0^2/2}\right], \quad (9.9) \end{aligned}$$

where SSR_g^z is as in the last section except based on the regressor matrix $\mathbf{X}_z$:

$$\mathrm{SSR}_g^z = \boldsymbol{y}^T(\mathbf{I} - \frac{g}{g+1}\mathbf{X}_z(\mathbf{X}_z^T\mathbf{X}_z)^{-1}\mathbf{X}_z^T)\boldsymbol{y} .$$

The part of Equation 9.9 that depends on γ is proportional to a gamma density, but in this case the normalizing constant is the part that we need:

$$\begin{aligned} &\gamma^{(\nu_0+n)/2-1}\exp[-\gamma \times (\nu_0\sigma_0^2 + \mathrm{SSR}_g^z)/2] = \\ &\frac{\Gamma([\nu_0+n]/2)}{([\nu_0\sigma_0^2 + \mathrm{SSR}_g^z]/2)^{(\nu_0+n)/2-1}} \times \mathrm{dgamma}[\gamma, (\nu_0+n)/2, (\nu_0\sigma_0^2 + \mathrm{SSR}_g^z)/2] . \end{aligned}$$

Since the gamma density integrates to 1, the integral of the left-hand side of the above equation must be equal to the constant on the right-hand side. Multiplying this constant by the other terms in Equation 9.9 gives the marginal probability we are interested in:

$$p(\boldsymbol{y}|\mathbf{X}, \boldsymbol{z}) = \pi^{-n/2}\frac{\Gamma([\nu_0+n]/2)}{\Gamma(\nu_0/2)}(1+g)^{-p_z/2}\frac{(\nu_0\sigma_0^2)^{\nu_0/2}}{(\nu_0\sigma_0^2 + \mathrm{SSR}_g^z)^{(\nu_0+n)/2}} .$$

Now suppose we set $g = n$ and use the unit information prior for $p(\sigma^2)$ for each model $\boldsymbol{z}$, so that $\nu_0 = 1$ for all $\boldsymbol{z}$, but σ_0^2 is the estimated residual variance under the least squares estimate for model $\boldsymbol{z}$. In this case, the ratio of the probabilities under any two models $\boldsymbol{z}_a$ and $\boldsymbol{z}_b$ is

$$\frac{p(\boldsymbol{y}|\mathbf{X},\boldsymbol{z}_a)}{p(\boldsymbol{y}|\mathbf{X},\boldsymbol{z}_b)} = (1+n)^{(p_{z_b}-p_{z_a})/2}\left(\frac{s^2_{z_a}}{s^2_{z_b}}\right)^{1/2} \times \left(\frac{s^2_{z_b}+\mathrm{SSR}_g^{z_b}}{s^2_{z_a}+\mathrm{SSR}_g^{z_a}}\right)^{(n+1)/2}.$$

Notice that the ratio of the marginal probabilities is essentially a balance between model complexity and goodness of fit: A large value of p_{z_b} compared to p_{z_a} penalizes model $\boldsymbol{z}_b$, although a large value of $\mathrm{SSR}_g^{z_a}$ compared to $\mathrm{SSR}_g^{z_b}$ penalizes model $\boldsymbol{z}_a$.

Oxygen uptake example

Recall our regression model for the oxygen uptake data:

$$\begin{aligned}\mathrm{E}[Y_i|\boldsymbol{\beta},\boldsymbol{x}_i] &= \beta_1 x_{i,1} + \beta_2 x_{i,2} + \beta_3 x_{i,3} + \beta_4 x_{i,4} \\ &= \beta_1 + \beta_2 \times \mathrm{group}_i + \beta_3 \times \mathrm{age}_i + \beta_4 \times \mathrm{group}_i \times \mathrm{age}_i\,.\end{aligned}$$

The question of whether or not there is an effect of group translates into the question of whether or not β_2 and β_4 are non-zero. Recall from our analyses in the previous sections that the estimated magnitudes of β_2 and β_4 were not large compared to their standard deviations, suggesting that maybe there is not an effect of group. However, we also noticed from the joint posterior distribution that β_2 and β_4 were negatively correlated, so whether or not β_2 is zero affects our information about β_4.

$\boldsymbol{z}$	model	$\log p(\boldsymbol{y}\|\mathbf{X},\boldsymbol{z})$	$p(\boldsymbol{z}\|\boldsymbol{y},\mathbf{X})$
(1,0,0,0)	β_1	-44.33	0.00
(1,1,0,0)	$\beta_1 + \beta_2 \times \mathrm{group}_i$	-42.35	0.00
(1,0,1,0)	$\beta_1 + \beta_3 \times \mathrm{age}_i$	-37.66	0.18
(1,1,1,0)	$\beta_1 + \beta_2 \times \mathrm{group}_i + \beta_3 \times \mathrm{age}_i$	-36.42	0.63
(1,1,1,1)	$\beta_1 + \beta_2 \times \mathrm{group}_i + \beta_3 \times \mathrm{age}_i + \beta_4 \times \mathrm{group}_i \times \mathrm{age}_i$	-37.60	0.19

Table 9.1. Marginal probabilities of the data under five different models.

We can formally evaluate whether β_2 or β_4 should be zero by computing the probability of the data under a variety of competing models. Table 9.1 lists five different regression models that we might like to consider for these data. Using the g-prior for $\boldsymbol{\beta}$ with $g = n$, and a unit information prior distribution for σ^2 for each value of $\boldsymbol{z}$, the values of $\log p(\boldsymbol{y}|\mathbf{X},\boldsymbol{z})$ can be computed for each of the five values of $\boldsymbol{z}$ we are considering. If we give each of these models equal prior weight, then posterior probabilities for each model can be computed as well. These calculations indicate that, among these five models, the most probable model is the one corresponding $\boldsymbol{z} = (1,1,1,0)$, having a slope for age with a separate intercept for each group. The evidence for an age effect is strong, as the posterior probabilities of the three models that include age essentially sum to 1. The evidence for an effect of group is weaker, as the combined probability for the three models with a group effect is 0.00+0.63+0.19=0.82. However, this probability is substantially higher than the corresponding prior probability of 0.20+0.20+0.20=0.60 for these three models.

9.3.2 Gibbs sampling and model averaging

If we allow each of the p regression coefficients to be either zero or non-zero, then there are 2^p different models to consider. If p is large, then it will be impractical for us to compute the marginal probability of each model. The diabetes data, for example, has $p = 64$ possible regressors, so the total number of models is $2^{64} \approx 1.8\times 10^{19}$. In these situations our data analysis goals become more modest: For example, we may be content with a decent estimate of $\boldsymbol{\beta}$ from which we can make predictions, or a list of relatively high-probability models. These items can be obtained with a Markov chain which searches through the space of models for values of $\boldsymbol{z}$ with high posterior probability. This can be done with a Gibbs sampler in which we iteratively sample each z_j from its full conditional distribution. Specifically, given a current value $\boldsymbol{z} = (z_1, \dots, z_p)$, a new value of z_j is generated by sampling from $p(z_j|\boldsymbol{y}, \mathbf{X}, \boldsymbol{z}_{-j})$, where $\boldsymbol{z}_{-j}$ refers to the values of $\boldsymbol{z}$ except the one corresponding to regressor j. The full conditional probability that z_j is 1 can be written as $o_j/(1+o_j)$, where o_j is the conditional odds that z_j is 1, given by

$$o_j = \frac{\Pr(z_j = 1|\boldsymbol{y}, \mathbf{X}, \boldsymbol{z}_{-j})}{\Pr(z_j = 0|\boldsymbol{y}, \mathbf{X}, \boldsymbol{z}_{-j})} = \frac{\Pr(z_j = 1)}{\Pr(z_j = 0)} \times \frac{p(\boldsymbol{y}|\mathbf{X}, \boldsymbol{z}_{-j}, z_j = 1)}{p(\boldsymbol{y}|\mathbf{X}, \boldsymbol{z}_{-j}, z_j = 0)}.$$

We may also want to obtain posterior samples of $\boldsymbol{\beta}$ and σ^2. Using the results of Section 9.2, values of these parameters can be sampled directly from their conditional distributions given $\boldsymbol{z}$, $\boldsymbol{y}$ and $\mathbf{X}$: For each $\boldsymbol{z}$ in our MCMC sample, we can construct the matrix $\mathbf{X}_z$ which consists of only those columns j corresponding to non-zero values of z_j. Using this matrix of regressors, a value of σ^2 can be sampled from $p(\sigma^2|\mathbf{X}, \boldsymbol{y}, \boldsymbol{z})$ (an inverse-gamma distribution) and then a value of $\boldsymbol{\beta}$ can be sampled from $p(\boldsymbol{\beta}|\mathbf{X}, \boldsymbol{y}, \boldsymbol{z}, \sigma^2)$ (a multivariate normal distribution). Our Gibbs sampling scheme therefore looks something like the following:

$$\begin{array}{ccccc} \boldsymbol{z}^{(s)} & \longrightarrow & \sigma^{2(s)} & \longrightarrow & \boldsymbol{\beta}^{(s)} \\ \downarrow & & & & \\ \boldsymbol{z}^{(s+1)} & \longrightarrow & \sigma^{2(s+1)} & \longrightarrow & \boldsymbol{\beta}^{(s+1)} \end{array}$$

More precisely, generating values of $\{\boldsymbol{z}^{(s+1)}, \sigma^{(s+1)}, \boldsymbol{\beta}^{(s+1)}\}$ from $\boldsymbol{z}^{(s)}$ is achieved with the following steps:

1. Set $\boldsymbol{z} = \boldsymbol{z}^{(s)}$;
2. For $j \in \{1, \dots, p\}$ in random order, replace z_j with a sample from $p(z_j|\boldsymbol{z}_{-j}, \boldsymbol{y}, \mathbf{X})$;
3. Set $\boldsymbol{z}^{(s+1)} = \boldsymbol{z}$;

4. Sample $\sigma^{2(s+1)} \sim p(\sigma^2|\boldsymbol{z}^{(s+1)}, \boldsymbol{y}, \mathbf{X})$;
5. Sample $\boldsymbol{\beta}^{(s+1)} \sim p(\boldsymbol{\beta}|\boldsymbol{z}^{(s+1)}, \sigma^{2(s+1)}, \boldsymbol{y}, \mathbf{X})$.

Note that the entries of $\boldsymbol{z}^{(s+1)}$ are not sampled from their full conditional distributions given $\sigma^{2(s)}$ and $\boldsymbol{\beta}^{(s)}$. This is not a problem: The Gibbs sampler for $\boldsymbol{z}$ ensures that the distribution of $\boldsymbol{z}^{(s)}$ converges to the target posterior distribution $p(\boldsymbol{z}|\boldsymbol{y}, \mathbf{X})$. Since $(\sigma^{2(s)}, \boldsymbol{\beta}^{(s)})$ are direct samples from $p(\sigma^2, \boldsymbol{\beta}|\boldsymbol{z}^{(s)}, \boldsymbol{y}, \mathbf{X})$, the distribution of $(\sigma^{2(s)}, \boldsymbol{\beta}^{(s)})$ converges to $p(\sigma^2, \boldsymbol{\beta}|\boldsymbol{y}, \mathbf{X})$. R-code to implement the Gibbs sampling algorithm in $\boldsymbol{z}$, along with a function `lpy.X` that calculates the log of $p(\boldsymbol{y}|\mathbf{X})$, is below. This code can be combined with the code in the previous section in order to generate samples of $\{\boldsymbol{z}, \sigma^2, \boldsymbol{\beta}\}$ from the joint posterior distribution.

```
##### a function to compute the marginal probability
lpy.X<-function(y,X,g=length(y),
   nu0=1,s20=try(summary(lm(y~-1+X))$sigma^2,silent=TRUE))
{
  n<-dim(X)[1] ; p<-dim(X)[2]
  if(p==0) { Hg<-0 ; s20<-mean(y^2) }
  if(p>0)  { Hg<-(g/(g+1)) * X%*%solve(t(X)%*%X)%*%t(X) }
  SSRg<- t(y)%*%( diag(1,nrow=n) - Hg )%*%y

  -.5*( n*log(pi)+p*log(1+g)+(nu0+n)*log(nu0*s20+SSRg)-
        nu0*log(nu0*s20) ) +
  lgamma( (nu0+n)/2 ) - lgamma(nu0/2)
}
#####

##### starting values and MCMC setup
z<-rep(1,dim(X)[2] )
lpy.c<-lpy.X(y,X[,z==1,drop=FALSE])
S<-10000
Z<-matrix(NA,S,dim(X)[2])
#####

##### Gibbs sampler
for(s in 1:S)
{
  for(j in sample(1:dim(X)[2]))
    {
      zp<-z ; zp[j]<-1-zp[j]
      lpy.p<-lpy.X(y,X[,zp==1,drop=FALSE])
      r<- (lpy.p - lpy.c)*(-1)^(zp[j]==0)
      z[j]<-rbinom(1,1,1/(1+exp(-r)))
      if(z[j]==zp[j]) {lpy.c<-lpy.p}
    }
  Z[s,]<-z
}
#####
```

Diabetes example

Using a uniform distribution on $\boldsymbol{z}$, the Gibbs sampling scheme described above was run for $S = 10,000$ iterations, generating 10,000 samples of $(\boldsymbol{z}, \sigma^2, \boldsymbol{\beta})$ which we can use to approximate the posterior distribution $p(\boldsymbol{z}, \sigma^2, \boldsymbol{\beta}|\boldsymbol{y}, \mathbf{X})$. How good will our approximation be? Recall that with $p = 64$ the total number of models, or possible values of $\boldsymbol{z}$, is $2^{64} \approx 10^{19}$, which is 10^{15} times as large as the number of approximate posterior samples we have. It should then not be too much of a surprise that, in the 10,000 MCMC samples of $\boldsymbol{z}$, only 32 of the possible models were sampled more than once: 28 models were sampled twice, two were sampled three times and two others were sampled five and six times. This means that, for large p, the Gibbs sampling scheme provides a poor approximation to the posterior distribution of $\boldsymbol{z}$. Nevertheless, in many situations where most of the regressors have no effect the Gibbs sampler can still provide reasonable estimates of the marginal posterior distributions of individual z_j's or β_j's. The first panel of Figure 9.7 shows the estimated posterior probabilities $\Pr(z_j = 1|\boldsymbol{y}, \mathbf{X})$ for each of the 64 regressors. There are six regression coefficients having a posterior probability higher than 0.5 of being non-zero. These six regressors are a subset of the 20 that remained after the backwards selection procedure described above. How well does this Bayesian approach do in terms of prediction? As usual, we can approximate the posterior mean of $\boldsymbol{\beta}$ with $\hat{\boldsymbol{\beta}}_{\text{bma}} = \sum_{s=1}^{S} \boldsymbol{\beta}^{(s)}/S$. This parameter estimate is sometimes called the (Bayesian) *model averaged* estimate of $\boldsymbol{\beta}$, because it is an average of regression parameters from different values of $\boldsymbol{z}$, i.e. over different regression models. This estimate, obtained by averaging the regression coefficients from several high-probability models, often performs better than the estimate of $\boldsymbol{\beta}$ obtained by considering only a single model. Returning to the problem of predicting data from the diabetes test set, we can compute model-averaged predicted values $\hat{\boldsymbol{y}}_{\text{test}} = \mathbf{X}\hat{\boldsymbol{\beta}}_{\text{bma}}$. These predicted values are plotted against the true values $\boldsymbol{y}_{\text{test}}$ in the second panel of Figure 9.7. These predictions have an average squared error of 0.452, which is better than the OLS estimates using either the full model or the one obtained from backwards elimination.

Finally, we evaluate the Bayesian model selection procedure when there is no relationship between Y and $\boldsymbol{x}$. Recall from above that when the backwards elimination procedure was applied to the permuted vector $\tilde{\boldsymbol{y}}$, which was constructed independently of $\mathbf{X}$, it erroneously returned 18 regressors. Running the Gibbs sampler above on the same dataset $(\tilde{y}, \mathbf{X})$ for 10,000 iterations provides approximated posterior probabilities $\Pr(z_j = 1|\boldsymbol{y}, \mathbf{X}) = \sum z_j^{(s)}/S$, all of which are less than 1/2, and all but two of which are less than 1/4. In contrast to the backwards selection procedure, for these data the Bayesian approach to model selection does not erroneously identify any regressors as having an effect on the distribution of $\tilde{\boldsymbol{y}}$.

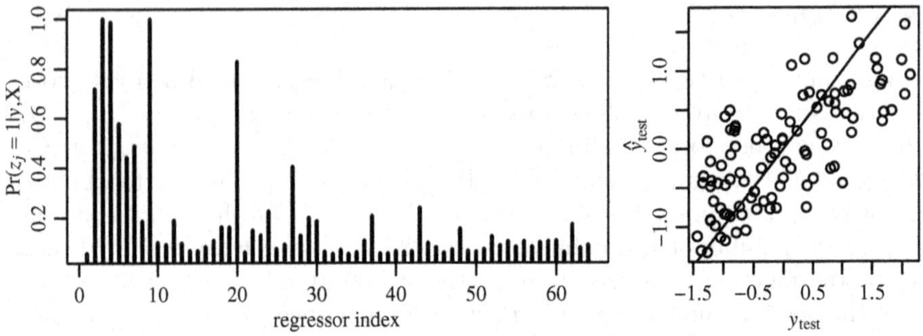

Fig. 9.7. The first panel shows posterior probabilities that each coefficient is non-zero. The second panel shows y_{test} versus predictions based on the model averaged estimate of $\boldsymbol{\beta}$.

9.4 Discussion and further references

There are many approaches to Bayesian model selection that use prior distributions allowing elements of $\boldsymbol{\beta}$ to be identically zero. George and McCulloch (1993) parameterize β_j as $b_j \times z_j$, where $z_j \in \{0, 1\}$, and use Gibbs sampling to do model selection. Liang et al (2008) review various types of g-priors in terms of two types of asymptotic consistency: model consistency and predictive consistency. The former is concerned with selecting the "true model," and the latter with making accurate posterior predictions. As pointed out by Leamer (1978), selecting a model and then acting as if it were true understates the uncertainty in the model selection process, and can result in suboptimal predictive performance. Predictive performance can be improved by Bayesian model averaging, i.e. averaging the predictive distributions under the different models according to their posterior probability (Madigan and Raftery, 1994; Raftery et al, 1997).

Many have argued that in most situations none of the regression models under consideration are actually true. Results of Bernardo and Smith (1994, Section 6.1.6) and Key et al (1999) indicate that in this situation, Bayesian model selection can still be meaningful in a decision-theoretic sense, where the task is to select the model with the best predictive performance. In this case, model selection proceeds using a modified Bayes factor that is similar to a cross-validation criterion.

10

Nonconjugate priors and Metropolis-Hastings algorithms

When conjugate or semiconjugate prior distributions are used, the posterior distribution can be approximated with the Monte Carlo method or the Gibbs sampler. In situations where a conjugate prior distribution is unavailable or undesirable, the full conditional distributions of the parameters do not have a standard form and the Gibbs sampler cannot be easily used. In this section we present the Metropolis-Hastings algorithm as a generic method of approximating the posterior distribution corresponding to any combination of prior distribution and sampling model. This section presents the algorithm in the context of two examples: The first involves Poisson regression, which is a type of generalized linear model. The second is a longitudinal regression model in which the observations are correlated over time.

10.1 Generalized linear models

Example: Song sparrow reproductive success

A sample from a population of 52 female song sparrows was studied over the course of a summer and their reproductive activities were recorded. In particular, the age and number of new offspring were recorded for each sparrow (Arcese et al, 1992). Figure 10.1 shows boxplots of the number of offspring versus age. The figure indicates that two-year-old birds in this population had the highest median reproductive success, with the number of offspring declining beyond two years of age. This is not surprising from a biological point of view: One-year-old birds are in their first mating season and are relatively inexperienced compared to two-year-old birds. As birds age beyond two years they experience a general decline in health and activity.

Suppose we wish to fit a probability model to these data, perhaps to understand the relationship between age and reproductive success, or to make population forecasts for this group of birds. Since the number of offspring for each bird is a non-negative integer $\{0,1,2,\ldots\}$, a simple probability model

P.D. Hoff, *A First Course in Bayesian Statistical Methods*,
Springer Texts in Statistics, DOI 10.1007/978-0-387-92407-6_10,

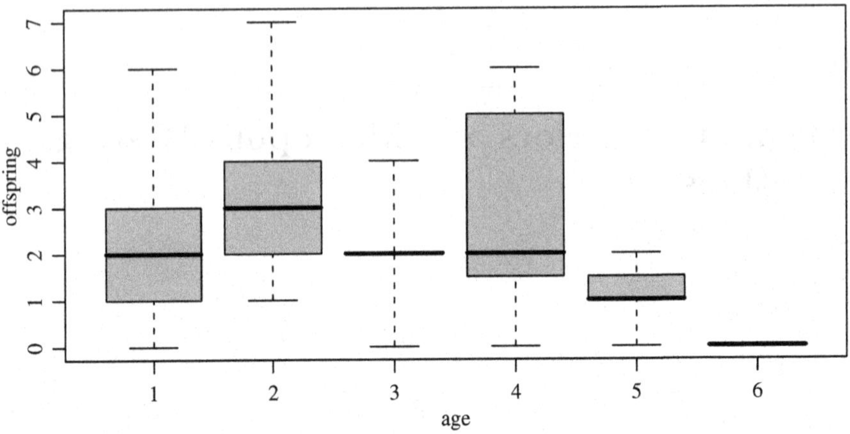

Fig. 10.1. Number of offspring versus age.

for Y=number of offspring conditional on x=age would be a Poisson model, $\{Y|x\} \sim \text{Poisson}(\theta_x)$. One possibility would be to estimate θ_x separately for each age group. However, the number of birds of each age is small and so the estimates of θ_x would be imprecise. To add stability to the estimation we will assume that the mean number of offspring is a smooth function of age. We will want to allow this function to be quadratic so that we can represent the increase in mean offspring while birds mature and the decline they experience thereafter. One possibility would be to express θ_x as $\theta_x = \beta_1 + \beta_2 x + \beta_3 x^2$. However, such a parameterization might allow some values of θ_x to be negative, which is not physically possible. As an alternative, we will model the log-mean of Y in terms of this regression, so that

$$\log \mathrm{E}[Y|x] = \log \theta_x = \beta_1 + \beta_2 x + \beta_3 x^2,$$

which means that $\mathrm{E}[Y|x] = \exp(\beta_1 + \beta_2 x + \beta_3 x^2)$, which is always greater than zero.

The resulting model, $\{Y|\boldsymbol{x}\} \sim \text{Poisson}(\exp[\boldsymbol{\beta}^T\boldsymbol{x}])$, is called a *Poisson regression model*. The term $\boldsymbol{\beta}^T\boldsymbol{x}$ is called the *linear predictor*. In this regression model the linear predictor is linked to $\mathrm{E}[Y|\boldsymbol{x}]$ via the log function, and so we say that this model has a *log link*. The Poisson regression model is a type of *generalized linear model*, a model which relates a function of the expectation to a linear predictor of the form $\boldsymbol{\beta}^T\boldsymbol{x}$. Another common generalized linear model is the *logistic regression model* for binary data. Writing $\Pr(Y = 1|\boldsymbol{x}) = \mathrm{E}[Y|\boldsymbol{x}] = \theta_x$, the logistic regression model parameterizes θ_x as

$$\theta_x = \frac{\exp(\boldsymbol{\beta}^T\boldsymbol{x})}{1 + \exp(\boldsymbol{\beta}^T\boldsymbol{x})}, \text{ so that}$$

$$\boldsymbol{\beta}^T\boldsymbol{x} = \log \frac{\theta_x}{1 - \theta_x}.$$

The function $\log \theta_x/(1-\theta_x)$ relating the mean to the linear predictor is called the *logit function*, so the logistic regression model could be described as a binary regression model with a logit link. Notice that the logit link forces θ_x to be between zero and one, even though $\boldsymbol{\beta}^T \boldsymbol{x}$ can range over the whole real line.

As in the case of ordinary regression, a natural class of prior distributions for $\boldsymbol{\beta}$ is the class of multivariate normal distributions. However, for neither the Poisson nor the logistic regression model would a prior distribution from this class result in a multivariate normal posterior distribution for $\boldsymbol{\beta}$. Furthermore, standard conjugate prior distributions for generalized linear models do not exist (except for the normal regression model).

One possible way to calculate the posterior distribution is to use a grid-based approximation, similar to the approach we used in Section 6.2: We can evaluate $p(\boldsymbol{y}|\mathbf{X}, \boldsymbol{\beta}) \times p(\boldsymbol{\beta})$ on a three-dimensional grid of $\boldsymbol{\beta}$-values, then normalize the result to obtain a discrete approximation to $p(\boldsymbol{\beta}|\mathbf{X}, \boldsymbol{y})$. Figure 10.2 shows approximate marginal and joint distributions of β_2 and β_3 based on the prior distribution $\boldsymbol{\beta} \sim$ multivariate normal$(\mathbf{0}, 100 \times \mathbf{I})$ and a grid having 100 values for each parameter. Computing these quantities for this three-parameter model required the calculation of $p(\boldsymbol{y}|\mathbf{X}, \boldsymbol{\beta}) \times p(\boldsymbol{\beta})$ at 1 million grid points. While feasible for this problem, a Poisson regression with only two more regressors and the same grid density would require 10 billion grid points, which is prohibitively large. Additionally, grid-based approximations can be very inefficient: The third panel of Figure 10.2 shows a strong negative posterior correlation between β_2 and β_3, which means that the probability mass is concentrated along a diagonal and so the vast majority of points of our cubical grid have essentially zero probability. In contrast, an approximation of $p(\boldsymbol{\beta}|\mathbf{X}, \boldsymbol{y})$ based on Monte Carlo samples could be stored in a computer much more efficiently, since our Monte Carlo sample would not include any points that have essentially zero posterior probability. Although independent Monte Carlo sampling from the posterior is not available for this Poisson regression model, the next section will show how to construct a Markov chain that can approximate $p(\boldsymbol{\beta}|\mathbf{X}, \boldsymbol{y})$ for any prior distribution $p(\boldsymbol{\beta})$.

10.2 The Metropolis algorithm

Let's consider a very generic situation where we have a sampling model $Y \sim p(y|\theta)$ and a prior distribution $p(\theta)$. Although in most problems $p(y|\theta)$ and $p(\theta)$ can be calculated for any values of y and θ, $p(\theta|y) = p(\theta)p(y|\theta)/\int p(\theta')p(y|\theta')\, d\theta'$ is often hard to calculate due to the integral in the denominator. If we were able to sample from $p(\theta|y)$, then we could generate $\theta^{(1)}, \ldots, \theta^{(S)} \sim$ i.i.d. $p(\theta|y)$ and obtain Monte Carlo approximations to posterior quantities, such as

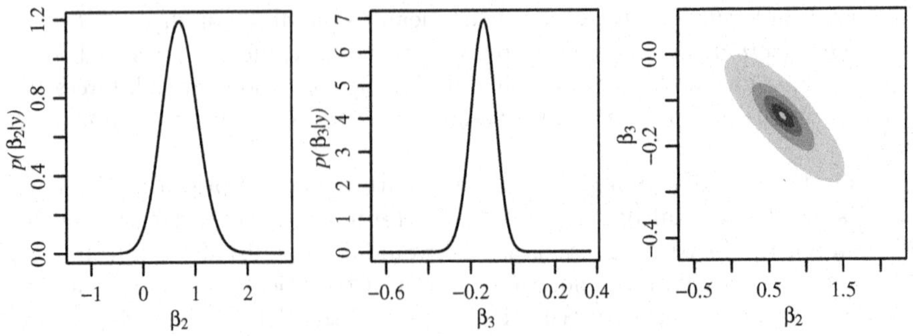

Fig. 10.2. Grid-based approximations to $p(\beta_2|\mathbf{X}, \boldsymbol{y})$, $p(\beta_3|\mathbf{X}, \boldsymbol{y})$ and $p(\beta_2, \beta_3|\mathbf{X}, \boldsymbol{y})$.

$$\mathrm{E}[g(\theta)|y] \approx \frac{1}{S}\sum_{s=1}^{S} g(\theta^{(s)}).$$

But what if we cannot sample directly from $p(\theta|y)$? In terms of approximating the posterior distribution, the critical thing is not that we have i.i.d. samples from $p(\theta|y)$ but rather that we are able to construct a large collection of θ-values, $\{\theta^{(1)}, \ldots, \theta^{(S)}\}$, whose empirical distribution approximates $p(\theta|y)$. Roughly speaking, for any two different values θ_a and θ_b we need

$$\frac{\#\{\theta^{(s)}\text{'s in the collection} = \theta_a\}}{\#\{\theta^{(s)}\text{'s in the collection} = \theta_b\}} \approx \frac{p(\theta_a|y)}{p(\theta_b|y)}.$$

Let's think intuitively about how we might construct such a collection. Suppose we have a working collection $\{\theta^{(1)}, \ldots, \theta^{(s)}\}$ to which we would like to add a new value $\theta^{(s+1)}$. Let's consider adding a value θ^* which is nearby $\theta^{(s)}$. Should we include θ^* in the set or not? If $p(\theta^*|y) > p(\theta^{(s)}|y)$ then we want more θ^*'s in the set than $\theta^{(s)}$'s. Since $\theta^{(s)}$ is already in the set, then it seems we should include θ^* as well. On the other hand, if $p(\theta^*|y) < p(\theta^{(s)}|y)$ then it seems we should not necessarily include θ^*. So perhaps our decision to include θ^* or not should be based on a comparison of $p(\theta^*|y)$ to $p(\theta^{(s)}|y)$. Fortunately, this comparison can be made even if we cannot compute $p(\theta|y)$:

$$r = \frac{p(\theta^*|y)}{p(\theta^{(s)}|y)} = \frac{p(y|\theta^*)p(\theta^*)}{p(y)} \frac{p(y)}{p(y|\theta^{(s)})p(\theta^{(s)})} = \frac{p(y|\theta^*)p(\theta^*)}{p(y|\theta^{(s)})p(\theta^{(s)})}. \tag{10.1}$$

Having computed r, how should we proceed?

If $r > 1$:

Intuition: Since $\theta^{(s)}$ is already in our set, we should include θ^* as it has a higher probability than $\theta^{(s)}$.

Procedure: Accept θ^* into our set, i.e. set $\theta^{(s+1)} = \theta^*$.

If $r < 1$:

> **Intuition:** The relative frequency of θ-values in our set equal to θ^* compared to those equal to $\theta^{(s)}$ should be $p(\theta^*|y)/p(\theta^{(s)}|y) = r$. This means that for every instance of $\theta^{(s)}$, we should have only a "fraction" of an instance of a θ^* value.
> **Procedure:** Set $\theta^{(s+1)}$ equal to either θ^* or $\theta^{(s)}$, with probability r and $1 - r$ respectively.

This is the basic intuition behind the famous *Metropolis algorithm*. The Metropolis algorithm proceeds by sampling a proposal value θ^* nearby the current value $\theta^{(s)}$ using a *symmetric proposal distribution* $J(\theta^*|\theta^{(s)})$. Symmetric here means that $J(\theta_b|\theta_a) = J(\theta_a|\theta_b)$, i.e. the probability of proposing $\theta^* = \theta_b$ given that $\theta^{(s)} = \theta_a$ is equal to the probability of proposing $\theta^* = \theta_a$ given that $\theta^{(s)} = \theta_b$. Usually $J(\theta^*|\theta^{(s)})$ is very simple, with samples from $J(\theta^*|\theta^{(s)})$ being near $\theta^{(s)}$ with high probability. Examples include

- $J(\theta^*|\theta^{(s)}) = \text{uniform}(\theta^{(s)} - \delta, \theta^{(s)} + \delta)$;
- $J(\theta^*|\theta^{(s)}) = \text{normal}(\theta^{(s)}, \delta^2)$.

The value of the parameter δ is generally chosen to make the approximation algorithm run efficiently, as will be discussed in more detail shortly.

Having obtained a proposal value θ^*, we add either it or a copy of $\theta^{(s)}$ to our set, depending on the ratio $r = p(\theta^*|y)/p(\theta^{(s)}|y)$. Specifically, given $\theta^{(s)}$, the Metropolis algorithm generates a value $\theta^{(s+1)}$ as follows:

1. Sample $\theta^* \sim J(\theta|\theta^{(s)})$;
2. Compute the acceptance ratio
$$r = \frac{p(\theta^*|y)}{p(\theta^{(s)}|y)} = \frac{p(y|\theta^*)p(\theta^*)}{p(y|\theta^{(s)})p(\theta^{(s)})} .$$
3. Let
$$\theta^{(s+1)} = \begin{cases} \theta^* & \text{with probability } \min(r,1) \\ \theta^{(s)} & \text{with probability } 1 - \min(r,1). \end{cases}$$

Step 3 can be accomplished by sampling $u \sim \text{uniform}(0,1)$ and setting $\theta^{(s+1)} = \theta^*$ if $u < r$ and setting $\theta^{(s+1)} = \theta^{(s)}$ otherwise.

Example: Normal distribution with known variance

Let's try out the Metropolis algorithm for the conjugate normal model with a known variance, a situation where we know the correct posterior distribution. Letting $\theta \sim \text{normal}(\mu, \tau^2)$ and $\{y_1, \ldots, y_n|\theta\} \sim \text{i.i.d. normal}(\theta, \sigma^2)$, the posterior distribution of θ is $\text{normal}(\mu_n, \tau_n^2)$ where

$$\mu_n = \bar{y}\frac{n/\sigma^2}{n/\sigma^2 + 1/\tau^2} + \mu\frac{1/\tau^2}{n/\sigma^2 + 1/\tau^2}$$
$$\tau_n^2 = 1/(n/\sigma^2 + 1/\tau^2).$$

Suppose $\sigma^2 = 1$, $\tau^2 = 10$, $\mu = 5$, $n = 5$ and $\boldsymbol{y} = (9.37, 10.18, 9.16, 11.60, 10.33)$. For these data, $\mu_n = 10.03$ and $\tau_n^2 = .20$, and so $p(\theta|\boldsymbol{y}) = \text{dnorm}(10.03, .44)$. Now suppose that for some reason we were unable to obtain the formula for this posterior distribution and needed to use the Metropolis algorithm to approximate it. Based on this model and prior distribution, the acceptance ratio comparing a proposed value θ^* to a current value $\theta^{(s)}$ is

$$r = \frac{p(\theta^*|\boldsymbol{y})}{p(\theta^{(s)}|\boldsymbol{y})} = \left(\frac{\prod_{i=1}^n \text{dnorm}(y_i, \theta^*, \sigma)}{\prod_{i=1}^n \text{dnorm}(y_i, \theta^{(s)}, \sigma)}\right) \times \left(\frac{\text{dnorm}(\theta^*, \mu, \tau)}{\text{dnorm}(\theta^{(s)}, \mu, \tau)}\right).$$

In many cases, computing the ratio r directly can be numerically unstable, a problem that often can be remedied by computing the logarithm of r:

$$\log r = \sum_{i=1}^n [\log \text{dnorm}(y_i, \theta^*, \sigma) - \log \text{dnorm}(y_i, \theta^{(s)}, \sigma)] +$$
$$\log \text{dnorm}(\theta^*, \mu, \tau) - \log \text{dnorm}(\theta^{(s)}, \mu, \tau).$$

Keeping things on the log scale, the proposal is accepted if $\log u < \log r$, where u is a sample from the uniform distribution on $(0, 1)$.

The R-code below generates 10,000 iterations of the Metropolis algorithm, starting at $\theta^{(0)} = 0$ and using a normal proposal distribution, $\theta^{(s+1)} \sim$ normal $(\theta^{(s)}, \delta^2)$ with $\delta^2 = 2$.

```
s2<-1 ; t2<-10 ; mu<-5
y<-c(9.37, 10.18, 9.16, 11.60, 10.33)
theta<-0 ; delta2<-2 ; S<-10000 ; THETA<-NULL ; set.seed(1)

for(s in 1:S)
{

  theta.star<-rnorm(1,theta,sqrt(delta2))

  log.r<-( sum(dnorm(y,theta.star,sqrt(s2),log=TRUE)) +
                dnorm(theta.star,mu,sqrt(t2),log=TRUE) ) -
         ( sum(dnorm(y,theta,sqrt(s2),log=TRUE)) +
                dnorm(theta,mu,sqrt(t2),log=TRUE) )

  if(log(runif(1))<log.r) { theta<-theta.star }

  THETA<-c(THETA,theta)

}
```

The first panel of Figure 10.3 plots these 10,000 simulated values as a function of iteration number. Although the value of θ starts nowhere near the posterior mean of 10.03, it quickly arrives there after a few iterations. The second panel gives a histogram of the 10,000 θ-values, and includes a plot of the normal(10.03, 0.20) density for comparison. Clearly the empirical distribution

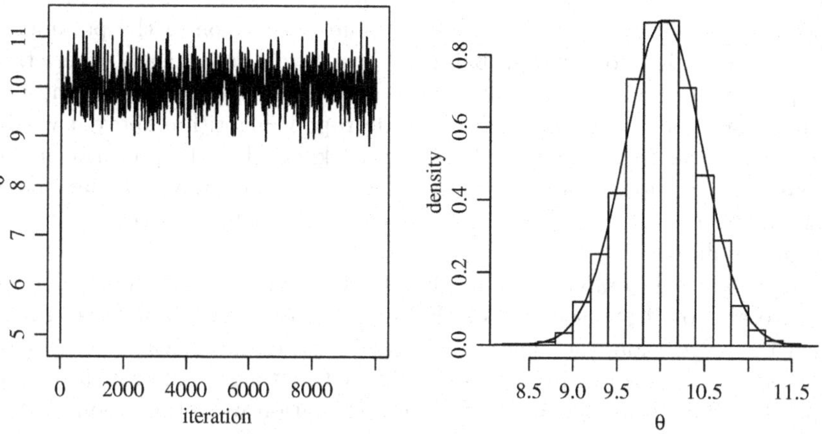

Fig. 10.3. Results from the Metropolis algorithm for the normal model.

of the simulated values is very close to the true posterior distribution. Will this similarity between $\{\theta^{(1)}, \ldots, \theta^{(S)}\}$ and $p(\theta|y)$ hold in general?

Output of the Metropolis algorithm

The Metropolis algorithm generates a dependent sequence $\{\theta^{(1)}, \theta^{(2)}, \ldots\}$ of θ-values. Since our procedure for generating $\theta^{(s+1)}$ depends only on $\theta^{(s)}$, the conditional distribution of $\theta^{(s+1)}$ given $\{\theta^{(1)}, \ldots, \theta^{(s)}\}$ also depends only on $\theta^{(s)}$ and so the sequence $\{\theta^{(1)}, \theta^{(2)}, \ldots\}$ is a Markov chain.

Under some mild conditions the marginal sampling distribution of $\theta^{(s)}$ is approximately $p(\theta|y)$ for large s. Additionally, for any given numerical value θ_a of θ,

$$\lim_{S\to\infty} \frac{\#\{\theta\text{'s in the sequence } < \theta_a\}}{S} = p(\theta < \theta_a|y).$$

Just as with the Gibbs sampler, this suggests we can approximate posterior means, quantiles and other posterior quantities of interest using the empirical distribution of $\{\theta^{(1)}, \ldots, \theta^{(S)}\}$. However, our approximation to these quantities will depend on how well our simulated sequence actually approximates $p(\theta|y)$. Results from probability theory say that, in the limit as $S \to \infty$, the approximation will be exact, but in practice we cannot run the Markov chain forever. Instead, the standard practice in MCMC approximation, using either the Metropolis algorithm or the Gibbs sampler, is as follows:

1. run algorithm until some iteration B for which it looks like the Markov chain has achieved stationarity;
2. run the algorithm S more times, generating $\{\theta^{(B+1)}, \ldots, \theta^{(B+S)}\}$;
3. discard $\{\theta^{(1)}, \ldots, \theta^{(B)}\}$ and use the empirical distribution of $\{\theta^{(B+1)}, \ldots, \theta^{(B+S)}\}$ to approximate $p(\theta|y)$.

The iterations up to and including B are called the "burn-in" period, in which the Markov chain moves from its initial value to a region of the parameter space that has high posterior probability. If we have a good idea of where this high probability region is, we can reduce the burn-in period by starting the Markov chain there. For example, in the Metropolis algorithm above it would have been better to start with $\theta^{(1)} = \bar{y}$ as we know that the posterior mode will be near $\bar{y}$. However, starting with $\theta^{(1)} = 0$ illustrates that the Metropolis algorithm is able to move from a low posterior probability region to one of high probability.

The θ-values generated from an MCMC algorithm are statistically dependent. Recall from the discussion of MCMC diagnostics in Chapter 6 that the higher the correlation, the longer it will take for the Markov chain to achieve stationarity and the more iterations it will take to get a good approximation to $p(\theta|y)$. Roughly speaking, the amount of information we obtain about $\mathrm{E}[\theta|y]$ from S positively correlated samples is less than the information we would obtain from S independent samples. The more correlated our Markov chain is, the less information we get per iteration (recall the notion of "effective sample size" from Section 6.6). In Gibbs sampling we do not have much control over the correlation of the Markov chain, but with the Metropolis algorithm the correlation can be adjusted by selecting an optimal value of δ in the proposal distribution. By selecting δ carefully, we can decrease the correlation in the Markov chain, leading to an increase in the rate of convergence, an increase in the effective sample size of the Markov chain and an improvement in the Monte Carlo approximation to the posterior distribution.

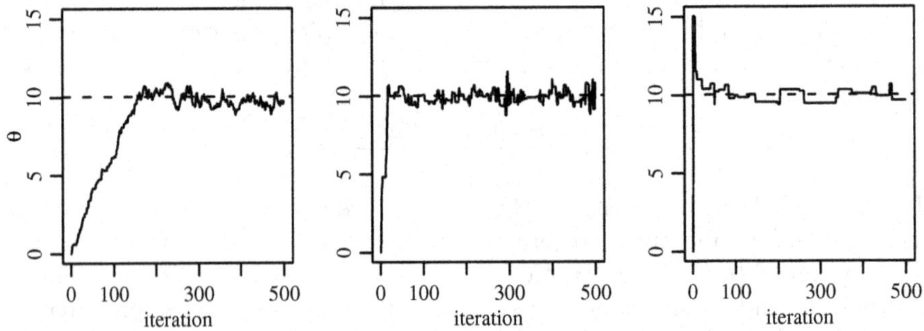

Fig. 10.4. Markov chains under three different proposal distributions. Going from left to right, the values of δ^2 are 1/32, 2 and 64 respectively.

To illustrate this, we can rerun the Metropolis algorithm for the one-sample normal problem using a range of δ values, including $\delta^2 \in \{1/32,\ 1/2,\ 2,\ 32,\ 64\ \}$. Doing so results in lag-1 autocorrelations of $(0.98, 0.77, 0.69, 0.84, 0.86)$ for these five different δ-values. Interestingly, the best δ-value among these five occurs in the middle of the set of values, and not at the extremes. The reason why can be understood by inspecting each sequence. Figure 10.4 plots the first 500 values for the sequences corresponding to $\delta^2 \in \{1/32, 2, 64\}$. In the first panel where $\delta^2 = 1/32$, the small proposal variance means that θ^* will be very close to $\theta^{(s)}$, and so $r \approx 1$ for most proposed values. As a result, θ^* is accepted as the value of $\theta^{(s+1)}$ for 87% of the iterations. Although this high acceptance rate keeps the chain moving, the moves are never very large and so the Markov chain is highly correlated. One consequence of this is that it takes a large number of iterations for the Markov chain to move from the starting value of zero to the posterior mode of 10.03. At the other extreme, the third plot in the figure shows the Markov chain for $\delta^2 = 64$. In this case the chain moves quickly to the posterior mode but once there it gets "stuck" for long periods. This is because the variance of the proposal distribution is so large that θ^* is frequently very far away from the posterior mode. Proposals in this Metropolis algorithm are accepted for only 5% of the iterations, and so $\theta^{(s+1)}$ is set equal to $\theta^{(s)}$ 95% of the time, resulting in a highly correlated Markov chain.

In order to construct a Markov chain with a low correlation we need a proposal variance that is large enough so that the Markov chain can quickly move around the parameter space, but not so large that the proposals end up getting rejected most of the time. Among the proposal variances considered for the data and normal model here, this balance was optimized with a δ^2 of 2, which gives an acceptance rate of 35%. In general, it is common practice to first select a proposal distribution by implementing several short runs of the Metropolis algorithm under different δ-values until one is found that gives an acceptance rate roughly between 20 and 50%. Once a reasonable value of δ is selected, a longer more efficient Markov chain can be run. Alternatively, modified versions of the Metropolis algorithm can be constructed that adaptively change the value of δ at the beginning of the chain in order to automatically find a good proposal distribution.

10.3 The Metropolis algorithm for Poisson regression

Let's implement the Metropolis algorithm for the Poisson regression model introduced at the beginning of the chapter. Recall that the model is that Y_i is a sample from a Poisson distribution with a log-mean given by $\log \mathrm{E}[Y_i|x_i] = \beta_1 + \beta_2 x_i + \beta_3 x_i^2$, where x_i is the age of the sparrow i. We will abuse notation slightly by writing $\boldsymbol{x}_i = (1, x_i, x_i^2)$ so that $\log \mathrm{E}[Y_i|x_i] = \boldsymbol{\beta}^T \boldsymbol{x}_i$. The prior distribution we used in Section 10.1 was that the regression coefficients were i.i.d. normal(0,100). Given a current value $\boldsymbol{\beta}^{(s)}$ and a value $\boldsymbol{\beta}^*$ generated from $J(\boldsymbol{\beta}^*|\boldsymbol{\beta}^{(s)})$, the acceptance ratio for the Metropolis algorithm is

$$r = \frac{p(\boldsymbol{\beta}^*|\mathbf{X}, \boldsymbol{y})}{p(\boldsymbol{\beta}^{(s)}|\mathbf{X}, \boldsymbol{y})}$$
$$= \frac{\prod_{i=1}^n \text{dpois}(y_i, \boldsymbol{x}_i^T\boldsymbol{\beta}^*)}{\prod_{i=1}^n \text{dpois}(y_i, \boldsymbol{x}_i^T\boldsymbol{\beta}^{(s)})} \times \frac{\prod_{j=1}^3 \text{dnorm}(\beta_j^*, 0, 10)}{\prod_{j=1}^3 \text{dnorm}(\beta_j^{(s)}, 0, 10)}.$$

All that remains to implement the algorithm is to specify the proposal distribution for θ^*. A convenient choice is a multivariate normal distribution with mean $\boldsymbol{\beta}^{(s)}$. In many problems, the posterior variance can be an efficient choice of a proposal variance. Although we do not know the posterior variance before we run the Metropolis algorithm, it is often sufficient just to use a rough approximation. In a normal regression problem, the posterior variance of $\boldsymbol{\beta}$ will be close to $\sigma^2(\mathbf{X}^T\mathbf{X})^{-1}$, where σ^2 is the variance of Y. In our Poisson regression, the model is that the log of Y has expectation equal to $\boldsymbol{\beta}^T\boldsymbol{x}$, so let's try a proposal variance of $\hat{\sigma}^2(\mathbf{X}^T\mathbf{X})^{-1}$ where $\hat{\sigma}^2$ is the sample variance of $\{\log(y_1+1/2), \ldots, \log(y_n+1/2)\}$ (we use $\log(y+1/2)$ instead of $\log y$ because the latter would be $-\infty$ if $y = 0$). If this results in an acceptance rate that is too high or too low, we can always adjust the proposal variance accordingly.

R-code to implement the Metropolis algorithm for a Poisson regression of $\boldsymbol{y}$ on $\mathbf{X}$ is as follows:

```
data(chapter10) ; y<-yX.sparrow[,1] ; X<-yX.sparrow[,-1]
n<-length(y) ; p<-dim(X)[2]

pmn.beta<-rep(0,p)        #prior expectation
psd.beta<-rep(10,p)       #prior var

var.prop<- var(log(y+1/2))*solve( t(X)%*%X ) #proposal var
S<-10000
beta<-rep(0,p) ; acs<-0
BETA<-matrix(0,nrow=S,ncol=p)
set.seed(1)

for(s in 1:S)
{
  beta.p<- t(rmvnorm(1, beta, var.prop ))

  lhr<- sum(dpois(y,exp(X%*%beta.p),log=T)) -
        sum(dpois(y,exp(X%*%beta),log=T)) +
        sum(dnorm(beta.p,pmn.beta,psd.beta,log=T)) -
        sum(dnorm(beta,pmn.beta,psd.beta,log=T))

  if( log(runif(1))< lhr ) { beta<-beta.p ; acs<-acs+1 }

  BETA[s,]<-beta
}
```

Applying this algorithm to the song sparrow data gives an acceptance rate of about 43%. A plot of β_3 versus iteration number appears in the first

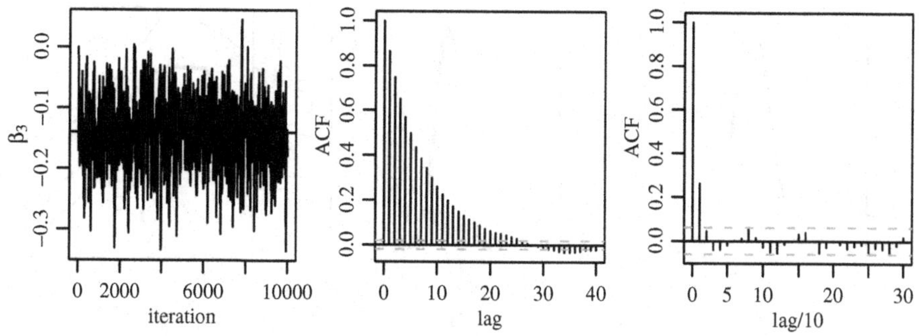

Fig. 10.5. Plot of the Markov chain in β_3 along with autocorrelation functions.

panel of Figure 10.5. The algorithm moves quickly from the starting value of $\beta_3 = 0$ to a region closer to the posterior mode. The second panel of the figure shows the autocorrelation function for β_3. We could possibly reduce the autocorrelation by modifying the proposal variance and obtaining a new Markov chain, although this Markov chain is perhaps sufficient to obtain a good approximation to $p(\boldsymbol{\beta}|\mathbf{X}, \boldsymbol{y})$. For example, the third panel of Figure 10.5 plots the autocorrelation function of every 10th value of β_3 from the Markov chain. This "thinned" subsequence contains 1,000 of the 10,000 β_3 values, but these 1,000 values are nearly independent. This suggests we have nearly the equivalent of 1,000 independent samples of β_3 with which to approximate the posterior distribution. To be more precise, we can calculate the effective sample size as described in Section 6.6. The effective sample sizes for β_1, β_2 and β_3 are 818, 778 and 726 respectively. The adequacy of this Markov chain is confirmed further in the first two panels of Figure 10.6, which plots the MCMC approximations to the marginal posterior densities of β_2 and β_3. These densities are nearly identical to the ones obtained from the grid-based approximation, which are shown in gray lines for comparison. Finally, the third panel of the figure plots posterior quantiles of $\mathrm{E}[Y|x]$ for each age x, which indicates the quadratic nature of reproductive success for this song sparrow population.

10.4 Metropolis, Metropolis-Hastings and Gibbs

Recall that a Markov chain is a sequentially generated sequence $\{x^{(1)}, x^{(2)}, \ldots\}$ such that the mechanism that generates $x^{(s+1)}$ can depend on the value of $x^{(s)}$ but not on $\{x^{(s-1)}, x^{(s-2)}, \ldots x^{(1)}\}$. A more poetic way of putting this is that for a Markov chain "the future depends on the present and not on the past." The Gibbs sampler and the Metropolis algorithm are both ways of generating Markov chains that approximate a target probability distribution $p_0(x)$ for a potentially vector-valued random variable x. In Bayesian analysis, x is typically a parameter or vector of parameters and $p_0(x)$ is a posterior

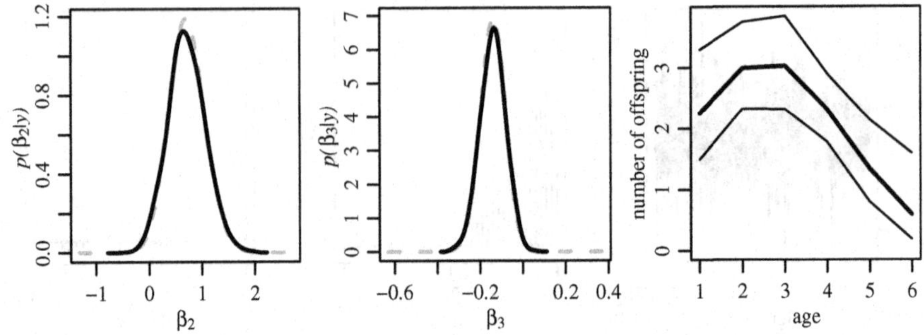

Fig. 10.6. The first two panels give the MCMC approximations to the posterior marginal distributions of β_2 and β_3 in black, with the grid-based approximations in gray. The third panel gives 2.5%, 50% and 97.5% posterior quantiles of $\exp(\boldsymbol{\beta}^T\boldsymbol{x})$.

distribution, but the Gibbs sampler and Metropolis algorithm are both used more broadly.

In this section we will show that these two algorithms are in fact special cases of a more general algorithm, called the *Metropolis-Hastings algorithm*. We will then describe why Markov chains generated by the Metropolis-Hastings algorithm are able to approximate a target probability distribution. Since the Gibbs and Metropolis algorithms are special cases of Metropolis-Hastings, this implies that these two algorithms are also valid ways to approximate probability distributions.

10.4.1 The Metropolis-Hastings algorithm

We'll first consider a simple example where our target probability distribution is $p_0(u, v)$, a bivariate distribution for two random variables U and V. In the one-sample normal problem, for example, we would have $U = \theta$, $V = \sigma^2$ and $p_0(u, v) = p(\theta, \sigma^2|\boldsymbol{y})$.

Recall that the Gibbs sampler proceeds by iteratively sampling values of U and V from their conditional distributions: Given $x^{(s)} = (u^{(s)}, v^{(s)})$, a new value of $x^{(s+1)}$ is generated as follows:

1. update U: sample $u^{(s+1)} \sim p_0(u|v^{(s)})$;
2. update V: sample $v^{(s+1)} \sim p_0(v|u^{(s+1)})$.

Alternatively, we could have first sampled $v^{(s+1)} \sim p_0(v|u^{(s)})$ and then $u^{(s+1)} \sim p_0(u|v^{(s+1)})$.

In contrast, the Metropolis algorithm proposes changes to $X = (U, V)$ and then accepts or rejects those changes based on p_0. In the Poisson regression example the proposed vector differed from its current value at each element of the vector, but this is not necessary. An alternative way to implement the Metropolis algorithm is to propose and then accept or reject changes to one element at a time:

1. update U:
   a) sample $u^* \sim J_u(u|u^{(s)})$;
   b) compute $r = p_O(u^*, v^{(s)})/p_O(u^{(s)}, v^{(s)})$;
   c) set $u^{(s+1)}$ to u^* or $u^{(s)}$ with probability $\min(1, r)$ and $\max(0, 1-r)$.
2. update V:
   a) sample $v^* \sim J_v(v|v^{(s)})$;
   b) compute $r = p_O(u^{(s+1)}, v^*)/p_O(u^{(s+1)}, v^{(s)})$;
   c) set $v^{(s+1)}$ to v^* or $v^{(s)}$ with probability $\min(1, r)$ and $\max(0, 1-r)$.

Here, J_u and J_v are separate symmetric proposal distributions for U and V.

This Metropolis algorithm generates proposals from J_u and J_v and accepts them with some probability $\min(1, r)$. Similarly, each step of the Gibbs sampler can be seen as generating a proposal from a full conditional distribution and then accepting it with probability 1. The Metropolis-Hastings algorithm generalizes both of these approaches by allowing arbitrary proposal distributions. The proposal distributions can be symmetric around the current values, full conditional distributions, or something else entirely. A Metropolis-Hastings algorithm for approximating $p_O(u, v)$ runs as follows:

1. update U:
   a) sample $u^* \sim J_u(u|u^{(s)}, v^{(s)})$;
   b) compute the acceptance ratio
   $$r = \frac{p_O(u^*, v^{(s)})}{p_O(u^{(s)}, v^{(s)})} \times \frac{J_u(u^{(s)}|u^*, v^{(s)})}{J_u(u^*|u^{(s)}, v^{(s)})};$$
   c) set $u^{(s+1)}$ to u^* or $u^{(s)}$ with probability $\min(1, r)$ and $\max(0, 1-r)$.
2. update V:
   a) sample $v^* \sim J_v(v|u^{(s+1)}, v^{(s)})$;
   b) compute the acceptance ratio
   $$r = \frac{p_O(u^{(s+1)}, v^*)}{p_O(u^{(s+1)}, v^{(s)})} \times \frac{J_v(v^{(s)}|u^{(s+1)}, v^*)}{J_v(v^*|u^{(s+1)}, v^{(s)})};$$
   c) set $v^{(s+1)}$ to v^* or $v^{(s)}$ with probability $\min(1, r)$ and $\max(0, 1-r)$.

In this algorithm the proposal distributions J_u and J_v are not required to be symmetric. In fact, the only requirement is that they do not depend on U or V values in our sequence previous to the most current values. This requirement ensures that the sequence is a Markov chain.

The Metropolis-Hastings algorithm looks a lot like the Metropolis algorithm, except that the acceptance ratio contains an extra factor, the ratio of the probability of generating the current value from the proposed to the probability of generating the proposed from the current. This can be viewed as a "correction factor:" If a value u^* is much more likely to be proposed than the current value $u^{(s)}$, then we must down-weight the probability of accepting u^* accordingly, otherwise the value u^* will be overrepresented in our sequence.

That the Metropolis algorithm is a special case of the Metropolis-Hastings algorithm is easy to see: If J_u is symmetric, meaning that $J(u_a|u_b, v) = J(u_b|u_a, v)$ for all possible u_a, u_b and v, then the correction factor in the Metropolis-Hastings acceptance ratio is equal to 1 and the acceptance probability is the same as in the Metropolis algorithm. That the Gibbs sampler is a type of Metropolis-Hastings algorithm is almost as easy to see. In the Gibbs sampler the proposal distribution for U is the full conditional distribution of U given $V = v$. If we use the full conditionals as our proposal distributions in the Metropolis-Hastings algorithm, we have $J_u(u^*|u^{(s)}, v^{(s)}) = p_0(u^*|v^{(s)})$. The Metropolis-Hastings acceptance ratio is then

$$\begin{aligned} r &= \frac{p_0(u^*, v^{(s)})}{p_0(u^{(s)}, v^{(s)})} \times \frac{J_u(u^{(s)}|u^*, v^{(s)})}{J_u(u^*|u^{(s)}, v^{(s)})} \\ &= \frac{p_0(u^*, v^{(s)})}{p_0(u^{(s)}, v^{(s)})} \frac{p_0(u^{(s)}|v^{(s)})}{p_0(u^*|v^{(s)})} \\ &= \frac{p_0(u^*|v^{(s)})p_0(v^{(s)})}{p_0(u^{(s)}|v^{(s)})p_0(v^{(s)})} \frac{p_0(u^{(s)}|v^{(s)})}{p_0(u^*|v^{(s)})} \\ &= \frac{p_0(v^{(s)})}{p_0(v^{(s)})} = 1, \end{aligned}$$

and so if we propose a value from the full conditional distribution the acceptance probability is 1, and the algorithm is equivalent to the Gibbs sampler.

10.4.2 Why does the Metropolis-Hastings algorithm work?

A more general form of the Metropolis-Hastings algorithm is as follows: Given a current value $x^{(s)}$ of X,

1. Generate x^* from $J_s(x^*|x^{(s)})$;
2. Compute the acceptance ratio

$$r = \frac{p_0(x^*)}{p_0(x^{(s)})} \times \frac{J_s(x^{(s)}|x^*)}{J_s(x^*|x^{(s)})};$$

3. Sample $u \sim$uniform$(0, 1)$. If $u < r$ set $x^{(s+1)} = x^*$, else set $x^{(s+1)} = x^{(s)}$.

Note that the proposal distribution may also depend on the iteration number s. For example, the Metropolis-Hastings algorithm presented in the last section can be equivalently described by steps 1, 2 and 3 above by setting J_s to be equal to J_u for odd values of s and equal to J_v for even values. This makes the algorithm alternately update values of U and V.

The primary restriction we place on $J_s(x^*|x^{(s)})$ is that it does not depend on values in the sequence previous to $x^{(s)}$. This restriction ensures that the algorithm generates a Markov chain. We also want to choose J_s so that the Markov chain is able to converge to the target distribution p_0. For example, we want to make sure that every value of x such that $p_0(x) > 0$ will eventually be proposed (and so accepted some fraction of the time), regardless of where we start the Markov chain. An example in which this is not the case is where the values of X having non-zero probability are the integers, and $J_s(x^*|x^{(s)})$ proposes $x^{(s)} \pm 2$ with equal probability. In this case the Metropolis-Hastings algorithm produces a Markov chain, but the chain will only generate even numbers if $x^{(1)}$ is even, and only odd number if $x^{(1)}$ is odd. This type of Markov chain is called *reducible*, as the set of possible X-values can be divided into non-overlapping sets (even and odd integers in this example), between which the algorithm is unable to move. In contrast, we want our Markov chain to be *irreducible*, that is, able to go from any one value of X to any other, eventually.

Additionally, we will want J_s to be such that the Markov chain is *aperiodic* and *recurrent*. A value x is periodic with period $k > 1$ in a Markov chain if it can only be visited every kth iteration. If x is periodic, then for every S there are an infinite number of iterations $s > S$ for which $\Pr(x^{(s)} = x) = 0$. Since we want the distribution of $x^{(s)}$ to converge to p_0, we should make sure that if $p_0(x) > 0$, then x is not periodic in our Markov chain. A Markov chain lacking any periodic states is called *aperiodic*. Finally, if $x^{(s)} = x$ for some iteration s, then this must mean that $p_0(x) > 0$. Therefore, we want our Markov chain to be able to return to x from time to time as we run our chain (otherwise the relative fraction of x's in the chain will go to zero, even though $p_0(x) > 0$). A value x is said to be recurrent if, when we continue to run the Markov chain from x, we are guaranteed to eventually return to x. Clearly we want all of the possible values of X to be recurrent in our Markov chain.

An irreducible, aperiodic and recurrent Markov chain is a very well behaved object. A theorem from probability theory says that the empirical distribution of samples generated from such a Markov chain will converge:

Theorem 2 *(Ergodic Theorem) If $\{x^{(1)}, x^{(2)}, \ldots\}$ is an irreducible, aperiodic and recurrent Markov chain, then there is a unique probability distribution π such that as $s \to \infty$,*

- $\Pr(x^{(s)} \in A) \to \pi(A)$ *for any set A;*
- $\frac{1}{S}\sum g(x^{(s)}) \to \int g(x)\pi(x)\, dx.$

The distribution π is called the *stationary distribution* of the Markov chain. It is called the stationary distribution because it has the following property:

If $x^{(s)} \sim \pi$,
and $x^{(s+1)}$ is generated from the Markov chain starting at $x^{(s)}$,
then $\Pr(x^{(s+1)} \in A) = \pi(A)$.

In other words, if you sample $x^{(s)}$ from π and then generate $x^{(s+1)}$ conditional on $x^{(s)}$ from the Markov chain, then the *unconditional* distribution of $x^{(s+1)}$ is π. Once you are sampling from the stationary distribution, you are always sampling from the stationary distribution.

In most problems it is not too hard to construct Metropolis-Hastings algorithms that generate Markov chains that are irreducible, aperiodic and recurrent. For example, if $p_0(x)$ is continuous, then using a normal proposal distribution centered around the current value guarantees that $\Pr(x^{(s+1)} \in A|x^{(s)} = x) > 0$ for every x, s and set A such that $p_0(A) > 0$. All of the Metropolis-Hastings algorithms in this book generate Markov chains that are irreducible, aperiodic and recurrent. As such, sequences of X-values generated from these algorithms can be used to approximate their stationary distributions. What is left to show is that the stationary distribution π for a Metropolis-Hastings algorithm is equal to the distribution p_0 we wish to approximate.

"Proof" that $\pi(x) = p_0(x)$

The theorem above says that the stationary distribution of the Metropolis-Hastings algorithm is unique, and so if we show that p_0 is a stationary distribution, we will have shown it is *the* stationary distribution. Our sketch of a proof follows closely a proof from Gelman et al (2004) for the Metropolis algorithm. In that proof and here, it is assumed for simplicity that X is a discrete random variable. Suppose $x^{(s)}$ is sampled from the target distribution p_0, and then $x^{(s+1)}$ is generated from $x^{(s)}$ using the Metropolis-Hastings algorithm. To show that p_0 is the stationary distribution we need to show that $\Pr(x^{(s+1)} = x) = p_0(x)$.

Let x_a and x_b be any two values of X such that $p_0(x_a)J_s(x_b|x_a) \geq p_0(x_b)J_s(x_a|x_b)$. Then under the Metropolis-Hastings algorithm the probability that $x^{(s)} = x_a$ and $x^{(s+1)} = x_b$ is equal to the probability of

1. sampling $x^{(s)} = x_a$ from p_0;
2. proposing $x^* = x_b$ from $J_s(x^*|x^{(s)})$;
3. accepting $x^{(s+1)} = x_b$.

The probability of these three things occurring is their product:

$$\begin{aligned}\Pr(x^{(s)} = x_a, x^{(s+1)} = x_b) &= p_0(x_a) \times J_s(x_b|x_a) \times \frac{p_0(x_b)}{p_0(x_a)} \frac{J_s(x_a|x_b)}{J_s(x_b|x_a)} \\ &= p_0(x_b)J_s(x_a|x_b) .\end{aligned}$$

On the other hand, the probability that $x^{(s)} = x_b$ and $x^{(s+1)} = x_a$ is the probability that x_b is sampled from p_0, that x_a is proposed from $J_s(x^*|x^{(s)})$ and that x_a is accepted as $x^{(s+1)}$. But in this case the acceptance probability is one because we assumed $p_0(x_a)J_s(x_b|x_a) \geq p_0(x_b)J_s(x_a|x_b)$. This means that $\Pr(x^{(s)} = x_b, x^{(s+1)} = x_a) = p_0(x_b)J_s(x_a|x_b)$.

The above two calculations have shown that the probability of observing $x^{(s)}$ and $x^{(s+1)}$ to be x_a and x_b, respectively, is the same as observing them to be x_b and x_a respectively, for any two values x_a and x_b. The final step of the proof is to use this fact to derive the marginal probability $\Pr(x^{(s+1)} = x)$:

$$\begin{aligned}
\Pr(x^{(s+1)} = x) &= \sum_{x_a} \Pr(x^{(s+1)} = x, x^{(s)} = x_a) \\
&= \sum_{x_a} \Pr(x^{(s+1)} = x_a, x^{(s)} = x) \\
&= \Pr(x^{(s)} = x)
\end{aligned}$$

This completes the proof that $\Pr(x^{(s+1)} = x) = p_0(x)$ if $\Pr(x^{(s)} = x) = p_0(x)$.

10.5 Combining the Metropolis and Gibbs algorithms

In complex models it is often the case that conditional distributions are available for some parameters but not for others. In these situations we can combine Gibbs and Metropolis-type proposal distributions to generate a Markov chain to approximate the joint posterior distribution of all of the parameters. In this section we do this in the context of estimating the parameters in a regression model for time-series data where the errors are temporally correlated. In this case, full conditional distributions are available for the regression parameters but not the parameter describing the dependence among the observations.

Example: Historical CO_2 *and temperature data*

Analyses of ice cores from East Antarctica have allowed scientists to deduce historical atmospheric conditions of the last few hundred thousand years (Petit et al, 1999). The first plot of Figure 10.7 plots time-series of temperature and carbon dioxide concentration on a standardized scale (centered and scaled to have a mean of zero and a variance of one). The data include 200 values of temperature measured at roughly equal time intervals, with time between consecutive measurements being approximately 2,000 years. For each value of temperature there is a CO_2 concentration value corresponding to a date that is roughly 1,000 years previous to the temperature value, on average. Temperature is recorded in terms of its difference from current present temperature in degrees Celsius, and CO_2 concentration is recorded in parts per million by volume.

The plot indicates that the temporal history of temperature and CO_2 follow very similar patterns. The second plot in Figure 10.7 indicates that CO_2 concentration at a given time point is predictive of temperature following that time point. One way to quantify this is by fitting a linear regression model for temperature (Y) as a function of CO_2 (x). Ordinary least squares regression gives an estimated model of $\hat{\mathrm{E}}[Y|x] = -23.02 + 0.08x$ with a nominal standard error of 0.0038 for the slope term. The validity of this standard error relies on the error terms in the regression model being independent and identically distributed, and standard confidence intervals further rely on the errors being normally distributed.

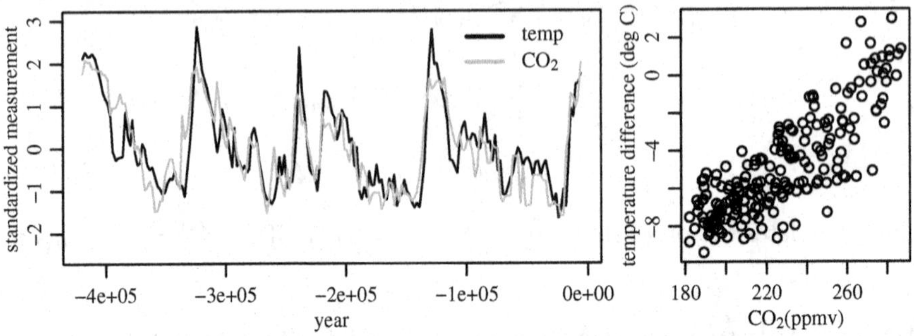

Fig. 10.7. Temperature and carbon dioxide data.

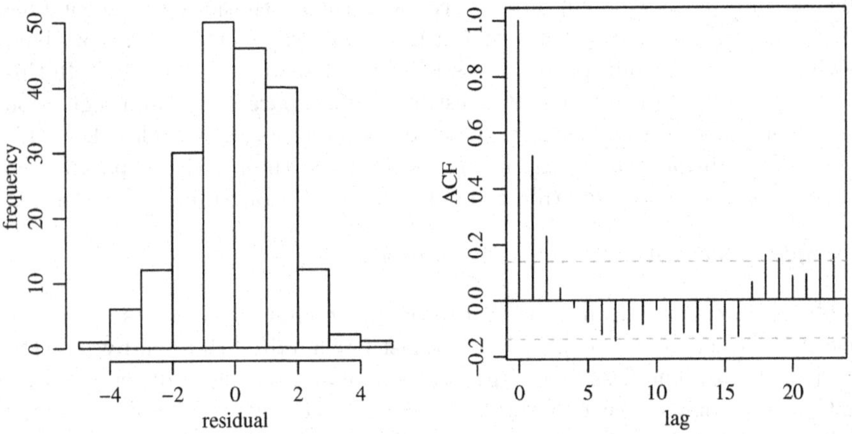

Fig. 10.8. Residual analysis for the least squares estimation.

These two assumptions are examined in the two residual diagnostic plots in Figure 10.8. The first plot, a histogram of the residuals, indicates no serious deviation from non-normality. The second plot gives the autocorrelation function of the residuals, and indicates a nontrivial correlation of 0.52 between residuals at consecutive time points. Such a positive correlation generally means that there is less information in the data, and less evidence for a relationship between the two variables, than is assumed by the ordinary least squares regression analysis.

10.5.1 A regression model with correlated errors

The ordinary regression model is

$$\boldsymbol{Y} = \begin{pmatrix} Y_1 \\ \vdots \\ Y_n \end{pmatrix} \sim \text{multivariate normal}(\mathbf{X}\boldsymbol{\beta}, \sigma^2\mathbf{I}).$$

The diagnostic plots suggest that a more appropriate model for the ice core data is one in which the error terms are not independent, but temporally correlated. This means we must replace the covariance matrix $\sigma^2\mathbf{I}$ in the ordinary regression model with a matrix Σ that can represent positive correlation between sequential observations. One simple, popular class of covariance matrices for temporally correlated data are those having *first-order autoregressive structure*:

$$\Sigma = \sigma^2\mathbf{C}_\rho = \sigma^2 \begin{pmatrix} 1 & \rho & \rho^2 & \cdots & \rho^{n-1} \\ \rho & 1 & \rho & \cdots & \rho^{n-2} \\ \rho^2 & \rho & 1 & & \\ \vdots & \vdots & & \ddots & \\ \rho^{n-1} & \rho^{n-2} & & & 1 \end{pmatrix}$$

Under this covariance matrix the variance of $\{Y_i|\boldsymbol{\beta}, \boldsymbol{x}_i\}$ is σ^2 but the correlation between Y_i and Y_{i+t} is ρ^t, which decreases to zero as the time difference t becomes larger.

Having observed $\boldsymbol{Y} = \boldsymbol{y}$, the parameters to estimate in this model include $\boldsymbol{\beta}$, σ^2 and ρ. Using the multivariate normal and inverse-gamma prior distributions of Section 9.2.1 for $\boldsymbol{\beta}$ and σ^2, it is left as an exercise to show that

$$\begin{aligned} \{\boldsymbol{\beta}|\mathbf{X}, \boldsymbol{y}, \sigma^2, \rho\} &\sim \text{multivariate normal}(\boldsymbol{\beta}_n, \Sigma_n)\ , \text{ where} \qquad (10.2)\\ \Sigma_n &= (\mathbf{X}^T\mathbf{C}_\rho^{-1}\mathbf{X}/\sigma^2 + \Sigma_0^{-1})^{-1} \\ \boldsymbol{\beta}_n &= \Sigma_n(\mathbf{X}^T\mathbf{C}_\rho^{-1}\boldsymbol{y}/\sigma^2 + \Sigma_0^{-1}\boldsymbol{\beta}_0)\ , \text{ and} \\ \{\sigma^2|\mathbf{X}, \boldsymbol{y}, \boldsymbol{\beta}, \rho\} &\sim \text{inverse-gamma}([\nu_0 + n]/2, [\nu_0\sigma_0^2 + \text{SSR}_\rho]/2)\ , \text{ where} \\ \text{SSR}_\rho &= (\boldsymbol{y} - \mathbf{X}\boldsymbol{\beta})^T\mathbf{C}_\rho^{-1}(\boldsymbol{y} - \mathbf{X}\boldsymbol{\beta}). \end{aligned}$$

If $\boldsymbol{\beta}_0 = \mathbf{0}$ and Σ_0 has large diagonal entries, then $\boldsymbol{\beta}_n$ is very close to $(\mathbf{X}^T\mathbf{C}_\rho^{-1}\mathbf{X})^{-1}\mathbf{X}^T\mathbf{C}_\rho^{-1}\boldsymbol{y}$. If ρ were known this would be the *generalized least squares* (GLS) estimate of $\boldsymbol{\beta}$, a type of weighted least squares estimate that is used when the error terms are not independent and identically distributed. In such situations, both OLS and GLS provide unbiased estimates of $\boldsymbol{\beta}$ but the GLS estimate has a lower variance. Bayesian analysis using a model that accounts for the correlated errors provides parameter estimates that are similar to those of GLS, so for convenience we will refer to our analysis as "Bayesian GLS."

If we knew the value of ρ we could use the Gibbs sampler to approximate $p(\boldsymbol{\beta}, \sigma^2|\mathbf{X}, \boldsymbol{y}, \rho)$ by iteratively sampling from the full conditional distributions given by the equations in 10.2. Of course ρ is unknown and so we will need to estimate it as well with our Markov chain. Unfortunately the full conditional distribution for ρ will be nonstandard for most prior distributions, suggesting that the Gibbs sampler is not applicable here and we may have to use a Metropolis algorithm (although a discrete approximation to $p(\rho|\mathbf{X}, \boldsymbol{y}, \boldsymbol{\beta}, \sigma^2)$ could be used).

It is in situations like this that the generality of the Metropolis-Hastings algorithm comes in handy. Recall that in this algorithm we are allowed to use different proposal distributions at each step. We can iteratively update $\boldsymbol{\beta}$, σ^2 and ρ at different steps, making proposals with full conditional distributions for $\boldsymbol{\beta}$ and σ^2 (Gibbs proposals) and a symmetric proposal distribution for ρ (a Metropolis proposal). Following the rules of the Metropolis-Hastings algorithm, we accept with probability 1 any proposal coming from a full conditional distribution, whereas we have to calculate an acceptance probability for proposals of ρ. Given $\{\boldsymbol{\beta}^{(s)}, \sigma^{2(s)}, \rho^{(s)}\}$, a Metropolis-Hastings algorithm to generate a new set of parameter values is as follows:

1. Update $\boldsymbol{\beta}$: Sample $\boldsymbol{\beta}^{(s+1)} \sim$ multivariate normal$(\boldsymbol{\beta}_n, \Sigma_n)$, where $\boldsymbol{\beta}_n$ and Σ_n depend on $\sigma^{2(s)}$ and $\rho^{(s)}$.
2. Update σ^2: Sample $\sigma^{2(s+1)} \sim$ inverse-gamma$([\nu_0+n]/2, [\nu_0\sigma_0^2+\text{SSR}_\rho]/2)$, where SSR_ρ depends on $\boldsymbol{\beta}^{(s+1)}$ and $\rho^{(s)}$.
3. Update ρ:
   a) Propose $\rho^* \sim$ uniform$(\rho^{(s)} - \delta, \rho^{(s)} + \delta)$. If $\rho^* < 0$ then reassign it to be $|\rho^*|$. If $\rho^* > 1$ reassign it to be $2 - \rho^*$.
   b) Compute the acceptance ratio

$$r = \frac{p(\boldsymbol{y}|\mathbf{X}, \boldsymbol{\beta}^{(s+1)}, \sigma^{2(s+1)}, \rho^*)p(\rho^*)}{p(\boldsymbol{y}|\mathbf{X}, \boldsymbol{\beta}^{(s+1)}, \sigma^{2(s+1)}, \rho^{(s)})p(\rho^{(s)})}$$

   and sample $u \sim$ uniform(0,1). If $u < r$ set $\rho^{(s+1)} = \rho^*$, otherwise set $\rho^{(s+1)} = \rho^{(s)}$.

The proposal distribution used in Step 3.a is called a *reflecting random walk*, which ensures that $0 < \rho < 1$. It is left as an exercise to show that this proposal distribution is symmetric. We also leave it as an exercise to show that the value of r given in Step 3.b is numerically equal to

$$\frac{p(\boldsymbol{\beta}^{(s+1)}, \sigma^{2(s+1)}, \rho^*|\boldsymbol{y}, \mathbf{X})}{p(\boldsymbol{\beta}^{(s+1)}, \sigma^{2(s+1)}, \rho^{(s)}|\boldsymbol{y}, \mathbf{X})},$$

the ratio as given in the definition of the Metropolis algorithm.

While technically the steps above constitute "three iterations" of the Metropolis-Hastings algorithm, it is convenient to group them together as one. A sequence of Metropolis-Hastings steps in which each parameter is updated is often referred to as a *scan* of the algorithm.

10.5.2 Analysis of the ice core data

We'll use diffuse prior distributions for the parameters, with $\boldsymbol{\beta}_0 = \mathbf{0}$, $\Sigma_0 = \text{diag}(1000)$, $\nu_0 = 1$ and $\sigma_0^2 = 1$. Our prior for ρ will be the uniform distribution on $(0, 1)$. The first panel of Figure 10.9 plots the first 1,000 values $\{\rho^{(1)}, \ldots, \rho^{(1000)}\}$ generated using the Metropolis-Hastings algorithm above. The acceptance rate for these values is 0.322 which seems good, but the autocorrelation of the sequence, shown in the second panel, is very high. The effective sample size for this correlated sequence of 1,000 ρ-values is only 23, indicating that we will need many more iterations of the algorithm to obtain a decent approximation to the posterior distribution.

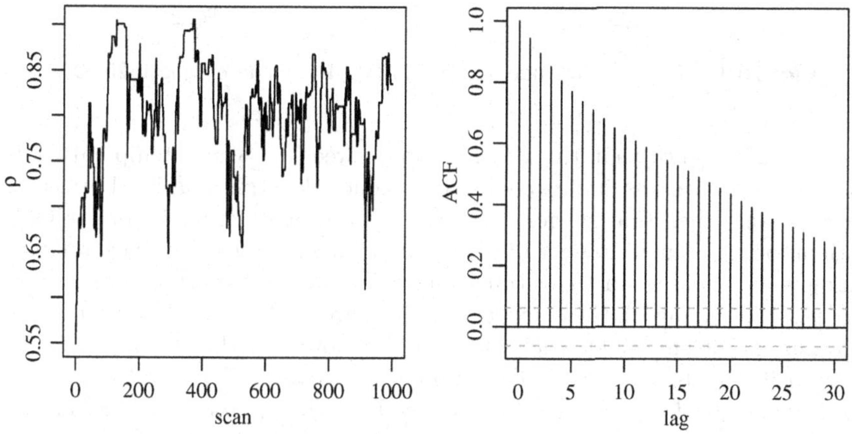

Fig. 10.9. The first 1,000 values of ρ generated from the Markov chain.

Suppose we were to generate 25,000 scans for a total of $25,000 \times 4 = 100,000$ parameter values. Storing and manipulating all of these values can be tedious and somewhat unnecessary: Since the Markov chain is so highly correlated, the values of $\rho^{(s)}$ and $\rho^{(s+1)}$ offer roughly the same information about the posterior distribution. With this in mind, for highly correlated Markov chains with moderate to large numbers of parameters we will often only save a fraction of the scans of the Markov chain. This practice of throwing away many iterations of a Markov chain is known as *thinning*. Figure 10.10 shows the thinned output of a 25,000-scan Markov chain for the ice core data, in which only every 25th scan was saved. Thinning the output reduces it down to a manageable 1,000 samples, having a much lower autocorrelation than 1,000 sequential samples from an unthinned Markov chain.

The Monte Carlo approximation to the posterior density of β_2, the slope parameter, appears in the first panel of Figure 10.11. The posterior mean of β_2 is 0.028 and a posterior 95% quantile-based confidence interval is $(0.01, 0.05)$,

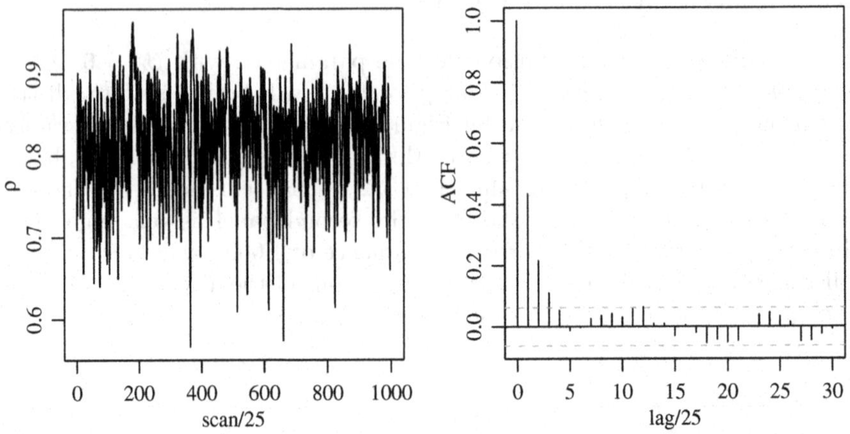

Fig. 10.10. Every 25th value of ρ from a Markov chain of length 25,000.

indicating evidence that the relationship between CO_2 and temperature is positive. However, as indicated in the second plot this relationship seems much weaker than that suggested by the OLS estimate of 0.08. For the OLS estimation, the small number of data points with high y-values have a larger amount of influence on the estimate of $\boldsymbol{\beta}$. In contrast, the GLS model recognizes that many of these extreme points are highly correlated with one another and down-weights their influence. We note that this "weaker" regression coefficient is a result of the temporally correlated data, and not of the particular prior distribution we used or the Bayesian approach in general. The reader is encouraged to repeat the analysis with different prior distributions, or to perform a non-Bayesian GLS estimation for comparison. In any case, the data analysis indicates evidence of a relationship between temperature measurements and the CO_2 measurements that precede them in time.

10.6 Discussion and further references

The Metropolis algorithm was introduced by Metropolis et al (1953) in an application to a problem in statistical physics. The algorithm was generalized by Hastings (1970), but it was not until the publication of Gelfand and Smith (1990) that MCMC became widely used in the statistics community. See Robert and Casella (2008) for a brief history of Monte Carlo and MCMC methods.

A number of modifications and extensions of MCMC methods have appeared since the 1990s. One technique that is broadly applicable is automatic, adaptive tuning of the proposal distribution in order to achieve good mixing (Gilks et al, 1998; Haario et al, 2001). Not all adaptive algorithms will result

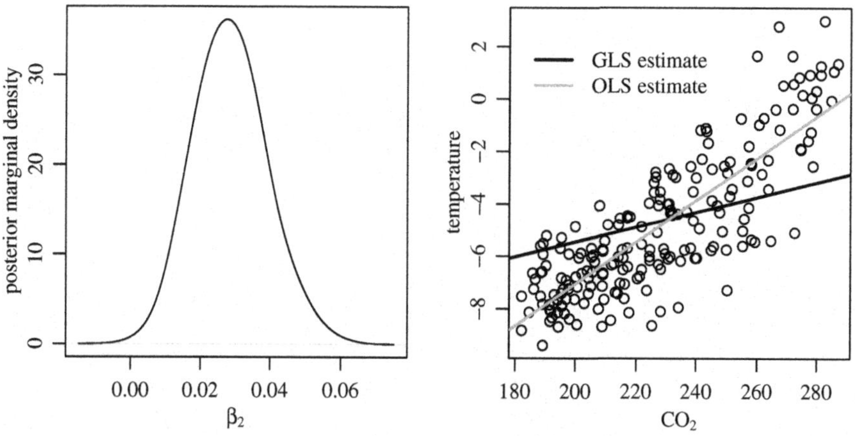

Fig. 10.11. Posterior distribution of the slope parameter β_2, along with the posterior mean regression line.

in chains that converge to the target distribution, but there are known conditions under which convergence is guaranteed (Atchadé and Rosenthal, 2005; Roberts and Rosenthal, 2007).

11

Linear and generalized linear mixed effects models

In Chapter 8 we learned about the concept of hierarchical modeling, a data analysis approach that is appropriate when we have multiple measurements within each of several groups. In that chapter, variation in the data was represented with a between-group sampling model for group-specific means, in addition to a within-group sampling model to represent heterogeneity of observations within a group. In this chapter we extend the hierarchical model to describe how relationships between variables may differ between groups. This can be done with a regression model to describe within-group variation, and a multivariate normal model to describe heterogeneity among regression coefficients across the groups. We also cover estimation for hierarchical generalized linear models, which are hierarchical models that have a generalized linear regression model representing within-group heterogeneity.

11.1 A hierarchical regression model

Let's return to the math score data described in Section 8.4, which included math scores of 10th grade children from 100 different large urban public high schools. In Chapter 8 we estimated school-specific expected math scores, as well as how these expected values varied from school to school. Now suppose we are interested in examining the relationship between math score and another variable, socioeconomic status (SES), which was calculated from parental income and education levels for each student in the dataset.

In Chapter 8 we quantified the between-school heterogeneity in expected math score with a hierarchical model. Given the amount of variation we observed it seems possible that the relationship between math score and SES might vary from school to school as well. A quick and easy way to assess this possibility is to fit a linear regression model of math score as a function of SES for each of the 100 schools in the dataset. To make the parameters more interpretable we will center the SES scores within each school separately, so that the sample average SES score within each school is zero. As a result, the

P.D. Hoff, *A First Course in Bayesian Statistical Methods*,
Springer Texts in Statistics, DOI 10.1007/978-0-387-92407-6_11,

intercept of the regression line can be interpreted as the school-level average math score.

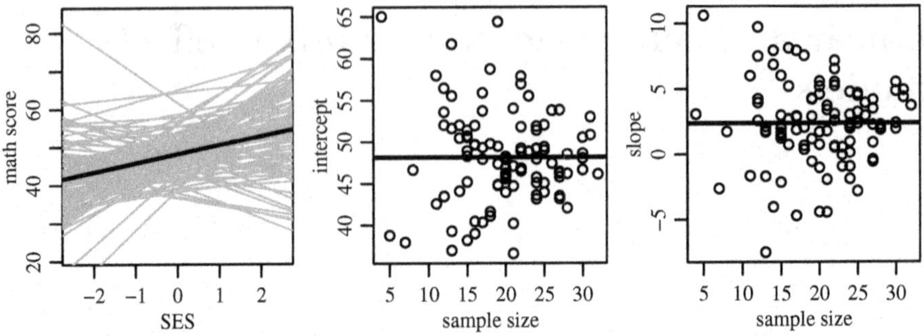

Fig. 11.1. Least squares regression lines for the math score data, and plots of estimates versus group sample size.

The first panel of Figure 11.1 plots least squares estimates of the regression lines for the 100 schools, along with an average of these lines in black. A large majority show an increase in expected math score with increasing SES, although a few show a negative relationship. The second and third panels of the figure relate the least squares estimates to sample size. Notice that schools with the highest sample sizes have regression coefficients that are generally close to the average, whereas schools with extreme coefficients are generally those with low sample sizes. This phenomenon is reminiscent of what we discussed in Section 8.4: The smaller the sample size for the group, the more probable that unrepresentative data are sampled and an extreme least squares estimate is produced. As in Chapter 8, our remedy to this problem will be to stabilize the estimates for small sample size schools by sharing information across groups, using a hierarchical model.

The hierarchical model in the linear regression setting is a conceptually straightforward generalization of the normal hierarchical model from Chapter 8. We use an ordinary regression model to describe within-group heterogeneity of observations, then describe between-group heterogeneity using a sampling model for the group-specific regression parameters. Expressed symbolically, our within-group sampling model is

$$Y_{i,j} = \boldsymbol{\beta}_j^T \boldsymbol{x}_{i,j} + \epsilon_{i,j} \ , \ \{\epsilon_{i,j}\} \sim \text{ i.i.d. normal}(0, \sigma^2), \tag{11.1}$$

where $\boldsymbol{x}_{i,j}$ is a $p \times 1$ vector of regressors for observation i in group j. Expressing $Y_{1,j}, \ldots, Y_{n_j,j}$ as a vector $\boldsymbol{Y}_j$ and combining $\boldsymbol{x}_{1,j}, \ldots, \boldsymbol{x}_{n_j,j}$ into an $n_j \times p$ matrix $\mathbf{X}_j$, the within-group sampling model can be expressed equivalently as $\boldsymbol{Y}_j \sim$ multivariate normal$(\mathbf{X}_j \boldsymbol{\beta}_j, \sigma^2 \mathbf{I})$, with the group-specific data vectors $\boldsymbol{Y}_1, \ldots, \boldsymbol{Y}_m$ being conditionally independent given $\boldsymbol{\beta}_1, \ldots, \boldsymbol{\beta}_m$ and σ^2.

The heterogeneity among the regression coefficients $\boldsymbol{\beta}_1, \ldots, \boldsymbol{\beta}_m$ will be described with a between-group sampling model. If we have no prior information distinguishing the different groups we can model them as being exchangeable, or (roughly) equivalently, as being i.i.d. from some distribution representing the sampling variability across groups. The *normal hierarchical regression model* describes the across-group heterogeneity with a multivariate normal model, so that

$$\boldsymbol{\beta}_1, \ldots, \boldsymbol{\beta}_m \sim \text{i.i.d. multivariate normal}(\boldsymbol{\theta}, \Sigma). \tag{11.2}$$

A graphical representation of the hierarchical model appears in Figure 11.2, which makes clear that the multivariate normal distribution for $\boldsymbol{\beta}_1, \ldots, \boldsymbol{\beta}_m$ is not a prior distribution representing uncertainty about a fixed but unknown quantity. Rather, it is a sampling distribution representing heterogeneity among a collection of objects. The values of $\boldsymbol{\theta}$ and Σ are fixed but unknown parameters to be estimated.

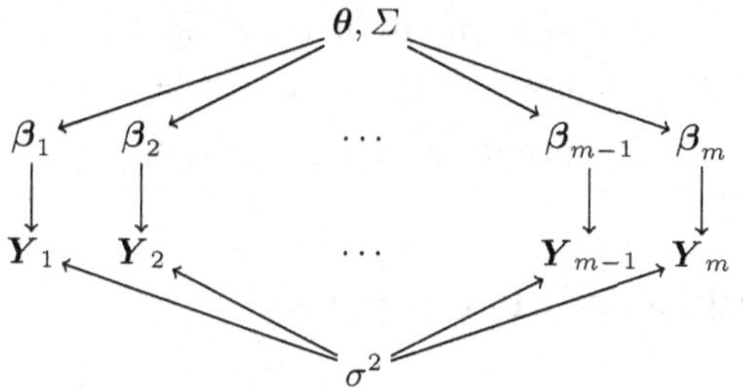

Fig. 11.2. A graphical representation of the hierarchical normal regression model.

This hierarchical regression model is sometimes called a *linear mixed effects model*. This name is motivated by an alternative parameterization of Equations 11.1 and 11.2. We can rewrite the between-group sampling model as

$$\begin{aligned} \boldsymbol{\beta}_j &= \boldsymbol{\theta} + \boldsymbol{\gamma}_j \\ \boldsymbol{\gamma}_1 \ldots, \boldsymbol{\gamma}_m &\sim \text{i.i.d. multivariate normal}(\mathbf{0}, \Sigma). \end{aligned}$$

Plugging this into our within-group regression model gives

$$\begin{aligned} Y_{i,j} &= \boldsymbol{\beta}_j^T \boldsymbol{x}_{i,j} + \epsilon_{i,j} \\ &= \boldsymbol{\theta}^T \boldsymbol{x}_{i,j} + \boldsymbol{\gamma}_j^T \boldsymbol{x}_{i,j} + \epsilon_{i,j}. \end{aligned}$$

In this parameterization $\boldsymbol{\theta}$ is referred to as a *fixed effect* as it is constant across groups, whereas $\boldsymbol{\gamma}_1, \ldots, \boldsymbol{\gamma}_m$ are called *random effects*, as they vary. The name "mixed effects model" comes from the fact that the regression model contains both fixed and random effects. Although for our particular example the regressors corresponding to the fixed and random effects are the same, this does not have to be the case. A more general model would be $Y_{i,j} = \boldsymbol{\theta}^T \boldsymbol{x}_{i,j} + \boldsymbol{\gamma}_j^T \boldsymbol{z}_{i,j} + \epsilon_{i,j}$, where $\boldsymbol{x}_{i,j}$ and $\boldsymbol{z}_{i,j}$ could be vectors of different lengths which may or may not contain overlapping variables. In particular, $\boldsymbol{x}_{i,j}$ might contain regressors that are group specific, that is, constant across all observations in the same group. Such variables are not generally included in $\boldsymbol{z}_{i,j}$, as there would be no information in the data with which to estimate the corresponding group-specific regression coefficients.

Given a prior distribution for $(\boldsymbol{\theta}, \Sigma, \sigma^2)$ and having observed $\boldsymbol{Y}_1 = \boldsymbol{y}_1, \ldots, \boldsymbol{Y}_m = \boldsymbol{y}_m$, a Bayesian analysis proceeds by computing the posterior distribution $p(\boldsymbol{\beta}_1, \ldots, \boldsymbol{\beta}_m, \boldsymbol{\theta}, \Sigma, \sigma^2 | \mathbf{X}_1, \ldots, \mathbf{X}_m, \boldsymbol{y}_1, \ldots, \boldsymbol{y}_m)$. If semiconjugate prior distributions are used for $\boldsymbol{\theta}, \Sigma$ and σ^2, then the posterior distribution can be approximated quite easily with Gibbs sampling. The classes of semiconjugate prior distributions for $\boldsymbol{\theta}$ and Σ are as in the multivariate normal model discussed in Chapter 7. The prior we will use for σ^2 is the usual inverse-gamma distribution.

$$
\begin{aligned}
\boldsymbol{\theta} &\sim \text{multivariate normal}(\boldsymbol{\mu}_0, \Lambda_0) \\
\Sigma &\sim \text{inverse-Wishart}(\eta_0, \mathbf{S}_0^{-1}) \\
\sigma^2 &\sim \text{inverse-gamma}(\nu_0/2, \nu_0\sigma_0^2/2)
\end{aligned}
$$

11.2 Full conditional distributions

While computing the posterior distribution for so many parameters may seem daunting, the calculations involved in computing the full conditional distributions have the same mathematical structure as models we have studied in previous chapters. Once we have the full conditional distributions we can iteratively sample from them to approximate the joint posterior distribution.

Full conditional distributions of $\boldsymbol{\beta}_1, \ldots, \boldsymbol{\beta}_m$

Our hierarchical regression model shares information across groups via the parameters $\boldsymbol{\theta}$, Σ and σ^2. As a result, conditional on $\boldsymbol{\theta}, \Sigma, \sigma^2$ the regression coefficients $\boldsymbol{\beta}_1, \ldots, \boldsymbol{\beta}_m$ are independent. Referring to the graph in Figure 11.2, from the perspective of a given $\boldsymbol{\beta}_j$ the model looks like an ordinary one-group regression problem where the prior mean and variance for $\boldsymbol{\beta}_j$ are $\boldsymbol{\theta}$ and Σ. This analogy is in fact correct, and the results of Section 9.2.1 show that $\{\boldsymbol{\beta}_j | \boldsymbol{y}_j, \mathbf{X}_j, \boldsymbol{\theta}, \Sigma, \sigma^2\}$ has a multivariate normal distribution with

$$
\begin{aligned}
\text{Var}[\boldsymbol{\beta}_j | \boldsymbol{y}_j, \mathbf{X}_j, \sigma^2, \boldsymbol{\theta}, \Sigma] &= (\Sigma^{-1} + \mathbf{X}_j^T \mathbf{X}_j / \sigma^2)^{-1} \\
\text{E}[\boldsymbol{\beta}_j | \boldsymbol{y}_j, \mathbf{X}_j, \sigma^2, \boldsymbol{\theta}, \Sigma] &= (\Sigma^{-1} + \mathbf{X}_j^T \mathbf{X}_j / \sigma^2)^{-1} (\Sigma^{-1} \boldsymbol{\theta} + \mathbf{X}_j^T \boldsymbol{y}_j / \sigma^2).
\end{aligned}
$$

Full conditional distributions of $\boldsymbol{\theta}$ and Σ

Our sampling model for the $\boldsymbol{\beta}_j$'s is that they are i.i.d. samples from a multivariate normal population with mean $\boldsymbol{\theta}$ and variance Σ. Inference for the population mean and variance of a multivariate normal population was covered in Chapter 7, in which we derived the full conditional distributions when semiconjugate priors are used. There, we saw that the full conditional distribution of a population mean is multivariate normal with expectation equal to a combination of the prior expectation and the sample mean, and precision equal to the sum of the prior and data precisions. In the context of the hierarchical regression model, given Σ and our sample of regression coefficients $\boldsymbol{\beta}_1, \ldots, \boldsymbol{\beta}_m$, the full conditional distribution of $\boldsymbol{\theta}$ is as follows:

$$\begin{aligned} \{\boldsymbol{\theta}|\boldsymbol{\beta}_1, \ldots, \boldsymbol{\beta}_m, \Sigma\} &\sim \text{multivariate normal}(\boldsymbol{\mu}_m, \Lambda_m) \text{ , where} \\ \Lambda_m &= (\Lambda_0^{-1} + m\Sigma^{-1})^{-1} \\ \boldsymbol{\mu}_m &= \Lambda_m(\Lambda_0^{-1}\boldsymbol{\mu}_0 + m\Sigma^{-1}\bar{\boldsymbol{\beta}}) \end{aligned}$$

where $\bar{\boldsymbol{\beta}}$ is the vector average $\frac{1}{m}\sum \boldsymbol{\beta}_j$. In Chapter 7 we also saw that the full conditional distribution of a covariance matrix is an inverse-Wishart distribution, with sum of squares matrix equal to the prior sum of squares $\mathbf{S}_0$ plus the sum of squares from the sample:

$$\begin{aligned} \{\Sigma|\boldsymbol{\theta}, \boldsymbol{\beta}_1, \ldots, \boldsymbol{\beta}_m\} &\sim \text{inverse-Wishart}(\eta_0 + m, [\mathbf{S}_0 + \mathbf{S}_\theta]^{-1}) \text{ , where} \\ \mathbf{S}_\theta &= \sum_{j=1}^{m} (\boldsymbol{\beta}_j - \boldsymbol{\theta})(\boldsymbol{\beta}_j - \boldsymbol{\theta})^T . \end{aligned}$$

Note that $\mathbf{S}_\theta$ depends on $\boldsymbol{\theta}$ and so must be recomputed each time $\boldsymbol{\theta}$ is updated in the Markov chain.

Full conditional distribution of σ^2

The parameter σ^2 represents the error variance, assumed to be common across all groups. As such, conditional on $\boldsymbol{\beta}_1, \ldots, \boldsymbol{\beta}_m$, the data provide information about σ^2 via the sum of squared residuals from each group:

$$\begin{aligned} \sigma^2 &\sim \text{inverse-gamma}([\nu_0 + \textstyle\sum n_j]/2, [\nu_0\sigma_0^2 + \text{SSR}]/2) \text{ , where} \\ \text{SSR} &= \sum_{j=1}^{m}\sum_{i=1}^{n_j} (y_{i,j} - \boldsymbol{\beta}_j^T \boldsymbol{x}_{i,j})^2 . \end{aligned}$$

It is important to remember that SSR depends on the value of $\boldsymbol{\beta}_j$, and so SSR must be recomputed in each scan of the Gibbs sampler before σ^2 is updated.

11.3 Posterior analysis of the math score data

To analyze the math score data we will use a prior distribution that is similar in spirit to the unit information priors that were discussed in Chapter 9. For example, we'll take $\boldsymbol{\mu}_0$, the prior expectation of $\boldsymbol{\theta}$, to be equal to the average of the ordinary least squares regression estimates and the prior variance Λ_0 to be their sample covariance. Such a prior distribution represents the information of someone with unbiased but weak prior information. For example, a 95% prior confidence interval for the slope parameter θ_2 under this prior is (-3.86,8.60), which is quite a large range when considering what the extremes of the interval imply in terms of average change in score per unit change in SES score. Similarly, we will take the prior sum of squares matrix $\mathbf{S}_0$ to be equal to the covariance of the least squares estimate, but we'll take the prior degrees of freedom η_0 to be $p+2=4$, so that the prior distribution of Σ is reasonably diffuse but has an expectation equal to the sample covariance of the least squares estimates. Finally, we'll take σ_0^2 to be the average of the within-group sample variance but set $\nu_0 = 1$.

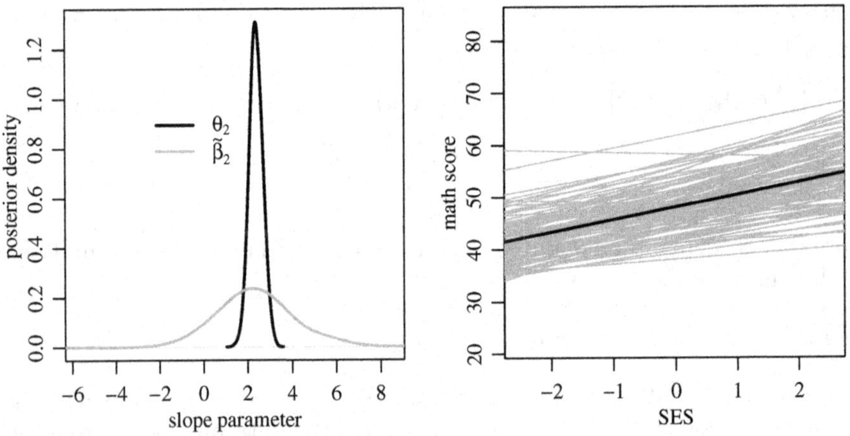

Fig. 11.3. Relationship between SES and math score. The first panel plots the posterior density of the expected slope θ_2 of a randomly sampled school, as well as the posterior predictive distribution of a randomly sampled slope. The second panel gives posterior expectations of the 100 school-specific regression lines, with the average line given in black.

Running a Gibbs sampler for 10,000 scans and saving every 10th scan produces a sequence of 1,000 values for each parameter, each sequence having a fairly low autocorrelation. For example, the lag-10 autocorrelations of θ_1 and θ_2 are -0.024 and 0.038. As usual, we can use these simulated values to make Monte Carlo approximations to various posterior quantities of interest. For example, the first plot in Figure 11.3 shows the posterior distribution of θ_2, the expected within-school slope parameter. A 95% quantile-based posterior confidence interval for this parameter is (1.83, 2.96), which, compared to our prior interval of (-3.86, 8.60), indicates a strong alteration in our information about θ_2.

The fact that θ_2 is extremely unlikely to be negative only indicates that the population average of school-level slopes is positive. It does not indicate that any given within-school slope cannot be negative. To clarify this distinction, the posterior predictive distribution of $\tilde{\beta}_2$, the slope for a to-be-sampled school, is plotted in the same figure. Samples from this distribution can be generated by sampling a value $\tilde{\boldsymbol{\beta}}^{(s)}$ from a multivariate normal$(\boldsymbol{\theta}^{(s)}, \Sigma^{(s)})$ distribution for each scan s of the Gibbs sampler. Notice that this posterior predictive distribution is much more spread out than the posterior distribution of θ_2, reflecting the heterogeneity in slopes across schools. Using the Monte Carlo approximation, we have $\Pr(\tilde{\beta}_2 < 0|\boldsymbol{y}_1, \ldots, \boldsymbol{y}_m, \mathbf{X}_1, \ldots, \mathbf{X}_m) \approx 0.07$, which is small but not negligible.

The second panel in Figure 11.3 plots posterior expectations of the 100 school-specific regression lines, with the line given by the posterior mean of $\boldsymbol{\theta}$ in black. Comparing this to the first panel of Figure 11.1 indicates how the hierarchical model is able to share information across groups, shrinking extreme regression lines towards the across-group average. In particular, hardly any of the slopes are negative when we share information across groups.

11.4 Generalized linear mixed effects models

As the name suggests, a generalized linear mixed effects model combines aspects of linear mixed effects models with those of generalized linear models described in Chapter 10. Such models are useful when we have a hierarchical data structure but the normal model for the within-group variation is not appropriate. For example, if the variable Y were binary or a count, then more appropriate models for within-group variation would be logistic or Poisson regression models, respectively.

A basic generalized linear mixed model is as follows:

$$\boldsymbol{\beta}_1, \ldots, \boldsymbol{\beta}_m \sim \text{i.i.d. multivariate normal}(\boldsymbol{\theta}, \Sigma)$$

$$p(\boldsymbol{y}_j|\mathbf{X}_j, \boldsymbol{\beta}_j, \gamma) = \prod_{i=1}^{n_j} p(y_{i,j}|\boldsymbol{\beta}_j^T \boldsymbol{x}_{i,j}, \gamma),$$

with observations from different groups also being conditionally independent. In this formulation $p(y|\boldsymbol{\beta}^T\boldsymbol{x}, \gamma)$ is a density whose mean depends on $\boldsymbol{\beta}^T\boldsymbol{x}$, and γ is an additional parameter often representing variance or scale. For example, in the normal model $p(y|\boldsymbol{\beta}^T\boldsymbol{x}, \gamma) = \text{dnorm}(y, \boldsymbol{\beta}^T\boldsymbol{x}, \gamma^{1/2})$ where γ represents the variance. In the Poisson model $p(y|\boldsymbol{\beta}^T\boldsymbol{x}) = \text{dpois}(\exp\{\boldsymbol{\beta}^T\boldsymbol{x}\})$, and there is no γ parameter.

11.4.1 A Metropolis-Gibbs algorithm for posterior approximation

Estimation for the linear mixed effects model was straightforward because the full conditional distribution of each parameter was standard, allowing for the easy implementation of a Gibbs sampling algorithm. In contrast, for non-normal generalized linear mixed models, typically only $\boldsymbol{\theta}$ and Σ have standard full conditional distributions. This suggests we use a Metropolis-Hastings algorithm to approximate the posterior distribution of the parameters, using a combination of Gibbs steps for updating $(\boldsymbol{\theta}, \Sigma)$ with a Metropolis step for updating each $\boldsymbol{\beta}_j$. In what follows we assume there is no γ parameter. If there is such a parameter, it can be updated using a Gibbs step if a full conditional distribution is available, and a Metropolis step if not.

Gibbs steps for $(\boldsymbol{\theta}, \Sigma)$

Just as in the linear mixed effects model, the full conditional distributions of $\boldsymbol{\theta}$ and Σ depend only on $\boldsymbol{\beta}_1, \dots, \boldsymbol{\beta}_m$. This means that the form of $p(y|\boldsymbol{\beta}^T\boldsymbol{x})$ has no effect on the full conditional distributions of $\boldsymbol{\theta}$ and Σ. Whether $p(y|\boldsymbol{\beta}^T\boldsymbol{x})$ is a normal model, a Poisson model, or some other generalized linear model, the full conditional distributions of $\boldsymbol{\theta}$ and Σ will be the multivariate normal and inverse-Wishart distributions described in Section 11.2.

Metropolis step for $\boldsymbol{\beta}_j$

Updating $\boldsymbol{\beta}_j$ in a Markov chain can proceed by proposing a new value of $\boldsymbol{\beta}_j^*$ based on the current parameter values and then accepting or rejecting it with the appropriate probability. A standard proposal distribution in this situation would be a multivariate normal distribution with mean equal to the current value $\boldsymbol{\beta}_j^{(s)}$ and with some proposal variance $V_j^{(s)}$. In this case the Metropolis step is as follows:

1. Sample $\boldsymbol{\beta}_j^* \sim$ multivariate normal$(\boldsymbol{\beta}_j^{(s)}, V_j^{(s)})$.
2. Compute the acceptance ratio

$$r = \frac{p(\boldsymbol{y}_j|\mathbf{X}_j, \boldsymbol{\beta}_j^*)p(\boldsymbol{\beta}_j^*|\boldsymbol{\theta}^{(s)}, \Sigma^{(s)})}{p(\boldsymbol{y}_j|\mathbf{X}_j, \boldsymbol{\beta}_j^{(s)})p(\boldsymbol{\beta}_j^{(s)}|\boldsymbol{\theta}^{(s)}, \Sigma^{(s)})}.$$

3. Sample $u \sim$ uniform(0,1). Set $\boldsymbol{\beta}_j^{(s+1)}$ to $\boldsymbol{\beta}_j^*$ if $u < r$ and to $\boldsymbol{\beta}_j^{(s)}$ if $u > r$.

In many cases, setting $V_j^{(s)}$ equal to a scaled version of $\Sigma^{(s)}$ produces a well-mixing Markov chain, although the task of finding the right scale might have to proceed by trial and error.

A Metropolis-Hastings approximation algorithm

Putting these steps together results in the following Metropolis-Hastings algorithm for approximating $p(\boldsymbol{\beta}_1, \ldots, \boldsymbol{\beta}_m, \boldsymbol{\theta}, \Sigma | \mathbf{X}_1, \ldots, \mathbf{X}_m, \boldsymbol{y}_1, \ldots, \boldsymbol{y}_m)$: Given current values at scan s of the Markov chain, we obtain new values as follows:

1. Sample $\boldsymbol{\theta}^{(s+1)}$ from its full conditional distribution.
2. Sample $\Sigma^{(s+1)}$ from its full conditional distribution.
3. For each $j \in \{1, \ldots, m\}$,
   a) propose a new value $\boldsymbol{\beta}_j^*$;
   b) set $\boldsymbol{\beta}_j^{(s+1)}$ equal to $\boldsymbol{\beta}_j^*$ or $\boldsymbol{\beta}_j^{(s)}$ with the appropriate probability.

11.4.2 Analysis of tumor location data

(From Haigis et al (2004)) A certain population of laboratory mice experiences a high rate of intestinal tumor growth. One item of interest to researchers is how the rate of tumor growth varies along the length of the intestine. To study this, the intestine of each of 21 sample mice was divided into 20 sections and the number of tumors occurring in each section was recorded. The first panel of Figure 11.4 shows 21 lines, each one representing the observed tumor counts of each of the mice plotted against the fraction of the length along their intestine. Although it is hard to tell from the figure, the lines for some mice are consistently below the average (given in black), and others are consistently above. This suggests that tumor counts are more similar within a mouse than between mice, and a hierarchical model with mouse-specific effects may be appropriate.

A natural model for count data such as these is a Poisson distribution with a log-link. Letting $Y_{x,j}$ be mouse j's tumor count at location x of their intestine, we will model $Y_{x,j}$ as Poisson($e^{f_j(x)}$), where f_j is a smoothly varying function of $x \in [0, 1]$. A simple way to parameterize f_j is as a polynomial, so that $f_j(x) = \beta_{1,j} + \beta_{2,j}x + \beta_{3,j}x^2 + \cdots + \beta_{p,j}x^{p-1}$ for some maximum degree $p - 1$. Such a parameterization allows us to represent each f_j as a regression on $(1, x, x^2, \ldots, x^{p-1})$.

Averaging across the 21 mice gives an observed average tumor count $\bar{y}_x$ at each of the 20 locations $x \in (.05, .10, \ldots, .95)$ along the intestine. This average curve is plotted in the first panel of Figure 11.4 in black, and the log of this curve is given in the second panel of the figure. Also in the second panel are approximations of this curve using polynomials of degree 2, 3 and 4. The second- and third-degree approximations indicate substantial lack-of-fit, whereas the fourth-degree polynomial fits the log average tumor count function rather well. For simplicity we'll model each f_j as a fourth-degree polynomial, although it is possible that a particular f_j may be better fit with a higher degree.

Our between-group sampling model for the $\boldsymbol{\beta}_j$'s will be as in the previous section, so that $\boldsymbol{\beta}_1, \ldots, \boldsymbol{\beta}_m \sim$ i.i.d. multivariate normal($\boldsymbol{\theta}, \Sigma$). Unconditional

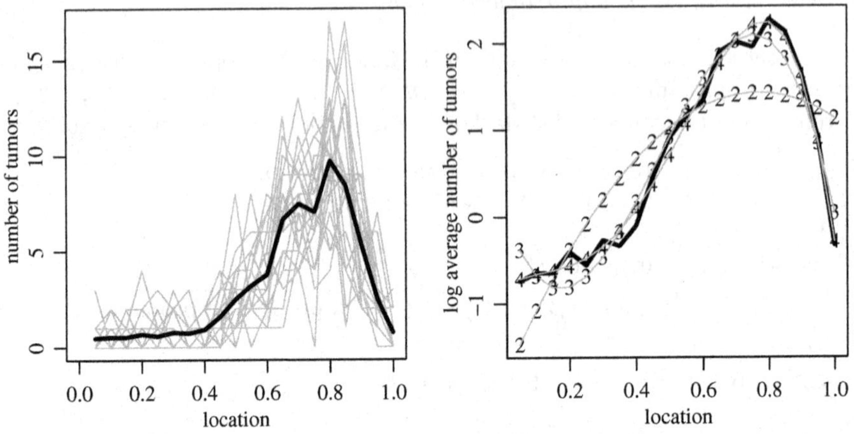

Fig. 11.4. Tumor count data. The first panel gives mouse-specific tumor counts as a function of location in gray, with a population average in black. The second panel gives quadratic, cubic and quartic polynomial fits to the log sample average tumor count.

on $\boldsymbol{\beta}_j$, the observations coming from a given mouse are statistically dependent as determined by Σ. Estimating Σ in this mixed effects model allows us to account for and describe potential within-mouse dependencies in the data. The unknown parameters in this model are $\boldsymbol{\theta}$ and Σ for which we need to specify prior distributions. Using conjugate normal and inverse-Wishart prior distributions, we need to specify $\boldsymbol{\mu}_0$ and Λ_0 for $p(\boldsymbol{\theta})$ and η_0 and $\mathbf{S}_0$ for $p(\Sigma)$. Specifying reasonable values for this many parameters can be difficult, especially in the absence of explicit prior data. As an alternative, we'll take an approach based on unit information priors, in which the prior distributions for the parameters are weakly centered around estimates derived from the observed data. As mentioned before, such prior distributions might represent the prior information of someone with a small amount of unbiased information.

Our unit information prior requires estimates of $\boldsymbol{\theta}$ and Σ, the population mean and covariance of the $\boldsymbol{\beta}_j$'s. For each mouse we can obtain a preliminary ad hoc estimate $\tilde{\boldsymbol{\beta}}_j$ by regressing $\{\log(y_{1,j} + 1/n), \ldots, \log(y_{n,j} + 1/n)\}$ on $\{\boldsymbol{x}_1, \ldots, \boldsymbol{x}_{20}\}$, where $\boldsymbol{x}_i = (1, x_i, x_i^2, x_i^3, x_i^4)^T$ for $x_i \in (.05, .10, \ldots, .95)$ (alternatively, we could obtain $\tilde{\boldsymbol{\beta}}_j$ using maximum likelihood estimates from a Poisson regression model). A unit-information type of prior distribution for $\boldsymbol{\theta}$ would be a multivariate normal distribution with an expectation of $\boldsymbol{\mu}_0 = \frac{1}{m}\sum_{j=1} \tilde{\boldsymbol{\beta}}_j$ and a prior covariance matrix equal to the sample covariance of the $\tilde{\boldsymbol{\beta}}_j$'s. We also set $\mathbf{S}_0$ equal to this sample covariance matrix, but set $\eta_0 = p + 2 = 7$, so that the prior expectation of Σ is equal to $\mathbf{S}_0$ but the prior distribution is relatively diffuse.

In terms of MCMC posterior approximation, recall from the steps outlined in the previous subsection that values of $\boldsymbol{\theta}$ and Σ can be sampled from their full conditional distributions. The full conditional distributions of the $\boldsymbol{\beta}_j$'s, however, are not standard and so we'll propose changes to these parameters from distributions centered around their current values. After a bit of trial and error, it turns out that a multivariate normal$(\boldsymbol{\beta}_j^{(s)}, \Sigma^{(s)}/2)$ proposal distribution yields an acceptance rate of about 31% and a reasonably well-mixing Markov chain. Running the Markov chain for 50,000 scans and saving the values every 10th scan gives 5,000 approximate posterior samples for each parameter. The effective sample sizes for the elements of Σ are all above 1,000 except for that of Σ_{11}, which was about 950. The effective sample sizes for the five $\boldsymbol{\theta}$ parameters are $(674, 1003, 1092, 1129, 1145)$. This roughly means that our Monte Carlo standard error in approximating $\mathrm{E}[\theta_1|\boldsymbol{y}_1, \ldots, \boldsymbol{y}_m, \mathbf{X}]$, for example, is $\sqrt{674} = 25.96$ times smaller than the posterior standard error of θ_1. Of course, we can reduce the Monte Carlo standard error to be as small as we want by running the Markov chain for more iterations. R-code for generating this Markov chain appears below:

```
## data
data(chapter11)
Y<-XY.tumor$Y ; X<-XY.tumor$X ; m<-dim(Y)[1] ; p<-dim(X)[2]

## priors
BETA<-NULL
for(j in 1:m)
{
  BETA<-rbind(BETA,lm(log(Y[j,]+1/20)~-1+X[,,j])$coef)
}

mu0<-apply(BETA,2,mean)
S0<-cov(BETA) ; eta0<-p+2
iL0<-iSigma<-solve(S0)

## MCMC
THETA.post<-NULL ; set.seed(1)
for(s in 1:50000)
{
  ##update theta
  Lm<-solve( iL0 + m*iSigma )
  mum<-Lm%*%( iL0%*%mu0 + iSigma%*%apply(BETA,2,sum) )
  theta<-t(rmvnorm(1,mum,Lm))
  ##

  ##update Sigma
  mtheta<-matrix(theta,m,p,byrow=TRUE)
  iSigma<-rwish(1,eta0+m,
          solve( S0+t(BETA-mtheta)%*%(BETA-mtheta)) )
  ##
```

```
 ##update beta
 Sigma<-solve(iSigma) ; dSigma<-det(Sigma)
 for(j in 1:m)
 {
   beta.p<-t(rmvnorm(1,BETA[j,],.5*Sigma))

   lr<-sum( dpois(Y[j,],exp(X[,,j]%*%beta.p),log=TRUE ) -
           dpois(Y[j,],exp(X[,,j]%*%BETA[j,]),log=TRUE ) ) +
       ldmvnorm( t(beta.p),theta,Sigma,
             iSigma=iSigma,dSigma=dSigma ) -
       ldmvnorm( t(BETA[j,]),theta,Sigma,
             iSigma=iSigma,dSigma=dSigma )

   if( log(runif(1))<lr ) { BETA[j,]<-beta.p }
  }
 ##

 ##store some output
 if(s%%10==0){THETA.post<-rbind(THETA.post,t(theta))}
 ##

}
```

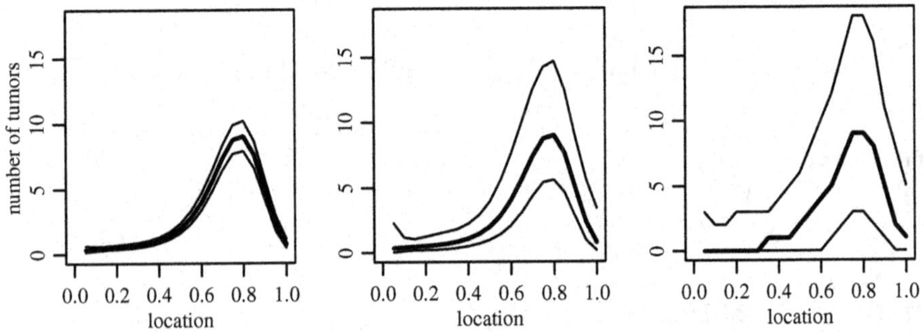

Fig. 11.5. 2.5, 50 and 97.5% quantiles for $\exp(\boldsymbol{\theta}^T\boldsymbol{x})$, $\exp(\boldsymbol{\beta}^T\boldsymbol{x})$ and $\{Y|\boldsymbol{x}\}$.

The three panels in Figure 11.5 show posterior distributions for a variety of quantities. The first panel gives 2.5%, 50% and 97.5% posterior quantiles for $\exp(\boldsymbol{\theta}^T\boldsymbol{x})$. The second panel gives the same quantiles for the posterior predictive distribution of $\exp(\boldsymbol{\beta}^T\boldsymbol{x})$. The difference in the width of the confidence bands is due to the estimated across-mouse heterogeneity. If there were no across-mouse heterogeneity then Σ would be zero, each $\boldsymbol{\beta}_j$ would be equal to $\boldsymbol{\theta}$ and the plots in the first two panels of the figure would be identical. Finally, the third panel gives 2.5%, 50% and 97.5% quantiles of the posterior predictive distribution of $\{Y|\boldsymbol{x}\}$ for each of the 20 values of $\boldsymbol{x}$. The difference between this plot and the one in the second panel is due to the variability of a Poisson random variable Y around its expected value $\exp(\boldsymbol{\beta}^T\boldsymbol{x})$. The widening confidence bands of the three plots in this figure describe cumulative sources of uncertainty: The first panel shows the uncertainty in the fixed but unknown value of $\boldsymbol{\theta}$. The second panel shows this uncertainty in addition to the uncertainty due to across-mouse heterogeneity. Finally, the third panel includes both of these sources of uncertainty as well as that due to fluctuations of a mouse's observed tumor counts around its own expected tumor count function. Understanding these different sources of uncertainty can be very relevant to inference and decision making: For example, if we want to predict the observed tumor count distribution of a new mouse, we should use the confidence bands in the third panel, whereas the bands in the first panel would be appropriate if we just wanted to describe the uncertainty in the fixed value of $\boldsymbol{\theta}$.

11.5 Discussion and further references

Posterior approximation via MCMC for hierarchical models can suffer from poor mixing. One reason for this is that many of the parameters in the model are highly correlated, and generating them one at a time in the Gibbs sampler can lead to a high degree of autocorrelation. For example, $\boldsymbol{\theta}$ and the $\boldsymbol{\beta}_j$'s are positively correlated, and so an extreme value of $\boldsymbol{\theta}$ at one iteration can lead to extreme values of the $\boldsymbol{\beta}_j$'s when they get updated, especially if the amount of within-group data is low. This in turn leads to an extreme value of $\boldsymbol{\theta}$ at the next iteration. Section 15.4 of Gelman et al (2004) provides a detailed discussion of several alternative Gibbs sampling strategies for improving mixing for hierarchical models. Improvements also can be made by careful reparameterizations of the model (Gelfand et al, 1995; Papaspiliopoulos et al, 2007).

12

Latent variable methods for ordinal data

Many datasets include variables whose distributions cannot be represented by the normal, binomial or Poisson distributions we have studied thus far. For example, distributions of common survey variables such as age, education level and income generally cannot be accurately described by any of the above-mentioned sampling models. Additionally, such variables are often binned into ordered categories, the number of which may vary from survey to survey. In such situations, interest often lies not in the scale of each individual variable, but rather in the associations between the variables: Is the relationship between two variables positive, negative or zero? What happens if we "account" for a third variable? For normally distributed data these types of questions can be addressed with the multivariate normal and linear regression models of Chapters 7 and 9. In this chapter we extend these models to situations where the data are not normal, by expressing non-normal random variables as functions of unobserved, "latent" normally distributed random variables. Multivariate normal and linear regression models then can be applied to the latent data.

12.1 Ordered probit regression and the rank likelihood

Suppose we are interested in describing the relationship between the educational attainment and number of children of individuals in a population. Additionally, we might suspect that an individual's educational attainment may be influenced by their parent's education level. The 1994 General Social Survey provides data on variables DEG, CHILD and PDEG for a sample of individuals in the United States, where DEG_i indicates the highest degree obtained by individual i, CHILD_i is their number of children and PDEG_i is the binary indicator of whether or not either parent of i obtained a college degree. Using these data, we might be tempted to investigate the relationship between the variables with a linear regression model:

P.D. Hoff, *A First Course in Bayesian Statistical Methods*,
Springer Texts in Statistics, DOI 10.1007/978-0-387-92407-6_12,

$$\text{DEG}_i = \beta_1 + \beta_2 \times \text{CHILD}_i + \beta_3 \times \text{PDEG}_i + \beta_4 \times \text{CHILD}_i \times \text{PDEG}_i + \epsilon_i,$$

where we assume that $\epsilon_1, \ldots, \epsilon_n \sim$ i.i.d. normal$(0, \sigma^2)$. However, such a model would be inappropriate for a couple of reasons. Empirical distributions of DEG and CHILD for a sample of 1,002 males in the 1994 workforce are shown in Figure 12.1. The value of DEG is recorded as taking a value in $\{1, 2, 3, 4, 5\}$ corresponding to the highest degree of the respondent being no degree, high school degree, associate's degree, bachelor's degree, or graduate degree.

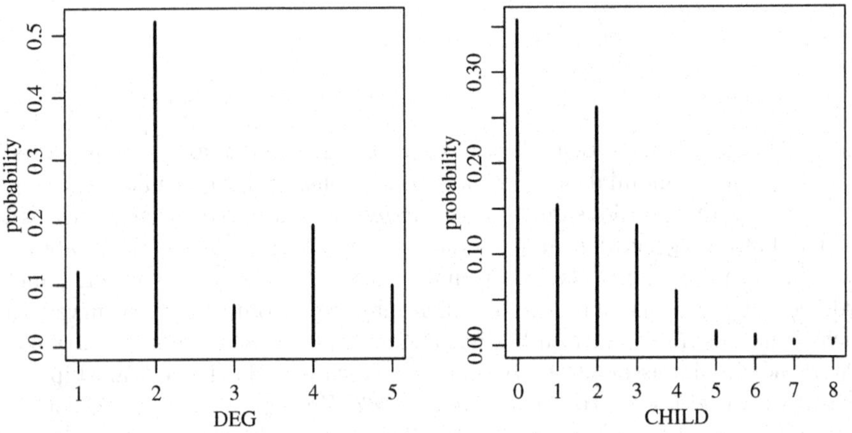

Fig. 12.1. Two ordinal variables having non-normal distributions.

Since the variable DEG takes on only a small set of discrete values, the normality assumption of the residuals will certainly be violated. But perhaps more importantly, the regression model imposes a numerical scale to the data that is not really present: A bachelor's degree is not "twice as much" as a high school degree, and an associate's degree is not "two less" than a graduate degree. There is an order to the categories in the sense that a graduate degree is "higher" than a bachelor's degree, but otherwise the scale of DEG is not meaningful.

Variables for which there is a logical ordering of the sample space are known as *ordinal variables*. With this definition, the discrete variables DEG and CHILD are ordinal variables, as are "continuous" variables like height or weight. However, CHILD, height and weight are variables that are measured on meaningful numerical scales, whereas DEG is not. In this chapter we will use the term "ordinal" to refer to any variable for which there is a logical ordering of the sample space. We will use the term "numeric" to refer to variables that have meaningful numerical scales, and "continuous" if a variable can have a value that is (roughly) any real number in an interval. For example,

DEG is ordinal but not numeric, whereas CHILD is ordinal, numeric and discrete. Variables like height or weight are ordinal, numeric and continuous.

12.1.1 Probit regression

Linear or generalized linear regression models, which assume a numeric scale to the data, may be appropriate for variables like CHILD, height or weight, but are not appropriate for non-numeric ordinal variables like DEG. However, it is natural to think of many ordinal, non-numeric variables as arising from some underlying numeric process. For example, the severity of a disease might be described "low", "moderate" or "high", although we imagine a patient's actual condition lies within a continuum. Similarly, the amount of effort a person puts into formal education may lie within a continuum, but a survey may only record a rough, categorized version of this variable, such as DEG. This idea motivates a modeling technique known as *ordered probit regression*, in which we relate a variable Y to a vector of predictors $\boldsymbol{x}$ via a regression in terms of a latent variable Z. More precisely, the model is

$$\begin{aligned} \epsilon_1, \ldots, \epsilon_n &\sim \text{i.i.d. normal}(0,1) \\ Z_i &= \boldsymbol{\beta}^T \boldsymbol{x}_i + \epsilon_i && (12.1) \\ Y_i &= g(Z_i), && (12.2) \end{aligned}$$

where $\boldsymbol{\beta}$ and g are unknown parameters. For example, to model the conditional distribution of DEG given CHILD and PDEG we would let Y_i be DEG_i and let $\boldsymbol{x}_i = (\text{CHILD}_i, \text{PDEG}_i, \text{CHILD}_i \times \text{PDEG}_i)$. The regression coefficients $\boldsymbol{\beta}$ describe the relationship between the explanatory variables and the unobserved latent variable Z, and the function g relates the value of Z to the observed variable Y. The function g is taken to be non-decreasing, so that we can interpret small and large values of Z as corresponding to small and large values of Y. This also means that the sign of a regression coefficient β_j indicates whether Y is increasing or decreasing in x_j.

Notice that in this probit regression model we have taken the variance of $\epsilon_1, \ldots, \epsilon_n$ to be one. This is because the scale of the distribution of Y can already be represented by g, as g is allowed to be any non-decreasing function. Similarly, g can represent the location of the distribution of Y, and so we do not need to include an intercept term in the model.

If the sample space for Y takes on K values, say $\{1, \ldots, K\}$, then the function g can be described with only $K-1$ ordered parameters $g_1 < g_2 < \cdots < g_{k-1}$ as follows:

$$\begin{aligned} y = g(z) &= 1 \text{ if } \; -\infty = g_0 < z < g_1 \\ &= 2 \text{ if } \quad g_1 < z < g_2 \\ &\;\vdots && (12.3) \\ &= K \text{ if } \quad g_{K-1} < z < g_K = \infty . \end{aligned}$$

The values $\{g_1, \ldots, g_{K-1}\}$ can be thought of as "thresholds," so that moving z past a threshold moves y into the next highest category. The unknown parameters in the model include the regression coefficients $\boldsymbol{\beta}$ and the thresholds $g_1, \ldots, g_{K-1}$. If we use normal prior distributions for these quantities, the joint posterior distribution of $\{\boldsymbol{\beta}, g_1, \ldots, g_{K-1}, Z_1, \ldots, Z_n\}$ given $\boldsymbol{Y} = \boldsymbol{y} = (y_1, \ldots, y_n)$ can be approximated using a Gibbs sampler.

Full conditional distribution of $\boldsymbol{\beta}$

Given $\boldsymbol{Y} = \boldsymbol{y}$, $\boldsymbol{Z} = \boldsymbol{z}$, and $\boldsymbol{g} = (g_1 \ldots, g_{K-1})$, the full conditional distribution of $\boldsymbol{\beta}$ depends only on $\boldsymbol{z}$ and satisfies $p(\boldsymbol{\beta}|\boldsymbol{y}, \boldsymbol{z}, \boldsymbol{g}) \propto p(\boldsymbol{\beta})p(\boldsymbol{z}|\boldsymbol{\beta})$. Just as in ordinary regression, a multivariate normal prior distribution for $\boldsymbol{\beta}$ gives a multivariate normal posterior distribution. For example, if we use $\boldsymbol{\beta} \sim$ multivariate normal$(\mathbf{0}, n(\mathbf{X}^T\mathbf{X})^{-1})$, then $p(\boldsymbol{\beta}|\boldsymbol{z})$ is multivariate normal with

$$\mathrm{Var}[\boldsymbol{\beta}|\boldsymbol{z}] = \frac{n}{n+1}(\mathbf{X}^T\mathbf{X})^{-1}, \text{ and}$$
$$\mathrm{E}[\boldsymbol{\beta}|\boldsymbol{z}] = \frac{n}{n+1}(\mathbf{X}^T\mathbf{X})^{-1}\mathbf{X}^T\boldsymbol{z}.$$

Full conditional distribution of $\boldsymbol{Z}$

The full conditional distributions of the Z_i's are only slightly more complicated. Under the sampling model, the conditional distribution of Z_i given $\boldsymbol{\beta}$ is $Z_i \sim$ normal$(\boldsymbol{\beta}^T\boldsymbol{x}_i, 1)$. Given g, observing $Y_i = y_i$ tells us that Z_i must lie in the interval (g_{y_i-1}, g_{y_i}). Letting $a = g_{y_i-1}$ and $b = g_{y_i}$, the full conditional distribution of Z_i given $\{\boldsymbol{\beta}, \boldsymbol{y}, \boldsymbol{g}\}$ is then

$$p(z_i|\boldsymbol{\beta}, \boldsymbol{y}, \boldsymbol{g}) \propto \mathrm{dnorm}(z_i, \boldsymbol{\beta}^T\boldsymbol{x}_i, 1) \times \delta_{(a,b)}(z_i).$$

This is the density of a *constrained normal distribution*. To sample a value x from a normal(μ, σ^2) distribution constrained to the interval (a, b), we perform the following two steps:

1. sample $u \sim$ uniform$(\Phi[(a-\mu)/\sigma], \Phi[(b-\mu)/\sigma])$
2. set $x = \mu + \sigma\Phi^{-1}(u)$

where Φ and Φ^{-1} are the cdf and inverse-cdf of the standard normal distribution (given by `pnorm` and `qnorm` in R). Code to sample from the full conditional distribution of Z_i is as follows:

```
ez<- t(beta)%*%X[i,]
a<-max(-Inf,g[y[i]-1],na.rm=TRUE)
b<-min(g[y[i]],Inf,na.rm=TRUE)

u<-runif(1, pnorm(a-ez),pnorm(b-ez) )
z[i]<- ez + qnorm(u)
```

The added complexity in assigning a and b in the above code is to deal with the special cases $g_0 = -\infty$ and $g_K = \infty$.

Full conditional distribution of $\boldsymbol{g}$

Suppose the prior distribution for $\boldsymbol{g}$ is some arbitrary density $p(\boldsymbol{g})$. Given $\boldsymbol{Y} = \boldsymbol{y}$ and $\boldsymbol{Z} = \boldsymbol{z}$, we know from Equation 12.3 that g_k must be higher than all z_i's for which $y_i = k$ and lower than all z_i's for which $y_i = k + 1$. Letting $a_k = \max\{z_i : y_i = k\}$ and $b_k = \min\{z_i : y_i = k + 1\}$ the full conditional distribution of $\boldsymbol{g}$ is then proportional to $p(\boldsymbol{g})$ but constrained to the set $\{\boldsymbol{g} : a_k < g_k < b_k\}$. For example, if $p(\boldsymbol{g})$ is proportional to the product $\prod_{k=1}^{K-1} \text{dnorm}(g_k, \mu_k, \sigma_k)$ but constrained so that $g_1 < \cdots < g_{k-1}$, then the full conditional density of g_k is a normal(μ_k, σ_k^2) density constrained to the interval (a_k, b_k). R-code to sample from the full conditional distribution of g_k is given below:

```
a<-max(z[y==k])
b<-min(z[y==k+1])

u<-runif(1,pnorm((a-mu[k])/sig[k]),pnorm((b-mu[k])/sig[k]) )
g[k]<- mu[k] + sig[k]*qnorm(u)
```

Example: Educational attainment

Some researchers suggest that having children reduces opportunities for educational attainment (Moore and Waite, 1977). Here we examine this hypothesis in a sample of males in the labor force (meaning not retired, not in school and not in an institution), obtained from the 1994 General Social Survey. For 959 of the 1,002 survey respondents we have complete data on the variables DEG, CHILD and PDEG described above. Letting $Y_i = \text{DEG}_i$

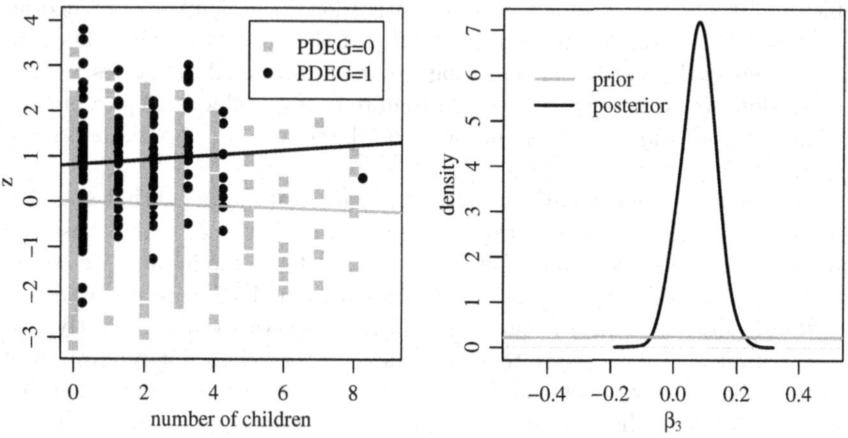

Fig. 12.2. Results from the probit regression analysis.

and $\boldsymbol{x}_i = (\text{CHILD}_i, \text{PDEG}_i, \text{CHILD}_i \times \text{PDEG}_i)$, we will estimate the parameters in the ordered probit regression model using prior distributions of $\boldsymbol{\beta} \sim$ multivariate normal$(\mathbf{0}, n(\mathbf{X}^T\mathbf{X})^{-1})$ and $p(\boldsymbol{g}) \propto \prod_{k=1}^{K-1} \text{dnorm}(g_k, 0, 100)$ but constrained so that $g_1 < \cdots < g_{K-1}$. We'll approximate the corresponding posterior distribution of $\{\boldsymbol{\beta}, \boldsymbol{Z}, \boldsymbol{g}\}$ with a Gibbs sampler consisting of 25,000 scans. Saving parameter values every 25th scan results in 1,000 values for each parameter with which to approximate the posterior distribution. The posterior mean regression line for people without a college-educated parent ($x_{i,2} = 0$) is $\text{E}[Z|\boldsymbol{y}, x_1, x_2 = 0] = -0.024 \times x_1$ while the regression line for people with a college-educated parent is $\text{E}[Z|\boldsymbol{y}, x_1, x_2 = 1] = 0.818 + 0.054 \times x_1$. These lines are shown in the first panel of Figure 12.2, along with the value of $\boldsymbol{Z}$ that was obtained in the last scan of the Gibbs sampler. The lines suggest that for people whose parents did not go to college, the number of children they have is indeed weakly negatively associated with their educational outcome. However, the opposite seems to be true among people whose parents went to college. The posterior distribution of β_3 is given in the second panel of the figure, along with the prior distribution for comparison. The 95% quantile-based posterior confidence interval for β_3 is (-0.026,0.178), which contains zero but still represents a reasonable amount of evidence that the slope for the $x_2 = 1$ group is larger than that of the $x_2 = 0$ group.

12.1.2 Transformation models and the rank likelihood

The analysis of the educational attainment data above required us to specify a prior distribution for $\boldsymbol{\beta}$ and the transformation $g(z)$, as specified by the vector $\boldsymbol{g}$ of $K-1$ threshold parameters. While simple default prior distributions for $\boldsymbol{\beta}$ exist (such as Zellner's g-prior), the same is not true for $\boldsymbol{g}$. Coming up with a prior distribution for $\boldsymbol{g}$ that represents actual prior information seems like a difficult task. Of course, this task is much harder if the number of categories K is large. For example, the incomes (INC) of the subjects in the 1994 GSS dataset were each recorded as belonging to one of 21 ordered categories, so that a regression in which $Y_i = \text{INC}_i$ would require that $\boldsymbol{g}$ includes 20 parameters. Estimation and prior specification for such a large number of parameters can be difficult.

Fortunately there is an alternative approach to estimating $\boldsymbol{\beta}$ that does not require us to estimate the function $g(z)$. Note that if the Z_i's were observed directly, then we could ignore Equation (12.2) of the model and we would be left with an ordinary regression problem without having to estimate the transformation $g(z)$. Unfortunately we do not observe the Z_i's directly, but there is information in the data about the Z_i's that does not require us to specify $g(z)$: Since we know that g is non-decreasing, we do know something about the *order* of the Z_i's. For example, if our observed data are such that $y_1 > y_2$, then since $y_i = g(Z_i)$, we know that $g(Z_1) > g(Z_2)$. Since g is non-decreasing, this means that we know $Z_1 > Z_2$. In other words, having observed $\boldsymbol{Y} = \boldsymbol{y}$, we know that the Z_i's must lie in the set

$$R(\boldsymbol{y}) = \{z \in \mathbb{R}^n : z_{i_1} < z_{i_2} \text{ if } y_{i_1} < y_{i_2}\}.$$

Since the distribution of the Z_i's does not depend on g, the probability that $\boldsymbol{Z} \in R(\boldsymbol{y})$ for a given $\boldsymbol{y}$ also does not depend on the unknown function g. This suggests that we base our posterior inference on the knowledge that $\boldsymbol{Z} \in R(\boldsymbol{y})$. Our posterior distribution for $\boldsymbol{\beta}$ in this case is given by

$$\begin{aligned} p(\boldsymbol{\beta}|\boldsymbol{Z} \in R(\boldsymbol{y})) &\propto p(\boldsymbol{\beta}) \times \Pr(\boldsymbol{Z} \in R(\boldsymbol{y})|\boldsymbol{\beta}) \\ &= p(\boldsymbol{\beta}) \times \int_{R(\boldsymbol{y})} \prod_{i=1}^{n} \text{dnorm}(z_i, \boldsymbol{\beta}^T \boldsymbol{x}_i, 1)\, dz_i . \end{aligned}$$

As a function of $\boldsymbol{\beta}$, the probability $\Pr(\boldsymbol{Z} \in R(\boldsymbol{y})|\boldsymbol{\beta})$ is known as the *rank likelihood*. For continuous y-variables this likelihood was introduced by Pettitt (1982) and its theoretical properties were studied by Bickel and Ritov (1997). It is called a rank likelihood because for continuous data it contains the same information about $\boldsymbol{y}$ as knowing the ranks of $\{y_1, \ldots, y_n\}$, i.e. which one has the highest value, which one has the second highest value, etc. If Y is discrete then observing $R(\boldsymbol{y})$ is not exactly the same as knowing the ranks, but for simplicity we will still refer to $\Pr(\boldsymbol{Z} \in R(\boldsymbol{y})|\beta)$ as the rank likelihood, whether or not Y is discrete or continuous. The important thing to note is that for any ordinal outcome variable Y (non-numeric, numeric, discrete or continuous), information about $\boldsymbol{\beta}$ can be obtained from $\Pr(\boldsymbol{Z} \in R(\boldsymbol{y})|\boldsymbol{\beta})$ without having to specify $g(z)$.

For any given $\boldsymbol{\beta}$ the value of $\Pr(\boldsymbol{Z} \in R(\boldsymbol{y})|\boldsymbol{\beta})$ involves a very complicated integral that is difficult to compute. However, by estimating $\boldsymbol{Z}$ simultaneously with $\boldsymbol{\beta}$ we can obtain an estimate of $\boldsymbol{\beta}$ without ever having to numerically compute $\Pr(\boldsymbol{Z} \in R(\boldsymbol{y})|\boldsymbol{\beta})$. The joint posterior distribution of $\{\boldsymbol{\beta}, \boldsymbol{Z}\}$ can be approximated by using Gibbs sampling, alternately sampling from full conditional distributions. The full conditional distribution of $\boldsymbol{\beta}$ is very easy: Given a current value $\boldsymbol{z}$ of $\boldsymbol{Z}$, the full conditional density $p(\boldsymbol{\beta}|\boldsymbol{Z} = \boldsymbol{z}, \boldsymbol{Z} \in R(\boldsymbol{y}))$ reduces to $p(\boldsymbol{\beta}|\boldsymbol{Z} = \boldsymbol{z})$ because knowing the value of $\boldsymbol{Z}$ is more informative than knowing just that $\boldsymbol{Z}$ lies in the set $R(\boldsymbol{y})$. A multivariate normal prior distribution for $\boldsymbol{\beta}$ then results in a multivariate normal full conditional distribution, as before. The full conditional distributions of the Z_i's are also straightforward to derive. Let's consider the full conditional distribution of Z_i given $\{\boldsymbol{\beta}, \boldsymbol{Z} \in R(\boldsymbol{y}), \boldsymbol{z}_{-i}\}$, where $\boldsymbol{z}_{-i}$ denotes the values of all of the Z's except Z_i. Conditional on $\boldsymbol{\beta}$, Z_i is normal$(\boldsymbol{\beta}^T \boldsymbol{x}_i, 1)$. Conditional on $\{\boldsymbol{\beta}, \boldsymbol{Z} \in R(\boldsymbol{y}), \boldsymbol{z}_{-i}\}$, the density of Z_i is proportional to a normal density but constrained by the fact that $\boldsymbol{Z} \in R(\boldsymbol{y})$. Let's recall the nature of this constraint: $y_i < y_j$ implies $Z_i < Z_j$, and $y_i > y_j$ implies $Z_i > Z_j$. This means that Z_i must lie in the following interval:

$$\max\{z_j : y_j < y_i\} < Z_i < \min\{z_j : y_i < y_j\} .$$

Letting a and b denote the numerical values of the lower and upper endpoints of this interval, the full conditional distribution of Z_i is then

$$p(z_i|\boldsymbol{\beta}, \boldsymbol{Z} \in R(\boldsymbol{y}), \boldsymbol{z}_{-i}) \propto \text{dnorm}(z_i, \boldsymbol{\beta}^T \boldsymbol{x}_i, 1) \times \delta_{(a,b)}(z_i).$$

This full conditional distribution is exactly the same as that of Z_i in the ordered probit model, except that now the constraints on Z_i are determined directly by the current value $\boldsymbol{Z}_{-i}$, instead of on the threshold variables. As such, sampling from this full conditional distribution is very similar to sampling from the analogous distribution in the probit regression model:

```
ez<- t(beta)%*%X[i,]
a<-max(z[y<y[i]])
b<-min(z[y[i]<y])

u<-runif(1, pnorm(a-ez),pnorm(b-ez) )
z[i]<- ez + qnorm(u)
```

Not surprisingly, for the educational attainment data the posterior distribution of $\boldsymbol{\beta}$ based on the rank likelihood is very similar to the one based on the full ordered probit model. The three panels of Figure 12.3 indicate that the marginal posterior densities of β_1, β_2 and β_3 are nearly identical under the two models. In general, if K is small and n is large, we expect the two methods to behave similarly. However, the rank likelihood approach is applicable to a wider array of datasets since with this approach, Y is allowed to be any type of ordinal variable, discrete or continuous. The drawback to using the rank likelihood is that it does not provide us with inference about $g(z)$, which describes the relationship between the latent and observed variables. If this parameter is of interest, then the rank likelihood is not appropriate, but if interest lies only in $\boldsymbol{\beta}$, then the rank likelihood provides a simple alternative to the ordered probit model.

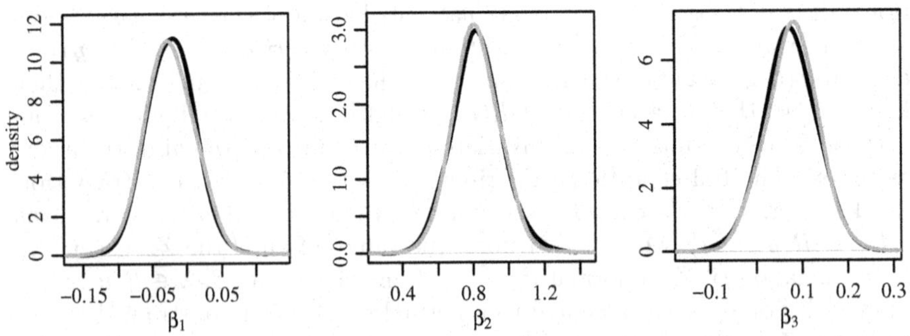

Fig. 12.3. Marginal posterior distributions of $(\beta_1, \beta_2, \beta_3)$, under the ordinal probit regression model (in gray) and the rank likelihood (in black).

12.2 The Gaussian copula model

The regression model above is somewhat limiting because it only describes the conditional distribution of one variable given the others. In general, we may be interested in the relationships among all of the variables in a dataset. If the variables were approximately jointly normally distributed, or at least were all measured on a meaningful numerical scale, then we could describe the relationships among the variables with the sample covariance matrix or a multivariate normal model. However, such a model is inappropriate for non-numeric ordinal variables like INC, DEG and PDEG. To accommodate variables such as these we can extend the ordered probit model above to a latent, multivariate normal model that is appropriate for all types of ordinal data, both numeric and non-numeric. Letting $\boldsymbol{Y}_1, \ldots, \boldsymbol{Y}_n$ be i.i.d. random samples from a p-variate population, our latent normal model is

$$\boldsymbol{Z}_1, \ldots, \boldsymbol{Z}_n \sim \text{i.i.d. multivariate normal}(\boldsymbol{0}, \Psi) \tag{12.4}$$

$$Y_{i,j} = g_j(Z_{i,j}), \tag{12.5}$$

where $g_1, \ldots, g_p$ are non-decreasing functions and Ψ is a correlation matrix, having diagonal entries equal to 1. In this model, the matrix Ψ represents the joint dependencies among the variables and the functions $g_1, \ldots, g_p$ represent their marginal distributions. To see how the g_j's represent the margins, let's calculate the marginal cdf $F_j(y)$ of a continuous random variable $Y_{i,j}$ under the model given by Equations 12.4 and 12.5. Recalling the definition of the cdf, we have

$$\begin{aligned} F_j(y) &= \Pr(Y_{i,j} \leq y) \\ &= \Pr(g_j(Z_{i,j}) \leq y) \text{ , since } Y_{i,j} = g_j(Z_{i,j}) \\ &= \Pr(Z_{i,j} \leq g_j^{-1}(y)) \\ &= \Phi(g_j^{-1}(y)), \end{aligned}$$

where $\Phi(z)$ is the cdf of the standard normal distribution. The last line holds because the diagonal entries of Ψ are all equal to 1, and so the *marginal* distribution of each $Z_{i,j}$ is a standard normal distribution with cdf $\Phi(z)$.

The above calculations show that $F_j(y) = \Phi(g_j^{-1}(y))$, indicating that the marginal distributions of the Y_j's are fully determined by the g_j's and do not depend on the matrix Ψ. A model having separate parameters for the univariate marginal distributions and the multivariate dependencies is generally called a *copula model*. The model given by Equations 12.4 and 12.5, where the dependence is described by a multivariate normal distribution, is called the *multivariate normal copula model*. The term "copula" refers to the method of "coupling" a model for multivariate dependence (such as the multivariate normal distribution) to a model for the marginal distributions of the data. As shown above, a copula model separates the parameters for the dependencies among the variables (Ψ) from the parameters describing their univariate marginal distributions $(g_1, \ldots, g_p)$. This separation comes in handy if we are primarily interested in the dependencies among the variables and not the univariate scales on which they were measured. In this case, the functions $g_1, \ldots, g_p$ are nuisance parameters and the parameter of interest is Ψ. Using an extension of the rank likelihood described in the previous section, we will be able to obtain a posterior distribution for Ψ without having to estimate or specify prior distributions for the nuisance parameters $g_1, \ldots, g_p$.

12.2.1 Rank likelihood for copula estimation

The unknown parameters in the copula model are the matrix Ψ and the non-decreasing functions $g_1, \ldots, g_p$. Bayesian inference for all of these parameters would require that we specify a prior for Ψ as well as p prior distributions over the complicated space of arbitrary non-decreasing functions. If we are not interested in $g_1, \ldots, g_p$ then we can use a version of the rank likelihood which quantifies information about $\boldsymbol{Z}_1, \ldots, \boldsymbol{Z}_n$ without having to specify these nuisance parameters. Recall that since each g_j is non-decreasing, observing the $n \times p$ data matrix $\mathbf{Y}$ tells us that the matrix of latent variables $\mathbf{Z}$ must lie in the set

$$R(\mathbf{Y}) = \{\mathbf{Z} : z_{i_1,j} < z_{i_2,j} \text{ if } y_{i_1,j} < y_{i_2,j}\}. \tag{12.6}$$

The probability of this event, $\Pr(\mathbf{Z} \in R(\mathbf{Y})|\Psi)$, does not depend on $g_1, \ldots, g_p$. As a function of Ψ, $\Pr(\mathbf{Z} \in R(\mathbf{Y})|\Psi)$ is called the rank likelihood for the multivariate normal copula model. Computing the likelihood for a given value of Ψ is very difficult, but as with the regression model in Section 12.1.2 we can make an MCMC approximation to $p(\Psi, \mathbf{Z}|\mathbf{Z} \in R(\mathbf{Y}))$ using Gibbs sampling, provided we use a prior for Ψ based on the inverse-Wishart distribution.

A parameter-expanded prior distribution for Ψ

Unfortunately there is no simple conjugate class of prior distributions for our correlation matrix Ψ. As an alternative, let's consider altering Equation 12.4 to be

$$\boldsymbol{Z}_1, \ldots, \boldsymbol{Z}_n \sim \text{i.i.d. multivariate normal}(\mathbf{0}, \Sigma),$$

where Σ is an arbitrary covariance matrix, not restricted to be a correlation matrix like Ψ. In this case a natural prior distribution for Σ would be an inverse-Wishart distribution, which would give an inverse-Wishart full conditional distribution and thus make posterior inference available via Gibbs sampling. However, careful inspection of the rank likelihood indicates that it does not provide us with a complete estimate of Σ. Specifically, the rank likelihood contains only information about the relative ordering among the $Z_{i,j}$'s, and no information about their scale. For example, if $Z_{1,j}$ and $Z_{2,j}$ are two i.i.d. samples from a normal$(0, \sigma_j^2)$ distribution, then the probability that $Z_{1,j} < Z_{2,j}$ does not depend on σ_j^2. For this reason we say that the diagonal entries of Σ are *non-identifiable* in this model, meaning that the rank likelihood provides no information about what the diagonal should be. In a Bayesian analysis, the posterior distribution of any non-identifiable parameter is determined by the prior distribution, and so in some sense the posterior distribution of such a parameter is not of interest. However, to each covariance matrix Σ there corresponds a unique correlation matrix Ψ, obtained by the function

$$\Psi = h(\Sigma) = \{\sigma_{i,j}/\sqrt{\sigma_i^2\sigma_j^2}\}.$$

The value of Ψ is identifiable from the rank likelihood, and so one estimation approach for the Gaussian copula model is to reparameterize the model in terms of a non-identifiable covariance matrix Σ, but focus our posterior inference on the identifiable correlation matrix $\Psi = h(\Sigma)$. This technique of modeling in terms of a non-identifiable parameter in order to simplify calculations is referred to as *parameter expansion* (Liu and Wu, 1999), and has been used in the context of modeling multivariate ordinal data by Hoff (2007) and Lawrence et al (2008).

To summarize, we will base our posterior distribution on

$$\begin{aligned} \Sigma &\sim \text{inverse-Wishart}(\nu_0, \mathbf{S}_0^{-1}) \\ \boldsymbol{Z}_1, \ldots, \boldsymbol{Z}_n &\sim \text{i.i.d. multivariate normal}(\mathbf{0}, \Sigma) \\ Y_{i,j} &= g_j(Z_{i,j}), \end{aligned} \tag{12.7}$$

but our estimation and inference will be restricted to $\Psi = h(\Sigma)$. Interestingly, the posterior distribution for Ψ obtained from this prior and model is exactly the same as that which would be obtained from the following:

$$\begin{aligned} \Sigma &\sim \text{inverse-Wishart}(\nu_0, \mathbf{S}_0^{-1}) \\ \Psi &= h(\Sigma) \\ \boldsymbol{Z}_1, \ldots, \boldsymbol{Z}_n &\sim \text{i.i.d. multivariate normal}(\mathbf{0}, \Psi) \\ Y_{i,j} &= g_j(Z_{i,j}). \end{aligned} \tag{12.8}$$

In other words, the non-identifiable model described in Equation 12.7 gives the same posterior distribution for Ψ as the identifiable model in Equation 12.8 in which the prior distribution for Ψ is defined by $\{\Sigma \sim \text{inverse-Wishart}(\nu_0, \mathbf{S}_0^{-1})\ , \Psi = h(\Sigma)\}$. The only difference is that the Gibbs sampling scheme for Equation 12.7 is easier to formulate. The equivalence of these two models relies on the scale invariance of the rank likelihood, and so will not generally hold for other types of models involving correlation matrices.

Full conditional distribution of Σ

If the prior distribution for Σ is inverse-Wishart$(\nu_0, \mathbf{S}_0^{-1})$, then, as described in Section 7.3, the full conditional distribution of Σ is inverse-Wishart as well. We review this fact here by first noting that the probability density of the $n \times p$ matrix $\mathbf{Z}$ can be written as

$$p(\mathbf{Z}|\Sigma) = \prod_{i=1}^{n} (2\pi)^{-p/2} |\Sigma|^{-1/2} \exp\{-\frac{1}{2} \boldsymbol{z}_i \Sigma^{-1} \boldsymbol{z}_i\}$$
$$= (2\pi)^{-np/2} |\Sigma|^{-n/2} \exp\{-\text{tr}(\mathbf{Z}^T \mathbf{Z} \Sigma^{-1})/2\},$$

where "tr($\mathbf{A}$)" stands for the trace of matrix $\mathbf{A}$, which is the sum of the diagonal elements of $\mathbf{A}$. The full conditional distribution $p(\Sigma|\mathbf{Z}, \mathbf{Z} \in R(\mathbf{Y})) = p(\Sigma|\mathbf{Z})$ is then given by

$$p(\Sigma|\mathbf{Z}) \propto p(\Sigma) \times p(\mathbf{Z}|\Sigma)$$
$$\propto |\Sigma|^{-(\nu_0+p+1)/2} \exp\{-\text{tr}(\mathbf{S}_0 \Sigma^{-1})/2\} \times |\Sigma|^{-n/2} \exp\{-\text{tr}(\mathbf{Z}^T \mathbf{Z} \Sigma^{-1})/2\}$$
$$= |\Sigma|^{-([\nu_0+n]+p+1)/2} \exp\{-\text{tr}([\mathbf{S}_0 + \mathbf{Z}^T \mathbf{Z}] \Sigma^{-1})/2\}$$

which is proportional to an inverse-Wishart$(\nu_0 + n, [\mathbf{S}_0 + \mathbf{Z}^T \mathbf{Z}]^{-1})$ density.

Full conditional distribution of $\mathbf{Z}$

Recall from Section 5 of Chapter 7 that if $\boldsymbol{Z}$ is a random multivariate normal$(\mathbf{0}, \Sigma)$ vector, then the conditional distribution of Z_j, given the other elements $\boldsymbol{Z}_{-j} = \boldsymbol{z}_{-j}$, is a univariate normal distribution with mean and variance given by

$$\text{E}[Z_j|\Sigma, \boldsymbol{z}_{-j}] = \Sigma_{j,-j} (\Sigma_{-j,-j})^{-1} \boldsymbol{z}_{-j}$$
$$\text{Var}[Z_j|\Sigma, \boldsymbol{z}_{-j}] = \Sigma_{j,j} - \Sigma_{j,-j} (\Sigma_{-j,-j})^{-1} \Sigma_{-j,j},$$

where $\Sigma_{j,-j}$ refers to the jth row of Σ with the jth column removed, and $\Sigma_{-j,-j}$ refers to Σ with both the jth row and column removed. If in addition we condition on the information that $\mathbf{Z} \in R(\mathbf{Y})$, then we know that

$$\max\{z_{k,j} : y_{k,j} < y_{i,j}\} < Z_{i,j} < \min\{z_{k,j} : y_{i,j} < y_{k,j}\}.$$

These two pieces of information imply that the full conditional distribution of $Z_{i,j}$ is a constrained normal distribution, which can be sampled from using the procedure described in the previous section and in the following R-code:

```
Sjc<- Sigma[j,-j]%*%solve(Sigma[-j,-j])
sz<- sqrt( Sigma[j,j] - Sjc%*%Sigma[-j,j])
ez<- Z[i,-j]%*%t(Sjc)

a<-max(Z[ Y[i,j]>Y[,j]  , j ], na.rm=TRUE)
b<-min(Z[ Y[i,j]<Y[,j]  , j ], na.rm=TRUE)

u<-runif(1, pnorm( (a-ez)/sz ),  pnorm( (b-ez)/sz )  )
Z[i,j]<- ez + sz*qnorm(u)
```

Missing data

The expression `na.rm=TRUE` in the above code allows for the possibility of missing data. Instead of throwing out the rows of the data matrix that contain some missing values, we would like to use all of the data we can. If the values are missing-at-random as described in Section 7.5, then they can simply be treated as unknown parameters and their values imputed using the Gibbs sampler. For the Gaussian copula model, this imputation happens at the level of the latent variables. For example, suppose that variable j for case i is not recorded, i.e. $y_{i,j}$ is not available. As described above, the full conditional distribution of $Z_{i,j}$ given $\boldsymbol{Z}_{i,-j}$ is normal. If $y_{i,j}$ were observed then the conditional distribution of $Z_{i,j}$ would be a constrained normal, as observing $y_{i,j}$ imposes a constraint on the allowable values of $Z_{i,j}$. But if $y_{i,j}$ is missing then no such constraint is imposed, and the full conditional distribution of $Z_{i,j}$ is simply the original unconstrained normal distribution. The R-code above handles this as follows: If $Y_{i,j}$ is missing, then `Z[ Y[i,j]>Y[,j] , j ]` is a vector of missing values. The option `na.rm=TRUE` removes all of these missing values, so `a` is the maximum of an empty set, which is defined to be $-\infty$. Similarly, `b` will be set to ∞.

Example: Social mobility data

The results of the probit regression of DEG on the variables PDEG and CHILD in the last section indicate that the educational level of an individual is related to that of their parents. In this section we analyze this further, by examining the joint relationships among respondent-specific variables DEG, CHILD, INC along with analogous parent-specific variables PDEG, PCHILD and PINC. In this data analysis PDEG is a five-level categorical variable with the same levels as DEG, recording the highest degree of the respondent's mother or father. PCHILD is the number of siblings of the respondent, and so is roughly the number of children of the respondent's parents. The variable PINC is a five-level ordered categorical variable recording the respondent's parent's financial status when the respondent was 16 years of age. Finally, we also include AGE, the respondent's age in years. Although not of primary interest, heterogeneity in a person's income, number of children and degree category is likely to be related to age.

Using an inverse-Wishart$(p+2,(p+2)\times\mathbf{I})$ prior distribution for Σ having a prior mean of $\mathrm{E}[\Sigma]=\mathbf{I}$, we can implement a Gibbs sampling algorithm using the full conditional distributions described above. Iterating the algorithm for 25,000 scans, saving parameter values every 25th scan, gives a total of 1,000 values of each parameter with which to approximate the posterior distribution of $\Psi=h(\Sigma)$. The Monte Carlo estimate of the posterior mean of Ψ is

$$\mathrm{E}[\Psi|\boldsymbol{y}_1,\ldots,\boldsymbol{y}_n] = \begin{pmatrix} 1.00 & 0.48 & 0.29 & 0.13 & 0.17 & -0.05 & 0.34 \\ 0.48 & 1.00 & -0.04 & 0.20 & 0.46 & -0.21 & 0.05 \\ 0.29 & -0.04 & 1.00 & -0.15 & -0.25 & 0.22 & 0.59 \\ 0.13 & 0.20 & -0.15 & 1.00 & 0.44 & -0.22 & -0.13 \\ 0.17 & 0.46 & -0.25 & 0.44 & 1.00 & -0.29 & -0.23 \\ -0.05 & -0.21 & 0.22 & -0.22 & -0.29 & 1.00 & 0.12 \\ 0.34 & 0.05 & 0.59 & -0.13 & -0.23 & 0.12 & 1.00 \end{pmatrix},$$

where the columns and rows are, in order, INC, DEG, CHILD, PINC, PDEG, PCHILD and AGE. We also may be interested in the "regression coefficients" $\boldsymbol{\beta}_{j|-j} = \Psi_{j,-j}(\Psi_{-j,-j})^{-1}$, which for each variable j is a vector of length $j-1$ that describes how the conditional mean of Z_j depends on the remaining variables Z_{-j}. Figure 12.4 summarizes the posterior distributions of each $\boldsymbol{\beta}_{j|-j}$ (except for that of AGE) as follows: A 95% quantile-based confidence interval is obtained for each $\boldsymbol{\beta}_{j,k}$. If the confidence interval does not contain zero, a line between variables j and k is drawn, with a "+" or a "−" indicating the sign of the posterior median. If the interval does contain zero, no line is drawn between the variables.

Such a graph is sometimes referred to as a *dependence graph*, which summarizes the conditional dependencies among the variables. Roughly speaking, two variables in the graph are conditionally independent given the other variables if there is no line between them. More precisely, the absence of a line indicates the lack of strong evidence of a conditional dependence. For example, although there is a positive marginal dependence between INC and PINC, the graph indicates that there is little evidence of any conditional dependence, given the other variables.

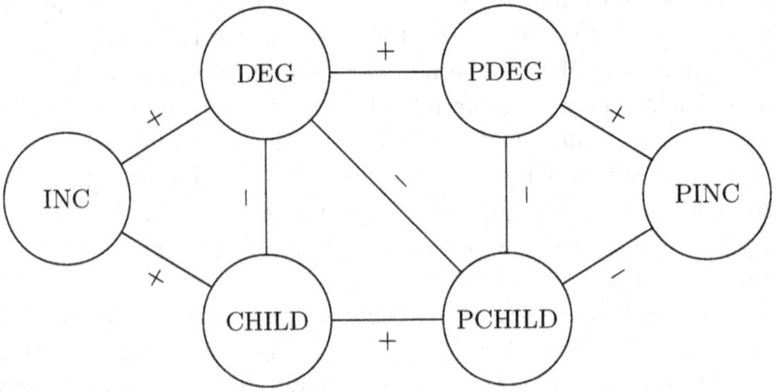

Fig. 12.4. Reduced conditional dependence graph for the GSS data.

12.3 Discussion and further references

Normally distributed latent variables are often used to induce dependence among a set of non-normal observed variables. For example, Chib and Winkelmann (2001) present a model for a vector of correlated count data in which each component is a Poisson random variable with a mean depending on a component-specific latent variable. Dependence among the count variables is induced by modeling the vector of latent variables with a multivariate normal distribution. Similar approaches are proposed by Dunson (2000) and described in Chapter 8 of Congdon (2003). Pitt et al (2006) discuss Bayesian inference for Gaussian copula models when the margins are known parametric families, and Quinn (2004) presents a factor analysis model for mixed continuous and discrete outcomes, in which the continuous variables are treated parametrically.

Pettitt (1982) develops the rank likelihood to estimate parameters in a latent normal regression model, allowing the transformation from the latent data to continuous observed data to be treated nonparametrically. Hoff (2007) extends this type of likelihood to accommodate both continuous and discrete ordinal data, and provides a Gibbs sampler for parameter estimation in a semiparametric Gaussian copula model.

The rank likelihood is based on the marginal distribution of the ranks, and so is called a *marginal* likelihood. Marginal likelihoods are typically constructed so that they use the information in the data that depends only on the parameters of interest, and do not use any information that depends on nuisance parameters. Marginal likelihoods do not generally provide efficient estimation, as they throw away part of the information in the data. However, they can turn a very difficult semiparametric estimation problem into essentially a parametric one. The use of marginal likelihoods in the context of Bayesian estimation is discussed in Monahan and Boos (1992).

Exercises

Chapter 2

2.1 Marginal and conditional probability: The social mobility data from Section 2.5 gives a joint probability distribution on (Y_1, Y_2)= (father's occupation, son's occupation). Using this joint distribution, calculate the following distributions:
  a) the marginal probability distribution of a father's occupation;
  b) the marginal probability distribution of a son's occupation;
  c) the conditional distribution of a son's occupation, given that the father is a farmer;
  d) the conditional distribution of a father's occupation, given that the son is a farmer.

2.2 Expectations and variances: Let Y_1 and Y_2 be two independent random variables, such that $\mathrm{E}[Y_i] = \mu_i$ and $\mathrm{Var}[Y_i] = \sigma_i^2$. Using the definition of expectation and variance, compute the following quantities, where a_1 and a_2 are given constants:
  a) $\mathrm{E}[a_1Y_1 + a_2Y_2]$, $\mathrm{Var}[a_1Y_1 + a_2Y_2]$;
  b) $\mathrm{E}[a_1Y_1 - a_2Y_2]$, $\mathrm{Var}[a_1Y_1 - a_2Y_2]$.

2.3 Full conditionals: Let X, Y, Z be random variables with joint density (discrete or continuous) $p(x, y, z) \propto f(x, z)g(y, z)h(z)$. Show that
  a) $p(x|y, z) \propto f(x, z)$, i.e. $p(x|y, z)$ is a function of x and z;
  b) $p(y|x, z) \propto g(y, z)$, i.e. $p(y|x, z)$ is a function of y and z;
  c) X and Y are conditionally independent given Z.

2.4 Symbolic manipulation: Prove the following form of Bayes' rule:

$$\Pr(H_j|E) = \frac{\Pr(E|H_j)\Pr(H_j)}{\sum_{k=1}^{K}\Pr(E|H_k)\Pr(H_k)}$$

where E is any event and $\{H_1, \ldots, H_K\}$ form a partition. Prove this using only axioms **P1-P3** from this chapter, by following steps a)-d) below:
  a) Show that $\Pr(H_j|E)\Pr(E) = \Pr(E|H_j)\Pr(H_j)$.

P.D. Hoff, *A First Course in Bayesian Statistical Methods*,
Springer Texts in Statistics, DOI 10.1007/978-0-387-92407-6,

b) Show that $\Pr(E) = \Pr(E \cap H_1) + \Pr(E \cap \{\cup_{k=2}^K H_k\})$.
c) Show that $\Pr(E) = \sum_{k=1}^K \Pr(E \cap H_k)$.
d) Put it all together to show Bayes' rule, as described above.

2.5 Urns: Suppose urn H is filled with 40% green balls and 60% red balls, and urn T is filled with 60% green balls and 40% red balls. Someone will flip a coin and then select a ball from urn H or urn T depending on whether the coin lands heads or tails, respectively. Let X be 1 or 0 if the coin lands heads or tails, and let Y be 1 or 0 if the ball is green or red.
a) Write out the joint distribution of X and Y in a table.
b) Find $\mathrm{E}[Y]$. What is the probability that the ball is green?
c) Find $\mathrm{Var}[Y|X=0]$, $\mathrm{Var}[Y|X=1]$ and $\mathrm{Var}[Y]$. Thinking of variance as measuring uncertainty, explain intuitively why one of these variances is larger than the others.
d) Suppose you see that the ball is green. What is the probability that the coin turned up tails?

2.6 Conditional independence: Suppose events A and B are conditionally independent given C, which is written $A \perp B|C$. Show that this implies that $A^c \perp B|C$, $A \perp B^c|C$, and $A^c \perp B^c|C$, where A^c means "not A." Find an example where $A \perp B|C$ holds but $A \perp B|C^c$ does not hold.

2.7 Coherence of bets: de Finetti thought of subjective probability as follows: Your probability $p(E)$ for event E is the amount you would be willing to pay or charge in exchange for a dollar on the occurrence of E. In other words, you must be willing to
- give $p(E)$ to someone, provided they give you \$1 if E occurs;
- take $p(E)$ from someone, and give them \$1 if E occurs.

Your probability for the event $E^c =$"not E" is defined similarly.
a) Show that it is a good idea to have $p(E) \leq 1$.
b) Show that it is a good idea to have $p(E) + p(E^c) = 1$.

2.8 Interpretations of probability: One abstract way to define probability is via measure theory, in that $\Pr(\cdot)$ is simply a "measure" that assigns mass to various events. For example, we can "measure" the number of times a particular event occurs in a potentially infinite sequence, or we can "measure" our information about the outcome of an unknown event. The above two types of measures are combined in de Finetti's theorem, which tells us that an exchangeable model for an infinite binary sequence $Y_1, Y_2, \ldots$ is equivalent to modeling the sequence as conditionally i.i.d. given a parameter θ, where $\Pr(\theta < c)$ represents our information that the long-run frequency of 1's is less than c. With this in mind, discuss the different ways in which probability could be interpreted in each of the following scenarios. Avoid using the word "probable" or "likely" when describing probability. Also discuss the different ways in which the events can be thought of as random.
a) The distribution of religions in Sri Lanka is 70% Buddhist, 15% Hindu, 8% Christian, and 7% Muslim. Suppose each person can be identified

by a number from 1 to K on a census roll. A number x is to be sampled from $\{1, \ldots, K\}$ using a pseudo-random number generator on a computer. Interpret the meaning of the following probabilities:
   i. Pr(person x is Hindu);
   ii. $\Pr(x = 6452859)$;
   iii. Pr(Person x is Hindu$|x$=6452859).

b) A quarter which you got as change is to be flipped many times. Interpret the meaning of the following probabilities:
   i. Pr(θ, the long-run relative frequency of heads, equals 1/3);
   ii. Pr(the first coin flip will result in a heads);
   iii. Pr(the first coin flip will result in a heads $| \ \theta = 1/3$).

c) The quarter above has been flipped, but you have not seen the outcome. Interpret Pr(the flip has resulted in a heads).

Chapter 3

3.1 Sample survey: Suppose we are going to sample 100 individuals from a county (of size much larger than 100) and ask each sampled person whether they support policy Z or not. Let $Y_i = 1$ if person i in the sample supports the policy, and $Y_i = 0$ otherwise.

a) Assume $Y_1, \ldots, Y_{100}$ are, conditional on θ, i.i.d. binary random variables with expectation θ. Write down the joint distribution of $\Pr(Y_1 = y_1, \ldots, Y_{100} = y_{100}|\theta)$ in a compact form. Also write down the form of $\Pr(\sum Y_i = y|\theta)$.

b) For the moment, suppose you believed that $\theta \in \{0.0, 0.1, \ldots, 0.9, 1.0\}$. Given that the results of the survey were $\sum_{i=1}^{100} Y_i = 57$, compute $\Pr(\sum Y_i = 57|\theta)$ for each of these 11 values of θ and plot these probabilities as a function of θ.

c) Now suppose you originally had no prior information to believe one of these θ-values over another, and so $\Pr(\theta = 0.0) = \Pr(\theta = 0.1) = \cdots = \Pr(\theta = 0.9) = \Pr(\theta = 1.0)$. Use Bayes' rule to compute $p(\theta|\sum_{i=1}^{n} Y_i = 57)$ for each θ-value. Make a plot of this posterior distribution as a function of θ.

d) Now suppose you allow θ to be any value in the interval $[0, 1]$. Using the uniform prior density for θ, so that $p(\theta) = 1$, plot $p(\theta) \times \Pr(\sum_{i=1}^{n} Y_i = 57|\theta)$ as a function of θ.

e) As discussed in this chapter, the posterior distribution of θ is beta$(1+57, 1+100-57)$. Plot the posterior density as a function of θ. Discuss the relationships among all of the plots you have made for this exercise.

3.2 Sensitivity analysis: It is sometimes useful to express the parameters a and b in a beta distribution in terms of $\theta_0 = a/(a + b)$ and $n_0 = a + b$, so that $a = \theta_0 n_0$ and $b = (1 - \theta_0)n_0$. Reconsidering the sample survey data in Exercise 3.1, for each combination of $\theta_0 \in \{0.1, 0.2, \ldots, 0.9\}$ and $n_0 \in \{1, 2, 8, 16, 32\}$ find the corresponding a, b values and compute $\Pr(\theta >$

$0.5|\sum Y_i = 57$) using a beta(a, b) prior distribution for θ. Display the results with a contour plot, and discuss how the plot could be used to explain to someone whether or not they should believe that $\theta > 0.5$, based on the data that $\sum_{i=1}^{100} Y_i = 57$.

3.3 Tumor counts: A cancer laboratory is estimating the rate of tumorigenesis in two strains of mice, A and B. They have tumor count data for 10 mice in strain A and 13 mice in strain B. Type A mice have been well studied, and information from other laboratories suggests that type A mice have tumor counts that are approximately Poisson-distributed with a mean of 12. Tumor count rates for type B mice are unknown, but type B mice are related to type A mice. The observed tumor counts for the two populations are

$$\boldsymbol{y}_A = (12, 9, 12, 14, 13, 13, 15, 8, 15, 6);$$

$$\boldsymbol{y}_B = (11, 11, 10, 9, 9, 8, 7, 10, 6, 8, 8, 9, 7).$$

a) Find the posterior distributions, means, variances and 95% quantile-based confidence intervals for θ_A and θ_B, assuming a Poisson sampling distribution for each group and the following prior distribution:

$$\theta_A \sim \text{gamma(120,10)},\ \theta_B \sim \text{gamma(12,1)},\ p(\theta_A, \theta_B) = p(\theta_A) \times p(\theta_B).$$

b) Compute and plot the posterior expectation of θ_B under the prior distribution $\theta_B \sim \text{gamma}(12\times n_0, n_0)$ for each value of $n_0 \in \{1, 2, \ldots, 50\}$. Describe what sort of prior beliefs about θ_B would be necessary in order for the posterior expectation of θ_B to be close to that of θ_A.
c) Should knowledge about population A tell us anything about population B? Discuss whether or not it makes sense to have $p(\theta_A, \theta_B) = p(\theta_A) \times p(\theta_B)$.

3.4 Mixtures of beta priors: Estimate the probability θ of teen recidivism based on a study in which there were $n = 43$ individuals released from incarceration and $y = 15$ re-offenders within 36 months.
a) Using a beta(2,8) prior for θ, plot $p(\theta)$, $p(y|\theta)$ and $p(\theta|y)$ as functions of θ. Find the posterior mean, mode, and standard deviation of θ. Find a 95% quantile-based confidence interval.
b) Repeat a), but using a beta(8,2) prior for θ.
c) Consider the following prior distribution for θ:

$$p(\theta) = \frac{1}{4}\frac{\Gamma(10)}{\Gamma(2)\Gamma(8)}[3\theta(1-\theta)^7 + \theta^7(1-\theta)],$$

which is a 75-25% mixture of a beta(2,8) and a beta(8,2) prior distribution. Plot this prior distribution and compare it to the priors in a) and b). Describe what sort of prior opinion this may represent.
d) For the prior in c):
   i. Write out mathematically $p(\theta) \times p(y|\theta)$ and simplify as much as possible.

ii. The posterior distribution is a mixture of two distributions you know. Identify these distributions.

iii. On a computer, calculate and plot $p(\theta) \times p(y|\theta)$ for a variety of θ values. Also find (approximately) the posterior mode, and discuss its relation to the modes in a) and b).

e) Find a general formula for the weights of the mixture distribution in d)ii, and provide an interpretation for their values.

3.5 Mixtures of conjugate priors: Let $p(y|\phi) = c(\phi)h(y)\exp\{\phi t(y)\}$ be an exponential family model and let $p_1(\phi), \ldots p_K(\phi)$ be K different members of the conjugate class of prior densities given in Section 3.3. A mixture of conjugate priors is given by $\tilde{p}(\theta) = \sum_{k=1}^{K} w_k p_k(\theta)$, where the w_k's are all greater than zero and $\sum w_k = 1$ (see also Diaconis and Ylvisaker (1985)).

a) Identify the general form of the posterior distribution of θ, based on n i.i.d. samples from $p(y|\theta)$ and the prior distribution given by $\tilde{p}$.

b) Repeat a) but in the special case that $p(y|\theta) = \text{dpois}(y, \theta)$ and $p_1, \ldots, p_K$ are gamma densities.

3.6 Exponential family expectations: Let $p(y|\phi) = c(\phi)h(y)\exp\{\phi t(y)\}$ be an exponential family model.

a) Take derivatives with respect to ϕ of both sides of the equation $\int p(y|\phi)\, dy = 1$ to show that $\text{E}[t(Y)|\phi] = -c'(\phi)/c(\phi)$.

b) Let $p(\phi) \propto c(\phi)^{n_0} e^{n_0 t_0 \phi}$ be the prior distribution for ϕ. Calculate $dp(\phi)/d\phi$ and, using the fundamental theorem of calculus, discuss what must be true so that $\text{E}[-c(\phi)/c(\phi)] = t_0$.

3.7 Posterior prediction: Consider a pilot study in which $n_1 = 15$ children enrolled in special education classes were randomly selected and tested for a certain type of learning disability. In the pilot study, $y_1 = 2$ children tested positive for the disability.

a) Using a uniform prior distribution, find the posterior distribution of θ, the fraction of students in special education classes who have the disability. Find the posterior mean, mode and standard deviation of θ, and plot the posterior density.

Researchers would like to recruit students with the disability to participate in a long-term study, but first they need to make sure they can recruit enough students. Let $n_2 = 278$ be the number of children in special education classes in this particular school district, and let Y_2 be the number of students with the disability.

b) Find $\Pr(Y_2 = y_2|Y_1 = 2)$, the posterior predictive distribution of Y_2, as follows:

i. Discuss what assumptions are needed about the joint distribution of (Y_1, Y_2) such that the following is true:

$$\Pr(Y_2 = y_2|Y_1 = 2) = \int_0^1 \Pr(Y_2 = y_2|\theta) p(\theta|Y_1 = 2)\, d\theta\,.$$

ii. Now plug in the forms for $\Pr(Y_2 = y_2|\theta)$ and $p(\theta|Y_1 = 2)$ in the above integral.

iii. Figure out what the above integral must be by using the calculus result discussed in Section 3.1.

c) Plot the function $\Pr(Y_2 = y_2|Y_1 = 2)$ as a function of y_2. Obtain the mean and standard deviation of Y_2, given $Y_1 = 2$.

d) The posterior mode and the MLE (maximum likelihood estimate; see Exercise 3.14) of θ, based on data from the pilot study, are both $\hat{\theta} = 2/15$. Plot the distribution $\Pr(Y_2 = y_2|\theta = \hat{\theta})$, and find the mean and standard deviation of Y_2 given $\theta = \hat{\theta}$. Compare these results to the plots and calculations in c) and discuss any differences. Which distribution for Y_2 would you use to make predictions, and why?

3.8 Coins: Diaconis and Ylvisaker (1985) suggest that coins spun on a flat surface display long-run frequencies of heads that vary from coin to coin. About 20% of the coins behave symmetrically, whereas the remaining coins tend to give frequencies of 1/3 or 2/3.

a) Based on the observations of Diaconis and Ylvisaker, use an appropriate mixture of beta distributions as a prior distribution for θ, the long-run frequency of heads for a particular coin. Plot your prior.

b) Choose a single coin and spin it at least 50 times. Record the number of heads obtained. Report the year and denomination of the coin.

c) Compute your posterior for θ, based on the information obtained in b).

d) Repeat b) and c) for a different coin, but possibly using a prior for θ that includes some information from the first coin. Your choice of a new prior may be informal, but needs to be justified. How the results from the first experiment influence your prior for the θ of the second coin may depend on whether or not the two coins have the same denomination, have a similar year, etc. Report the year and denomination of this coin.

3.9 Galenshore distribution: An unknown quantity Y has a Galenshore(a, θ) distribution if its density is given by

$$p(y|\theta) = \frac{2}{\Gamma(a)}\theta^{2a}y^{2a-1}e^{-\theta^2 y^2}$$

for $y > 0$, $\theta > 0$ and $a > 0$. Assume for now that a is known. For this density,

$$\mathrm{E}[Y|\theta] = \frac{\Gamma(a+1/2)}{\theta\Gamma(a)}, \quad \mathrm{E}[Y^2|\theta] = \frac{a}{\theta^2}.$$

a) Identify a class of conjugate prior densities for θ. Plot a few members of this class of densities.

b) Let $Y_1, \ldots, Y_n \sim$ i.i.d. Galenshore(a, θ). Find the posterior distribution of θ given $Y_1, \ldots, Y_n$, using a prior from your conjugate class.

c) Write down $p(\theta_a|Y_1, \ldots, Y_n)/p(\theta_b|Y_1, \ldots, Y_n)$ and simplify. Identify a sufficient statistic.

d) Determine $\mathrm{E}[\theta|y_1, \ldots, y_n]$.

e) Determine the form of the posterior predictive density $p(\tilde{y}|y_1 \ldots, y_n)$.

3.10 Change of variables: Let $\psi = g(\theta)$, where g is a monotone function of θ, and let h be the inverse of g so that $\theta = h(\psi)$. If $p_\theta(\theta)$ is the probability density of θ, then the probability density of ψ induced by p_θ is given by $p_\psi(\psi) = p_\theta(h(\psi)) \times |\frac{dh}{d\psi}|$.

a) Let $\theta \sim \text{beta}(a, b)$ and let $\psi = \log[\theta/(1-\theta)]$. Obtain the form of p_ψ and plot it for the case that $a = b = 1$.

b) Let $\theta \sim \text{gamma}(a, b)$ and let $\psi = \log \theta$. Obtain the form of p_ψ and plot it for the case that $a = b = 1$.

3.12 Jeffreys' prior: Jeffreys (1961) suggested a default rule for generating a prior distribution of a parameter θ in a sampling model $p(y|\theta)$. Jeffreys' prior is given by $p_J(\theta) \propto \sqrt{I(\theta)}$, where $I(\theta) = -\text{E}[\partial^2 \log p(Y|\theta)/\partial\theta^2|\theta]$ is the *Fisher information*.

a) Let $Y \sim \text{binomial}(n, \theta)$. Obtain Jeffreys' prior distribution $p_J(\theta)$ for this model.

b) Reparameterize the binomial sampling model with $\psi = \log \theta/(1-\theta)$, so that $p(y|\psi) = \binom{n}{y} e^{\psi y}(1+e^\psi)^{-n}$. Obtain Jeffreys' prior distribution $p_J(\psi)$ for this model.

c) Take the prior distribution from a) and apply the change of variables formula from Exercise 3.10 to obtain the induced prior density on ψ. This density should be the same as the one derived in part b) of this exercise. This consistency under reparameterization is the defining characteristic of Jeffrey's' prior.

3.13 Improper Jeffreys' prior: Let $Y \sim \text{Poisson}(\theta)$.

a) Apply Jeffreys' procedure to this model, and compare the result to the family of gamma densities. Does Jeffreys' procedure produce an actual probability density for θ? In other words, can $\sqrt{I(\theta)}$ be proportional to an actual probability density for $\theta \in (0, \infty)$?

b) Obtain the form of the function $f(\theta, y) = \sqrt{I(\theta)} \times p(y|\theta)$. What probability density for θ is $f(\theta, y)$ proportional to? Can we think of $f(\theta, y)/\int f(\theta, y)d\theta$ as a posterior density of θ given $Y = y$?

3.14 Unit information prior: Let $Y_1, \ldots, Y_n \sim$ i.i.d. $p(y|\theta)$. Having observed the values $Y_1 = y_1, \ldots, Y_n = y_n$, the *log likelihood* is given by $l(\theta|\boldsymbol{y}) = \sum \log p(y_i|\theta)$, and the value $\hat{\theta}$ of θ that maximizes $l(\theta|\boldsymbol{y})$ is called the *maximum likelihood estimator*. The negative of the curvature of the log-likelihood, $J(\theta) = -\partial^2 l(\theta|\boldsymbol{y})/\partial\theta^2$, describes the precision of the MLE $\hat{\theta}$ and is called the *observed Fisher information*. For situations in which it is difficult to quantify prior information in terms of a probability distribution, some have suggested that the "prior" distribution be based on the likelihood, for example, by centering the prior distribution around the MLE $\hat{\theta}$. To deal with the fact that the MLE is not really prior information, the curvature of the prior is chosen so that it has only "one nth" as much information as the likelihood, so that $-\partial^2 \log p(\theta)/\partial\theta^2 = J(\theta)/n$. Such a prior is called a *unit information prior* (Kass and Wasserman, 1995; Kass and Raftery, 1995), as it has as much information as the average amount of information from a single observation. The unit information prior is not really a prior distribution, as it is computed from the observed data. However, it can be roughly viewed as the prior information of someone with weak but accurate prior information.

a) Let $Y_1, \ldots, Y_n \sim$ i.i.d. binary(θ). Obtain the MLE $\hat{\theta}$ and $J(\hat{\theta})/n$.
b) Find a probability density $p_U(\theta)$ such that $\log p_U(\theta) = l(\theta|\boldsymbol{y})/n + c$, where c is a constant that does not depend on θ. Compute the information $-\partial^2 \log p_U(\theta)/\partial\theta^2$ of this density.
c) Obtain a probability density for θ that is proportional to $p_U(\theta) \times p(y_1, \ldots, y_n|\theta)$. Can this be considered a posterior distribution for θ?
d) Repeat a), b) and c) but with $p(y|\theta)$ being the Poisson distribution.

Chapter 4

4.1 Posterior comparisons: Reconsider the sample survey in Exercise 3.1. Suppose you are interested in comparing the rate of support in that county to the rate in another county. Suppose that a survey of sample size 50 was done in the second county, and the total number of people in the sample who supported the policy was 30. Identify the posterior distribution of θ_2 assuming a uniform prior. Sample 5,000 values of each of θ_1 and θ_2 from their posterior distributions and estimate $\Pr(\theta_1 < \theta_2|\text{the data and prior})$.

4.2 Tumor count comparisons: Reconsider the tumor count data in Exercise 3.3:
a) For the prior distribution given in part a) of that exercise, obtain $\Pr(\theta_B < \theta_A|\boldsymbol{y}_A, \boldsymbol{y}_B)$ via Monte Carlo sampling.
b) For a range of values of n_0, obtain $\Pr(\theta_B < \theta_A|\boldsymbol{y}_A, \boldsymbol{y}_B)$ for $\theta_A \sim$ gamma(120, 10) and $\theta_B \sim$ gamma($12 \times n_0, n_0$). Describe how sensitive the conclusions about the event $\{\theta_B < \theta_A\}$ are to the prior distribution on θ_B.
c) Repeat parts a) and b), replacing the event $\{\theta_B < \theta_A\}$ with the event $\{\tilde{Y}_B < \tilde{Y}_A\}$, where $\tilde{Y}_A$ and $\tilde{Y}_B$ are samples from the posterior predictive distribution.

4.3 Posterior predictive checks: Let's investigate the adequacy of the Poisson model for the tumor count data. Following the example in Section 4.4, generate posterior predictive datasets $\boldsymbol{y}_A^{(1)}, \ldots, \boldsymbol{y}_A^{(1000)}$. Each $\boldsymbol{y}_A^{(s)}$ is a sample of size $n_A = 10$ from the Poisson distribution with parameter $\theta_A^{(s)}$, $\theta_A^{(s)}$ is itself a sample from the posterior distribution $p(\theta_A|\boldsymbol{y}_A)$, and $\boldsymbol{y}_A$ is the observed data.
a) For each s, let $t^{(s)}$ be the sample average of the 10 values of $\boldsymbol{y}_A^{(s)}$, divided by the sample standard deviation of $\boldsymbol{y}_A^{(s)}$. Make a histogram of $t^{(s)}$ and compare to the observed value of this statistic. Based on this statistic, assess the fit of the Poisson model for these data.

b) Repeat the above goodness of fit evaluation for the data in population B.

4.4 Mixtures of conjugate priors: For the posterior density from Exercise 3.4:

a) Make a plot of $p(\theta|y)$ or $p(y|\theta)p(\theta)$ using the mixture prior distribution and a dense sequence of θ-values. Can you think of a way to obtain a 95% quantile-based posterior confidence interval for θ? You might want to try some sort of discrete approximation.

b) To sample a random variable z from the mixture distribution $wp_1(z) + (1-w)p_0(z)$, first toss a w-coin and let x be the outcome (this can be done in R with `x<-rbinom(1,1,w)`). Then if $x = 1$ sample z from p_1, and if $x = 0$ sample z from p_0. Using this technique, obtain a Monte Carlo approximation of the posterior distribution $p(\theta|y)$ and a 95% quantile-based confidence interval, and compare them to the results in part a).

4.5 Cancer deaths: Suppose for a set of counties $i \in \{1, \ldots, n\}$ we have information on the population size X_i = number of people in 10,000s, and Y_i = number of cancer fatalities. One model for the distribution of cancer fatalities is that, given the cancer rate θ, they are independently distributed with $Y_i \sim \text{Poisson}(\theta X_i)$.

a) Identify the posterior distribution of θ given data $(Y_1, X_1), \ldots, (Y_n, X_n)$ and a gamma(a, b) prior distribution.

The file `cancer_react.dat` contains 1990 population sizes (in 10,000s) and number of cancer fatalities for 10 counties in a Midwestern state that are near nuclear reactors. The file `cancer_noreact.dat` contains the same data on counties in the same state that are not near nuclear reactors. Consider these data as samples from two populations of counties: one is the population of counties with no neighboring reactors and a fatality rate of θ_1 deaths per 10,000, and the other is a population of counties having nearby reactors and a fatality rate of θ_2. In this exercise we will model beliefs about the rates as independent and such that $\theta_1 \sim \text{gamma}(a_1, b_1)$ and $\theta_2 \sim \text{gamma}(a_2, b_2)$.

b) Using the numerical values of the data, identify the posterior distributions for θ_1 and θ_2 for any values of (a_1, b_1, a_2, b_2).

c) Suppose cancer rates from previous years have been roughly $\tilde{\theta} = 2.2$ per 10,000 (and note that most counties are not near reactors). For each of the following three prior opinions, compute $\text{E}[\theta_1|\text{data}]$, $\text{E}[\theta_2|\text{data}]$, 95% quantile-based posterior intervals for θ_1 and θ_2, and $\Pr(\theta_2 > \theta_1|\text{data})$. Also plot the posterior densities (try to put $p(\theta_1|\text{data})$ and $p(\theta_2|\text{data})$ on the same plot). Comment on the differences across posterior opinions.

   i. Opinion 1: $(a_1 = a_2 = 2.2 \times 100, b_1 = b_2 = 100)$. Cancer rates for both types of counties are similar to the average rates across all counties from previous years.

ii. Opinion 2: ($a_1 = 2.2 \times 100, b_1 = 100, a_2 = 2.2, b_2 = 1$). Cancer rates in this year for nonreactor counties are similar to rates in previous years in nonreactor counties. We don't have much information on reactor counties, but perhaps the rates are close to those observed previously in nonreactor counties.
iii. Opinion 3: ($a_1 = a_2 = 2.2, b_1 = b_2 = 1$). Cancer rates in this year could be different from rates in previous years, for both reactor and nonreactor counties.

d) In the above analysis we assumed that population size gives no information about fatality rate. Is this reasonable? How would the analysis have to change if this is not reasonable?
e) We encoded our beliefs about θ_1 and θ_2 such that they gave no information about each other (they were *a priori* independent). Think about why and how you might encode beliefs such that they were *a priori* dependent.

4.6 Non-informative prior distributions: Suppose for a binary sampling problem we plan on using a uniform, or beta(1,1), prior for the population proportion θ. Perhaps our reasoning is that this represents "no prior information about θ." However, some people like to look at proportions on the log-odds scale, that is, they are interested in $\gamma = \log \frac{\theta}{1-\theta}$. Via Monte Carlo sampling or otherwise, find the prior distribution for γ that is induced by the uniform prior for θ. Is the prior informative about γ?

4.7 Mixture models: After a posterior analysis on data from a population of squash plants, it was determined that the total vegetable weight of a given plant could be modeled with the following distribution:

$$p(y|\theta,\sigma^2) = .31\text{dnorm}(y,\theta,\sigma) + .46\text{dnorm}(2\theta_1, 2\sigma) + .23\text{dnorm}(y, 3\theta_1, 3\sigma)$$

where the posterior distributions of the parameters have been calculated as $1/\sigma^2 \sim \text{gamma}(10, 2.5)$, and $\theta|\sigma^2 \sim \text{normal}(4.1, \sigma^2/20)$.

a) Sample at least 5,000 y values from the posterior predictive distribution.
b) Form a 75% quantile-based confidence interval for a new value of Y.
c) Form a 75% HPD region for a new Y as follows:
   i. Compute estimates of the posterior density of Y using the `density` command in R, and then normalize the density values so they sum to 1.
   ii. Sort these discrete probabilities in decreasing order.
   iii. Find the first probability value such that the cumulative sum of the sorted values exceeds 0.75. Your HPD region includes all values of y which have a discretized probability greater than this cutoff. Describe your HPD region, and compare it to your quantile-based region.
d) Can you think of a physical justification for the mixture sampling distribution of Y?

4.8 More posterior predictive checks: Let θ_A and θ_B be the average number of children of men in their 30s with and without bachelor's degrees, respectively.
  a) Using a Poisson sampling model, a gamma(2,1) prior for each θ and the data in the files `menchild30bach.dat` and `menchild30nobach.dat`, obtain 5,000 samples of $\tilde{Y}_A$ and $\tilde{Y}_B$ from the posterior predictive distribution of the two samples. Plot the Monte Carlo approximations to these two posterior predictive distributions.
  b) Find 95% quantile-based posterior confidence intervals for $\theta_B - \theta_A$ and $\tilde{Y}_B - \tilde{Y}_A$. Describe in words the differences between the two populations using these quantities and the plots in a), along with any other results that may be of interest to you.
  c) Obtain the empirical distribution of the data in group B. Compare this to the Poisson distribution with mean $\hat{\theta} = 1.4$. Do you think the Poisson model is a good fit? Why or why not?
  d) For each of the 5,000 θ_B-values you sampled, sample $n_B = 218$ Poisson random variables and count the number of 0s and the number of 1s in each of the 5,000 simulated datasets. You should now have two sequences of length 5,000 each, one sequence counting the number of people having zero children for each of the 5,000 posterior predictive datasets, the other counting the number of people with one child. Plot the two sequences against one another (one on the x-axis, one on the y-axis). Add to the plot a point marking how many people in the observed dataset had zero children and one child. Using this plot, describe the adequacy of the Poisson model.

Chapter 5

5.1 Studying: The files `school1.dat`, `school2.dat` and `school3.dat` contain data on the amount of time students from three high schools spent on studying or homework during an exam period. Analyze data from each of these schools separately, using the normal model with a conjugate prior distribution, in which $\{\mu_0 = 5, \sigma_0^2 = 4, \kappa_0 = 1, \nu_0 = 2\}$ and compute or approximate the following:
  a) posterior means and 95% confidence intervals for the mean θ and standard deviation σ from each school;
  b) the posterior probability that $\theta_i < \theta_j < \theta_k$ for all six permutations $\{i, j, k\}$ of $\{1, 2, 3\}$;
  c) the posterior probability that $\tilde{Y}_i < \tilde{Y}_j < \tilde{Y}_k$ for all six permutations $\{i, j, k\}$ of $\{1, 2, 3\}$, where $\tilde{Y}_i$ is a sample from the posterior predictive distribution of school i.
  d) Compute the posterior probability that θ_1 is bigger than both θ_2 and θ_3, and the posterior probability that $\tilde{Y}_1$ is bigger than both $\tilde{Y}_2$ and $\tilde{Y}_3$.

5.2 Sensitivity analysis: Thirty-two students in a science classroom were randomly assigned to one of two study methods, A and B, so that $n_A = n_B = 16$ students were assigned to each method. After several weeks of study, students were examined on the course material with an exam designed to give an average score of 75 with a standard deviation of 10. The scores for the two groups are summarized by $\{\bar{y}_A = 75.2, s_A = 7.3\}$ and $\{\bar{y}_B = 77.5, s_B = 8.1\}$. Consider independent, conjugate normal prior distributions for each of θ_A and θ_B, with $\mu_0 = 75$ and $\sigma_0^2 = 100$ for both groups. For each $(\kappa_0, \nu_0) \in \{(1,1),(2,2),(4,4),(8,8),(16,16),(32,32)\}$ (or more values), obtain $\Pr(\theta_A < \theta_B | \boldsymbol{y}_A, \boldsymbol{y}_B)$ via Monte Carlo sampling. Plot this probability as a function of $(\kappa_0 = \nu_0)$. Describe how you might use this plot to convey the evidence that $\theta_A < \theta_B$ to people of a variety of prior opinions.

5.3 Marginal distributions: Given observations $Y_1, \dots, Y_n \sim$ i.i.d. normal(θ, σ^2) and using the conjugate prior distribution for θ and σ^2, derive the formula for $p(\theta|y_1, \dots, y_n)$, the marginal posterior distribution of θ, conditional on the data but marginal over σ^2. Check your work by comparing your formula to a Monte Carlo estimate of the marginal distribution, using some values of $Y_1, \dots, Y_n$, μ_0, σ_0^2, ν_0 and κ_0 that you choose. Also derive $p(\tilde{\sigma}^2|y_1, \dots, y_n)$, where $\tilde{\sigma}^2 = 1/\sigma^2$ is the precision.

5.4 Jeffreys' prior: For sampling models expressed in terms of a p-dimensional vector $\boldsymbol{\psi}$, Jeffreys' prior (Exercise 3.11) is defined as $p_J(\boldsymbol{\psi}) \propto \sqrt{|I(\boldsymbol{\psi})|}$, where $|I(\boldsymbol{\psi})|$ is the determinant of the $p \times p$ matrix $I(\boldsymbol{\psi})$ having entries $I(\boldsymbol{\psi})_{k,l} = -\mathrm{E}[\partial^2 \log p(Y|\boldsymbol{\psi})/\partial\psi_k \partial\psi_l]$.

a) Show that Jeffreys' prior for the normal model is $p_J(\theta, \sigma^2) \propto (\sigma^2)^{-3/2}$.
b) Let $\boldsymbol{y} = (y_1, \dots, y_n)$ be the observed values of an i.i.d. sample from a normal(θ, σ^2) population. Find a probability density $p_J(\theta, \sigma^2|\boldsymbol{y})$ such that $p_J(\theta, \sigma^2|\boldsymbol{y}) \propto p_J(\theta, \sigma^2)p(\boldsymbol{y}|\theta, \sigma^2)$. It may be convenient to write this joint density as $p_J(\theta|\sigma^2, \boldsymbol{y}) \times p_J(\sigma^2|\boldsymbol{y})$. Can this joint density be considered a posterior density?

5.5 Unit information prior: Obtain a unit information prior for the normal model as follows:

a) Reparameterize the normal model as $p(y|\theta, \psi)$, where $\psi = 1/\sigma^2$. Write out the log likelihood $l(\theta, \psi|\boldsymbol{y}) = \sum \log p(y_i|\theta, \psi)$ in terms of θ and ψ.
b) Find a probability density $p_U(\theta, \psi)$ so that $\log p_U(\theta, \psi) = l(\theta, \psi|\boldsymbol{y})/n + c$, where c is a constant that does not depend on θ or ψ. Hint: Write $\sum(y_i - \theta)^2$ as $\sum(y_i - \bar{y} + \bar{y} - \theta)^2 = \sum(y_i - \bar{y})^2 + n(\theta - \bar{y})^2$, and recall that $\log p_U(\theta, \psi) = \log p_U(\theta|\psi) + \log p_U(\psi)$.
c) Find a probability density $p_U(\theta, \psi|\boldsymbol{y})$ that is proportional to $p_U(\theta, \psi) \times p(y_1, \dots, y_n|\theta, \psi)$. It may be convenient to write this joint density as $p_U(\theta|\psi, \boldsymbol{y}) \times p_U(\psi|\boldsymbol{y})$. Can this joint density be considered a posterior density?

Chapter 6

6.1 Poisson population comparisons: Let's reconsider the number of children data of Exercise 4.8. We'll assume Poisson sampling models for the two groups as before, but now we'll parameterize θ_A and θ_B as $\theta_A = \theta$, $\theta_B = \theta \times \gamma$. In this parameterization, γ represents the relative rate θ_B/θ_A. Let $\theta \sim \text{gamma}(a_\theta, b_\theta)$ and let $\gamma \sim \text{gamma}(a_\gamma, b_\gamma)$.

a) Are θ_A and θ_B independent or dependent under this prior distribution? In what situations is such a joint prior distribution justified?

b) Obtain the form of the full conditional distribution of θ given $\boldsymbol{y}_A$, $\boldsymbol{y}_B$ and γ.

c) Obtain the form of the full conditional distribution of γ given $\boldsymbol{y}_A$, $\boldsymbol{y}_B$ and θ.

d) Set $a_\theta = 2$ and $b_\theta = 1$. Let $a_\gamma = b_\gamma \in \{8, 16, 32, 64, 128\}$. For each of these five values, run a Gibbs sampler of at least 5,000 iterations and obtain $\text{E}[\theta_B - \theta_A | \boldsymbol{y}_A, \boldsymbol{y}_B]$. Describe the effects of the prior distribution for γ on the results.

6.2 Mixture model: The file `glucose.dat` contains the plasma glucose concentration of 532 females from a study on diabetes (see Exercise 7.6).

a) Make a histogram or kernel density estimate of the data. Describe how this empirical distribution deviates from the shape of a normal distribution.

b) Consider the following mixture model for these data: For each study participant there is an unobserved group membership variable X_i which is equal to 1 or 2 with probability p and $1 - p$. If $X_i = 1$ then $Y_i \sim \text{normal}(\theta_1, \sigma_1^2)$, and if $X_i = 2$ then $Y_i \sim \text{normal}(\theta_2, \sigma_2^2)$. Let $p \sim \text{beta}(a, b)$, $\theta_j \sim \text{normal}(\mu_0, \tau_0^2)$ and $1/\sigma_j \sim \text{gamma}(\nu_0/2, \nu_0\sigma_0^2/2)$ for both $j = 1$ and $j = 2$. Obtain the full conditional distributions of $(X_1, \ldots, X_n)$, p, θ_1, θ_2, σ_1^2 and σ_2^2.

c) Setting $a = b = 1$, $\mu_0 = 120$, $\tau_0^2 = 200$, $\sigma_0^2 = 1000$ and $\nu_0 = 10$, implement the Gibbs sampler for at least 10,000 iterations. Let $\theta_{(1)}^{(s)} = \min\{\theta_1^{(s)}, \theta_2^{(s)}\}$ and $\theta_{(2)}^{(s)} = \max\{\theta_1^{(s)}, \theta_2^{(s)}\}$. Compute and plot the autocorrelation functions of $\theta_{(1)}^{(s)}$ and $\theta_{(2)}^{(s)}$, as well as their effective sample sizes.

d) For each iteration s of the Gibbs sampler, sample a value $x \sim \text{binary}(p^{(s)})$, then sample $\tilde{Y}^{(s)} \sim \text{normal}(\theta_x^{(s)}, \sigma_x^{2(s)})$. Plot a histogram or kernel density estimate for the empirical distribution of $\tilde{Y}^{(1)}, \ldots, \tilde{Y}^{(S)}$, and compare to the distribution in part a). Discuss the adequacy of this two-component mixture model for the glucose data.

6.3 Probit regression: A panel study followed 25 married couples over a period of five years. One item of interest is the relationship between divorce rates and the various characteristics of the couples. For example, the researchers would like to model the probability of divorce as a function of age differential, recorded as the man's age minus the woman's age. The data can be found in the file `divorce.dat`. We will model these data with probit regression, in which a binary variable Y_i is described in terms of an explanatory variable x_i via the following latent variable model:

$$Z_i = \beta x_i + \epsilon_i$$
$$Y_i = \delta_{(c,\infty)}(Z_i),$$

where β and c are unknown coefficients, $\epsilon_1, \ldots, \epsilon_n \sim$ i.i.d. normal$(0, 1)$ and $\delta_{(c,\infty)}(z) = 1$ if $z > c$ and equals zero otherwise.

a) Assuming $\beta \sim$ normal$(0, \tau_\beta^2)$ obtain the full conditional distribution $p(\beta|\boldsymbol{y}, \boldsymbol{x}, \boldsymbol{z}, c)$.
b) Assuming $c \sim$ normal$(0, \tau_c^2)$, show that $p(c|\boldsymbol{y}, \boldsymbol{x}, \boldsymbol{z}, \beta)$ is a constrained normal density, i.e. proportional to a normal density but constrained to lie in an interval. Similarly, show that $p(z_i|\boldsymbol{y}, \boldsymbol{x}, \boldsymbol{z}_{-i}, \beta, c)$ is proportional to a normal density but constrained to be either above c or below c, depending on y_i.
c) Letting $\tau_\beta^2 = \tau_c^2 = 16$, implement a Gibbs sampling scheme that approximates the joint posterior distribution of $\boldsymbol{Z}$, β, and c (a method for sampling from constrained normal distributions is outlined in Section 12.1.1). Run the Gibbs sampler long enough so that the effective sample sizes of all unknown parameters are greater than 1,000 (including the Z_i's). Compute the autocorrelation function of the parameters and discuss the mixing of the Markov chain.
d) Obtain a 95% posterior confidence interval for β, as well as $\Pr(\beta > 0|\boldsymbol{y}, \boldsymbol{x})$.

Chapter 7

7.1 Jeffreys' prior: For the multivariate normal model, Jeffreys' rule for generating a prior distribution on $(\boldsymbol{\theta}, \Sigma)$ gives $p_J(\boldsymbol{\theta}, \Sigma) \propto |\Sigma|^{-(p+2)/2}$.
a) Explain why the function p_J cannot actually be a probability density for $(\boldsymbol{\theta}, \Sigma)$.
b) Let $p_J(\boldsymbol{\theta}, \Sigma|\boldsymbol{y}_1, \ldots, \boldsymbol{y}_n)$ be the probability density that is proportional to $p_J(\boldsymbol{\theta}, \Sigma) \times p(\boldsymbol{y}_1, \ldots, \boldsymbol{y}_n|\boldsymbol{\theta}, \Sigma)$. Obtain the form of $p_J(\boldsymbol{\theta}, \Sigma|\boldsymbol{y}_1, \ldots, \boldsymbol{y}_n)$, $p_J(\boldsymbol{\theta}|\Sigma, \boldsymbol{y}_1, \ldots, \boldsymbol{y}_n)$ and $p_J(\Sigma|\boldsymbol{y}_1, \ldots, \boldsymbol{y}_n)$.

7.2 Unit information prior: Letting $\Psi = \Sigma^{-1}$, show that a unit information prior for $(\boldsymbol{\theta}, \Psi)$ is given by $\boldsymbol{\theta}|\Psi \sim$ multivariate normal$(\bar{\boldsymbol{y}}, \Psi^{-1})$ and $\Psi \sim$ Wishart$(p+1, \mathbf{S}^{-1})$, where $\mathbf{S} = \sum(\boldsymbol{y}_i - \bar{\boldsymbol{y}})(\boldsymbol{y}_i - \bar{\boldsymbol{y}})^T/n$. This can be done by mimicking the procedure outlined in Exercise 5.5 as follows:
a) Reparameterize the multivariate normal model in terms of the precision matrix $\Psi = \Sigma^{-1}$. Write out the resulting log likelihood, and find a probability density $p_U(\boldsymbol{\theta}, \Psi) = p_U(\boldsymbol{\theta}|\Psi)p_U(\Psi)$ such that $\log p(\boldsymbol{\theta}, \Psi) = l(\boldsymbol{\theta}, \Psi|\mathbf{Y})/n + c$, where c does not depend on $\boldsymbol{\theta}$ or Ψ.

Hint: Write $(\boldsymbol{y}_i - \boldsymbol{\theta})$ as $(\boldsymbol{y}_i - \bar{\boldsymbol{y}} + \bar{\boldsymbol{y}} - \boldsymbol{\theta})$, and note that $\sum \boldsymbol{a}_i^T \mathbf{B} \boldsymbol{a}_i$ can be written as $\mathrm{tr}(\mathbf{AB})$, where $\mathbf{A} = \sum \boldsymbol{a}_i \boldsymbol{a}_i^T$.

b) Let $p_U(\Sigma)$ be the inverse-Wishart density induced by $p_U(\Psi)$. Obtain a density $p_U(\boldsymbol{\theta}, \Sigma | \boldsymbol{y}_1, \ldots, \boldsymbol{y}_n) \propto p_U(\boldsymbol{\theta}|\Sigma) p_U(\Sigma) p(\boldsymbol{y}_1, \ldots, \boldsymbol{y}_n | \boldsymbol{\theta}, \Sigma)$. Can this be interpreted as a posterior distribution for θ and Σ?

7.3 Australian crab data: The files `bluecrab.dat` and `orangecrab.dat` contain measurements of body depth (Y_1) and rear width (Y_2), in millimeters, made on 50 male crabs from each of two species, blue and orange. We will model these data using a bivariate normal distribution.

a) For each of the two species, obtain posterior distributions of the population mean $\boldsymbol{\theta}$ and covariance matrix Σ as follows: Using the semiconjugate prior distributions for $\boldsymbol{\theta}$ and Σ, set $\boldsymbol{\mu}_0$ equal to the sample mean of the data, Λ_0 and $\mathbf{S}_0$ equal to the sample covariance matrix and $\nu_0 = 4$. Obtain 10,000 posterior samples of $\boldsymbol{\theta}$ and Σ. Note that this "prior" distribution loosely centers the parameters around empirical estimates based on the observed data (and is very similar to the unit information prior described in the previous exercise). It cannot be considered as our true prior distribution, as it was derived from the observed data. However, it can be roughly considered as the prior distribution of someone with weak but unbiased information.

b) Plot values of $\boldsymbol{\theta} = (\theta_1, \theta_2)'$ for each group and compare. Describe any size differences between the two groups.

c) From each covariance matrix obtained from the Gibbs sampler, obtain the corresponding correlation coefficient. From these values, plot posterior densities of the correlations ρ_{blue} and ρ_{orange} for the two groups. Evaluate differences between the two species by comparing these posterior distributions. In particular, obtain an approximation to $\Pr(\rho_{\text{blue}} < \rho_{\text{orange}} | \boldsymbol{y}_{\text{blue}}, \boldsymbol{y}_{\text{orange}})$. What do the results suggest about differences between the two populations?

7.4 Marriage data: The file `agehw.dat` contains data on the ages of 100 married couples sampled from the U.S. population.

a) Before you look at the data, use your own knowledge to formulate a semiconjugate prior distribution for $\boldsymbol{\theta} = (\theta_h, \theta_w)^T$ and Σ, where θ_h, θ_w are mean husband and wife ages, and Σ is the covariance matrix.

b) Generate a *prior predictive dataset* of size $n = 100$, by sampling $(\boldsymbol{\theta}, \Sigma)$ from your prior distribution and then simulating $\boldsymbol{Y}_1, \ldots, \boldsymbol{Y}_n \sim$ i.i.d. multivariate normal$(\boldsymbol{\theta}, \Sigma)$. Generate several such datasets, make bivariate scatterplots for each dataset, and make sure they roughly represent your prior beliefs about what such a dataset would actually look like. If your prior predictive datasets do not conform to your beliefs, go back to part a) and formulate a new prior. Report the prior that you eventually decide upon, and provide scatterplots for at least three prior predictive datasets.

c) Using your prior distribution and the 100 values in the dataset, obtain an MCMC approximation to $p(\boldsymbol{\theta}, \Sigma | \boldsymbol{y}_1, \ldots, \boldsymbol{y}_{100})$. Plot the joint posterior distribution of θ_h and θ_w, and also the marginal posterior density of the correlation between Y_h and Y_w, the ages of a husband and wife. Obtain 95% posterior confidence intervals for θ_h, θ_w and the correlation coefficient.

d) Obtain 95% posterior confidence intervals for θ_h, θ_w and the correlation coefficient using the following prior distributions:
   i. Jeffreys' prior, described in Exercise 7.1;
   ii. the unit information prior, described in Exercise 7.2;
   iii. a "diffuse prior" with $\mu_0 = \mathbf{0}$, $\Lambda_0 = 10^5 \times \mathbf{I}$, $\mathbf{S}_0 = 1000 \times \mathbf{I}$ and $\nu_0 = 3$.

e) Compare the confidence intervals from d) to those obtained in c). Discuss whether or not you think that your prior information is helpful in estimating $\boldsymbol{\theta}$ and Σ, or if you think one of the alternatives in d) is preferable. What about if the sample size were much smaller, say $n = 25$?

7.5 Imputation: The file `interexp.dat` contains data from an experiment that was interrupted before all the data could be gathered. Of interest was the difference in reaction times of experimental subjects when they were given stimulus A versus stimulus B. Each subject is tested under one of the two stimuli on their first day of participation in the study, and is tested under the other stimulus at some later date. Unfortunately the experiment was interrupted before it was finished, leaving the researchers with 26 subjects with both A and B responses, 15 subjects with only A responses and 17 subjects with only B responses.

a) Calculate empirical estimates of θ_A, θ_B, ρ, σ_A^2, σ_B^2 from the data using the commands `mean`, `cor` and `var`. Use *all* the A responses to get $\hat{\theta}_A$ and $\hat{\sigma}_A^2$, and use *all* the B responses to get $\hat{\theta}_B$ and $\hat{\sigma}_B^2$. Use only the complete data cases to get $\hat{\rho}$.

b) For each person i with only an A response, impute a B response as

$$\hat{y}_{i,B} = \hat{\theta}_B + (y_{i,A} - \hat{\theta}_A)\hat{\rho}\sqrt{\hat{\sigma}_B^2/\hat{\sigma}_A^2}.$$

For each person i with only a B response, impute an A response as

$$\hat{y}_{i,A} = \hat{\theta}_A + (y_{i,B} - \hat{\theta}_B)\hat{\rho}\sqrt{\hat{\sigma}_A^2/\hat{\sigma}_B^2}.$$

You now have two "observations" for each individual. Do a paired sample t-test and obtain a 95% confidence interval for $\theta_A - \theta_B$.

c) Using either Jeffreys' prior or a unit information prior distribution for the parameters, implement a Gibbs sampler that approximates the joint distribution of the parameters and the missing data. Compute a posterior mean for $\theta_A - \theta_B$ as well as a 95% posterior confidence interval for $\theta_A - \theta_B$. Compare these results with the results from b) and discuss.

7.6 Diabetes data: A population of 532 women living near Phoenix, Arizona were tested for diabetes. Other information was gathered from these women at the time of testing, including number of pregnancies, glucose level, blood pressure, skin fold thickness, body mass index, diabetes pedigree and age. This information appears in the file `azdiabetes.dat`. Model the joint distribution of these variables for the diabetics and non-diabetics separately, using a multivariate normal distribution:

a) For both groups separately, use the following type of unit information prior, where $\hat{\Sigma}$ is the sample covariance matrix.
   i. $\boldsymbol{\mu}_0 = \bar{\boldsymbol{y}}$, $\Lambda_0 = \hat{\Sigma}$;
   ii. $\mathbf{S}_0 = \hat{\Sigma}$, $\nu_0 = p + 2 = 9$.

   Generate at least 10,000 Monte Carlo samples for $\{\boldsymbol{\theta}_d, \Sigma_d\}$ and $\{\boldsymbol{\theta}_n, \Sigma_n\}$, the model parameters for diabetics and non-diabetics respectively. For each of the seven variables $j \in \{1, \ldots, 7\}$, compare the marginal posterior distributions of $\theta_{d,j}$ and $\theta_{n,j}$. Which variables seem to differ between the two groups? Also obtain $\Pr(\theta_{d,j} > \theta_{n,j} | \mathbf{Y})$ for each $j \in \{1, \ldots, 7\}$.

b) Obtain the posterior means of Σ_d and Σ_n, and plot the entries versus each other. What are the main differences, if any?

Chapter 8

8.1 Components of variance: Consider the hierarchical model where

$$\theta_1, \ldots, \theta_m | \mu, \tau^2 \sim \text{i.i.d. normal}(\mu, \tau^2)$$
$$y_{1,j}, \ldots, y_{n_j,j} | \theta_j, \sigma^2 \sim \text{i.i.d. normal}(\theta_j, \sigma^2) .$$

For this problem, we will eventually compute the following:
$\text{Var}[y_{i,j}|\theta_j, \sigma^2]$, $\text{Var}[\bar{y}_{\cdot,j}|\theta_j, \sigma^2]$, $\text{Cov}[y_{i_1,j}, y_{i_2,j}|\theta_j, \sigma^2]$
$\text{Var}[y_{i,j}|\mu, \tau^2]$, $\text{Var}[\bar{y}_{\cdot,j}|\mu, \tau^2]$, $\text{Cov}[y_{i_1,j}, y_{i_2,j}|\mu, \tau^2]$
First, lets use our intuition to guess at the answers:

a) Which do you think is bigger, $\text{Var}[y_{i,j}|\theta_j, \sigma^2]$ or $\text{Var}[y_{i,j}|\mu, \tau^2]$? To guide your intuition, you can interpret the first as the variability of the Y's when sampling from a fixed group, and the second as the variability in first sampling a group, then sampling a unit from within the group.
b) Do you think $\text{Cov}[y_{i_1,j}, y_{i_2,j}|\theta_j, \sigma^2]$ is negative, positive, or zero? Answer the same for $\text{Cov}[y_{i_1,j}, y_{i_2,j}|\mu, \tau]$. You may want to think about what $y_{i_2,j}$ tells you about $y_{i_1,j}$ if θ_j is known, and what it tells you when θ_j is unknown.
c) Now compute each of the six quantities above and compare to your answers in a) and b).
d) Now assume we have a prior $p(\mu)$ for μ. Using Bayes' rule, show that

$$p(\mu|\theta_1, \ldots, \theta_m, \sigma^2, \tau^2, \boldsymbol{y}_1, \ldots, \boldsymbol{y}_m) = p(\mu|\theta_1, \ldots, \theta_m, \tau^2).$$

Interpret in words what this means.

8.2 Sensitivity analysis: In this exercise we will revisit the study from Exercise 5.2, in which 32 students in a science classroom were randomly assigned to one of two study methods, A and B, with $n_A = n_B = 16$. After several weeks of study, students were examined on the course material, and the scores are summarized by $\{\bar{y}_A = 75.2, s_A = 7.3\}$, $\{\bar{y}_B = 77.5, s_B = 8.1\}$. We will estimate $\theta_A = \mu + \delta$ and $\theta_B = \mu - \delta$ using the two-sample model and prior distributions of Section 8.1.

a) Let $\mu \sim \text{normal}(75, 100)$, $1/\sigma^2 \sim \text{gamma}(1, 100)$ and $\delta \sim \text{normal}(\delta_0, \tau_0^2)$. For each combination of $\delta_0 \in \{-4, -2, 0, 2, 4\}$ and $\tau_0^2 \in \{10, 50, 100, 500\}$, obtain the posterior distribution of μ, δ and σ^2 and compute
   i. $\Pr(\delta < 0|\mathbf{Y})$;
   ii. a 95% posterior confidence interval for δ;
   iii. the prior and posterior correlation of θ_A and θ_B.

b) Describe how you might use these results to convey evidence that $\theta_A < \theta_B$ to people of a variety of prior opinions.

8.3 Hierarchical modeling: The files `school1.dat` through `school8.dat` give weekly hours spent on homework for students sampled from eight different schools. Obtain posterior distributions for the true means for the eight different schools using a hierarchical normal model with the following prior parameters:

$$\mu_0 = 7, \gamma_0^2 = 5, \quad \tau_0^2 = 10, \eta_0 = 2, \quad \sigma_0^2 = 15, \nu_0 = 2\,.$$

a) Run a Gibbs sampling algorithm to approximate the posterior distribution of $\{\boldsymbol{\theta}, \sigma^2, \mu, \tau^2\}$. Assess the convergence of the Markov chain, and find the effective sample size for $\{\sigma^2, \mu, \tau^2\}$. Run the chain long enough so that the effective sample sizes are all above 1,000.

b) Compute posterior means and 95% confidence regions for $\{\sigma^2, \mu, \tau^2\}$. Also, compare the posterior densities to the prior densities, and discuss what was learned from the data.

c) Plot the posterior density of $R = \frac{\tau^2}{\sigma^2+\tau^2}$ and compare it to a plot of the prior density of R. Describe the evidence for between-school variation.

d) Obtain the posterior probability that θ_7 is smaller than θ_6, as well as the posterior probability that θ_7 is the smallest of all the θ's.

e) Plot the sample averages $\bar{y}_1, \ldots, \bar{y}_8$ against the posterior expectations of $\theta_1, \ldots, \theta_8$, and describe the relationship. Also compute the sample mean of all observations and compare it to the posterior mean of μ.

Chapter 9

9.1 Extrapolation: The file `swim.dat` contains data on the amount of time, in seconds, it takes each of four high school swimmers to swim 50 yards. Each swimmer has six times, taken on a biweekly basis.

a) Perform the following data analysis for each swimmer separately:
   i. Fit a linear regression model of swimming time as the response and week as the explanatory variable. To formulate your prior, use the information that competitive times for this age group generally range from 22 to 24 seconds.
   ii. For each swimmer j, obtain a posterior predictive distribution for Y_j^*, their time if they were to swim two weeks from the last recorded time.

b) The coach of the team has to decide which of the four swimmers will compete in a swimming meet in two weeks. Using your predictive distributions, compute $\Pr(Y_j^* = \max\{Y_1^*, \ldots, Y_4^*\}|\mathbf{Y}))$ for each swimmer j, and based on this make a recommendation to the coach.

9.2 Model selection: As described in Exercise 7.6, The file `azdiabetes.dat` contains data on health-related variables of a population of 532 women. In this exercise we will be modeling the conditional distribution of glucose level (`glu`) as a linear combination of the other variables, excluding the variable `diabetes`.
   a) Fit a regression model using the g-prior with $g = n$, $\nu_0 = 2$ and $\sigma_0^2 = 1$. Obtain posterior confidence intervals for all of the parameters.
   b) Perform the model selection and averaging procedure described in Section 9.3. Obtain $\Pr(\beta_j \neq 0|\boldsymbol{y})$, as well as posterior confidence intervals for all of the parameters. Compare to the results in part a).

9.3 Crime: The file `crime.dat` contains crime rates and data on 15 explanatory variables for 47 U.S. states, in which both the crime rates and the explanatory variables have been centered and scaled to have variance 1. A description of the variables can be obtained by typing library(MASS);?UScrime in R.
   a) Fit a regression model $\boldsymbol{y} = \mathbf{X}\boldsymbol{\beta} + \boldsymbol{\epsilon}$ using the g-prior with $g = n$, $\nu_0 = 2$ and $\sigma_0^2 = 1$. Obtain marginal posterior means and 95% confidence intervals for $\boldsymbol{\beta}$, and compare to the least squares estimates. Describe the relationships between crime and the explanatory variables. Which variables seem strongly predictive of crime rates?
   b) Lets see how well regression models can predict crime rates based on the $\mathbf{X}$-variables. Randomly divide the crime roughly in half, into a training set $\{\boldsymbol{y}_{\text{tr}}, \mathbf{X}_{\text{tr}}\}$ and a test set $\{\boldsymbol{y}_{\text{te}}, \mathbf{X}_{\text{te}}\}$
      i. Using only the training set, obtain least squares regression coefficients $\hat{\boldsymbol{\beta}}_{\text{ols}}$. Obtain predicted values for the test data by computing $\hat{\boldsymbol{y}}_{\text{ols}} = \mathbf{X}_{\text{te}}\hat{\boldsymbol{\beta}}_{\text{ols}}$. Plot $\hat{\boldsymbol{y}}_{\text{ols}}$ versus $\boldsymbol{y}_{\text{te}}$ and compute the prediction error $\frac{1}{n_{\text{te}}}\sum(y_{i,\text{te}} - \hat{y}_{i,\text{ols}})^2$.
      ii. Now obtain the posterior mean $\hat{\boldsymbol{\beta}}_{\text{Bayes}} = \mathrm{E}[\boldsymbol{\beta}|\boldsymbol{y}_{\text{tr}}]$ using the g-prior described above and the training data only. Obtain predictions for the test set $\hat{\boldsymbol{y}}_{\text{Bayes}} = \mathbf{X}_{\text{test}}\hat{\boldsymbol{\beta}}_{\text{Bayes}}$. Plot versus the test data, compute the prediction error, and compare to the OLS prediction error. Explain the results.

c) Repeat the procedures in b) many times with different randomly generated test and training sets. Compute the average prediction error for both the OLS and Bayesian methods.

Chapter 10

10.1 Reflecting random walks: It is often useful in MCMC to have a proposal distribution which is both symmetric and has support only on a certain region. For example, if we know $\theta > 0$, we would like our proposal distribution $J(\theta_1|\theta_0)$ to have support on positive θ values. Consider the following proposal algorithm:
- sample $\tilde{\theta} \sim \text{uniform}(\theta_0 - \delta, \theta_0 + \delta)$;
- if $\tilde{\theta} < 0$, set $\theta_1 = -\tilde{\theta}$;
- if $\tilde{\theta} \geq 0$, set $\theta_1 = \tilde{\theta}$.

In other words, $\theta_1 = |\tilde{\theta}|$. Show that the above algorithm draws samples from a symmetric proposal distribution which has support on positive values of θ. It may be helpful to write out the associated proposal density $J(\theta_1|\theta_0)$ under the two conditions $\theta_0 \leq \delta$ and $\theta_0 > \delta$ separately.

10.2 Nesting success: Younger male sparrows may or may not nest during a mating season, perhaps depending on their physical characteristics. Researchers have recorded the nesting success of 43 young male sparrows of the same age, as well as their wingspan, and the data appear in the file `msparrownest.dat`. Let Y_i be the binary indicator that sparrow i successfully nests, and let x_i denote their wingspan. Our model for Y_i is $\text{logit}\Pr(Y_i = 1|\alpha, \beta, x_i) = \alpha + \beta x_i$, where the logit function is given by $\text{logit}\,\theta = \log[\theta/(1-\theta)]$.

a) Write out the joint sampling distribution $\prod_{i=1}^n p(y_i|\alpha, \beta, x_i)$ and simplify as much as possible.

b) Formulate a prior probability distribution over α and β by considering the range of $\Pr(Y = 1|\alpha, \beta, x)$ as x ranges over 10 to 15, the approximate range of the observed wingspans.

c) Implement a Metropolis algorithm that approximates $p(\alpha, \beta|\boldsymbol{y}, \boldsymbol{x})$. Adjust the proposal distribution to achieve a reasonable acceptance rate, and run the algorithm long enough so that the effective sample size is at least 1,000 for each parameter.

d) Compare the posterior densities of α and β to their prior densities.

e) Using output from the Metropolis algorithm, come up with a way to make a confidence band for the following *function* $f_{\alpha\beta}(x)$ of wingspan:

$$f_{\alpha\beta}(x) = \frac{e^{\alpha+\beta x}}{1 + e^{\alpha+\beta x}},$$

where α and β are the parameters in your sampling model. Make a plot of such a band.

10.3 Tomato plants: The file `tplant.dat` contains data on the heights of ten tomato plants, grown under a variety of soil pH conditions. Each plant was measured twice. During the first measurement, each plant's height was recorded and a reading of soil pH was taken. During the second measurement only plant height was measured, although it is assumed that pH levels did not vary much from measurement to measurement.

a) Using ordinary least squares, fit a linear regression to the data, modeling plant height as a function of time (measurement period) and pH level. Interpret your model parameters.

b) Perform model diagnostics. In particular, carefully analyze the residuals and comment on possible violations of assumptions. In particular, assess (graphically or otherwise) whether or not the residuals within a plant are independent. What parts of your ordinary linear regression model do you think are sensitive to any violations of assumptions you may have detected?

c) Hypothesize a new model for your data which allows for observations within a plant to be correlated. Fit the model using a MCMC approximation to the posterior distribution, and present diagnostics for your approximation.

d) Discuss the results of your data analysis. In particular, discuss similarities and differences between the ordinary linear regression and the model fit with correlated responses. Are the conclusions different?

10.4 Gibbs sampling: Consider the general Gibbs sampler for a vector of parameters $\boldsymbol{\phi}$. Suppose $\boldsymbol{\phi}^{(s)}$ is sampled from the target distribution $p(\boldsymbol{\phi})$ and then $\boldsymbol{\phi}^{(s+1)}$ is generated using the Gibbs sampler by iteratively updating each component of the parameter vector. Show that the marginal probability $\Pr(\boldsymbol{\phi}^{(s+1)} \in A)$ equals the target distribution $\int_A p(\boldsymbol{\phi})\, d\boldsymbol{\phi}$.

10.5 Logistic regression variable selection: Consider a logistic regression model for predicting diabetes as a function of x_1 = number of pregnancies, x_2 = blood pressure, x_3 = body mass index, x_4 = diabetes pedigree and x_5 = age. Using the data in `azdiabetes.dat`, center and scale each of the x-variables by subtracting the sample average and dividing by the sample standard deviation for each variable. Consider a logistic regression model of the form $\Pr(Y_i = 1|\boldsymbol{x}_i, \boldsymbol{\beta}, \boldsymbol{z}) = e^{\theta_i}/(1 + e^{\theta_i})$ where

$$\theta_i = \beta_0 + \beta_1\gamma_1 x_{i,1} + \beta_2\gamma_2 x_{i,2} + \beta_3\gamma_3 x_{i,3} + \beta_4\gamma_4 x_{i,4} + \beta_5\gamma_5 x_{i,5}.$$

In this model, each γ_j is either 0 or 1, indicating whether or not variable j is a predictor of diabetes. For example, if it were the case that $\boldsymbol{\gamma} = (1, 1, 0, 0, 0)$, then $\theta_i = \beta_0 + \beta_1 x_{i,1} + \beta_2 x_{i,2}$. Obtain posterior distributions for $\boldsymbol{\beta}$ and $\boldsymbol{\gamma}$, using independent prior distributions for the parameters, such that $\gamma_j \sim \text{binary}(1/2)$, $\beta_0 \sim \text{normal}(0, 16)$ and $\beta_j \sim \text{normal}(0, 4)$ for each $j > 0$.

a) Implement a Metropolis-Hastings algorithm for approximating the posterior distribution of $\boldsymbol{\beta}$ and $\boldsymbol{\gamma}$. Examine the sequences $\beta_j^{(s)}$ and $\beta_j^{(s)} \times \gamma_j^{(s)}$ for each j and discuss the mixing of the chain.

b) Approximate the posterior probability of the top five most frequently occurring values of $\boldsymbol{\gamma}$. How good do you think the MCMC estimates of these posterior probabilities are?

c) For each j, plot posterior densities and obtain posterior means for $\beta_j \gamma_j$. Also obtain $\Pr(\gamma_j = 1|\boldsymbol{x}, \boldsymbol{y})$.

Chapter 11

11.1 Full conditionals: Derive formally the full conditional distributions of $\boldsymbol{\theta}, \Sigma, \sigma^2$ and the $\boldsymbol{\beta}_j$'s as given in Section 11.2.

11.2 Randomized block design: Researchers interested in identifying the optimal planting density for a type of perennial grass performed the following randomized experiment: Ten different plots of land were each divided into eight subplots, and planting densities of 2, 4, 6 and 8 plants per square meter were randomly assigned to the subplots, so that there are two subplots at each density in each plot. At the end of the growing season the amount of plant matter yield was recorded in metric tons per hectare. These data appear in the file `pdensity.dat`. The researchers want to fit a model like $y = \beta_1 + \beta_2 x + \beta_3 x^2 + \epsilon$, where y is yield and x is planting density, but worry that since soil conditions vary across plots they should allow for some across-plot heterogeneity in this relationship. To accommodate this possibility we will analyze these data using the hierarchical linear model described in Section 11.1.

a) Before we do a Bayesian analysis we will get some ad hoc estimates of these parameters via least squares regression. Fit the model $y = \beta_1 + \beta_2 x + \beta_3 x^2 + \epsilon$ using OLS for each group, and make a plot showing the heterogeneity of the least squares regression lines. From the least squares coefficients find ad hoc estimates of $\boldsymbol{\theta}$ and Σ. Also obtain an estimate of σ^2 by combining the information from the residuals across the groups.

b) Now we will perform an analysis of the data using the following distributions as prior distributions:

$$\Sigma^{-1} \sim \text{Wishart}(4, \hat{\Sigma}^{-1})$$
$$\boldsymbol{\theta} \sim \text{multivariate normal}(\hat{\boldsymbol{\theta}}, \hat{\Sigma})$$
$$\sigma^2 \sim \text{inverse} - \text{gamma}(1, \hat{\sigma}^2)$$

where $\hat{\boldsymbol{\theta}}, \hat{\Sigma}, \hat{\sigma}^2$ are the estimates you obtained in a). Note that this analysis is not combining prior information with information from the data, as the"prior" distribution is based on the observed data.

However, such an analysis can be roughly interpreted as the Bayesian analysis of an individual who has weak but unbiased prior information.

c) Use a Gibbs sampler to approximate posterior expectations of $\boldsymbol{\beta}$ for each group j, and plot the resulting regression lines. Compare to the regression lines in a) above and describe why you see any differences between the two sets of regression lines.

d) From your posterior samples, plot marginal posterior and prior densities of $\boldsymbol{\theta}$ and the elements of Σ. Discuss the evidence that the slopes or intercepts vary across groups.

e) Suppose we want to identify the planting density that maximizes average yield over a random sample of plots. Find the value $x_{\max}$ of x that maximizes expected yield, and provide a 95% posterior predictive interval for the yield of a randomly sampled plot having planting density $x_{\max}$.

11.3 Hierarchical variances:. The researchers in Exercise 11.2 are worried that the plots are not just heterogeneous in their regression lines, but also in their variances. In this exercise we will consider the same hierarchical model as above except that the sampling variability within a group is given by $y_{i,j} \sim \text{normal}(\beta_{1,j} + \beta_{2,j}x_{i,j} + \beta_{3,j}x_{i,j}^2, \sigma_j^2)$, that is, the variances are allowed to differ across groups. As in Section 8.5, we will model $\sigma_1^2, \ldots, \sigma_m^2 \sim$ i.i.d. inverse gamma$(\nu_0/2, \nu_0\sigma_0^2/2)$, with $\sigma_0^2 \sim$ gamma$(2,2)$ and $p(\nu_0)$ uniform on the integers $\{1, 2, \ldots, 100\}$.

a) Obtain the full conditional distribution of σ_0^2.

b) Obtain the full conditional distribution of σ_j^2.

c) Obtain the full conditional distribution of $\boldsymbol{\beta}_j$.

d) For two values $\nu_0^{(s)}$ and ν_0^* of ν_0, obtain the ratio $p(\nu_0^*|\sigma_0^2, \sigma_1^2, \ldots, \sigma_m^2)$ divided by $p(\nu_0^{(s)}|\sigma_0^2, \sigma_1^2, \ldots, \sigma_m^2)$, and simplify as much as possible.

e) Implement a Metropolis-Hastings algorithm for obtaining the joint posterior distribution of all of the unknown parameters. Plot values of σ_0^2 and ν_0 versus iteration number and describe the mixing of the Markov chain in terms of these parameters.

f) Compare the prior and posterior distributions of ν_0. Comment on any evidence there is that the variances differ across the groups.

11.4 Hierarchical logistic regression: The Washington Assessment of Student Learning (WASL) is a standardized test given to students in the state of Washington. Letting j index the counties within the state of Washington and i index schools within counties, the file `mathstandard.dat` includes data on the following variables:

$y_{i,j}$ = the indicator that more than half the 10th graders in school i, j passed the WASL math exam;

$x_{i,j}$ = the percentage of teachers in school i, j who have a masters degree.

In this exercise we will construct an algorithm to approximate the posterior distribution of the parameters in a generalized linear mixed-effects model for these data. The model is a mixed effects version of logistic regression:

$y_{i,j} \sim \text{binomial}(e^{\gamma_{i,j}}/[1+e^{\gamma_{i,j}}])$, where $\gamma_{i,j} = \beta_{0,j} + \beta_{1,j}x_{i,j}$
$\boldsymbol{\beta}_1, \ldots, \boldsymbol{\beta}_J \sim$ i.i.d. multivariate normal $(\boldsymbol{\theta}, \Sigma)$, where $\boldsymbol{\beta}_j = (\beta_{0,j}, \beta_{1,j})$

a) The unknown parameters in the model include population-level parameters $\{\boldsymbol{\theta}, \Sigma\}$ and the group-level parameters $\{\boldsymbol{\beta}_1, \ldots, \boldsymbol{\beta}_m\}$. Draw a diagram that describes the relationships between these parameters, the data $\{y_{i,j}, x_{i,j}, i = 1\ldots, n_j, j = 1, \ldots, m\}$, and prior distributions.

b) Before we do a Bayesian analysis, we will get some ad hoc estimates of these parameters via maximum likelihood: Fit a separate logistic regression model for each group, possibly using the `glm` command in R via `beta.j <- glm(y.j~X.j,family=binomial)$coef`. Explain any problems you have with obtaining estimates for each county. Plot $\exp\{\hat{\beta}_{0,j} + \hat{\beta}_{1,j}x\}/(1+\exp\{\hat{\beta}_{0,j} + \hat{\beta}_{1,j}x\})$ as a function of x for each county and describe what you see. Using maximum likelihood estimates only from those counties with 10 or more schools, obtain ad hoc estimates $\hat{\boldsymbol{\theta}}$ and $\hat{\Sigma}$ of $\boldsymbol{\theta}$ and Σ. Note that these estimates may not be representative of patterns from schools with small sample sizes.

c) Formulate a unit information prior distribution for $\boldsymbol{\theta}$ and Σ based on the observed data. Specifically, let $\boldsymbol{\theta} \sim$ multivariate normal$(\hat{\boldsymbol{\theta}}, \hat{\Sigma})$ and let $\Sigma^{-1} \sim$ Wishart$(4, \hat{\Sigma}^{-1})$. Use a Metropolis-Hastings algorithm to approximate the joint posterior distribution of all parameters.

d) Make plots of the samples of $\boldsymbol{\theta}$ and Σ (5 parameters) versus MCMC iteration number. Make sure you run the chain long enough so that your MCMC samples are likely to be a reasonable approximation to the posterior distribution.

e) Obtain posterior expectations of $\boldsymbol{\beta}_j$ for each group j, plot $\text{E}[\beta_{0,j}|\boldsymbol{y}] + \text{E}[\beta_{1,j}|\boldsymbol{y}]x$ as a function of x for each county, compare to the plot in b) and describe why you see any differences between the two sets of regression lines.

f) From your posterior samples, plot marginal posterior and prior densities of $\boldsymbol{\theta}$ and the elements of Σ. Include your ad hoc estimates from b) in the plots. Discuss the evidence that the slopes or intercepts vary across groups.

11.5 Disease rates: The number of occurrences of a rare, nongenetic birth defect in a five-year period for six neighboring counties is $\boldsymbol{y} = (1, 3, 2, 12, 1, 1)$. The counties have populations of $\boldsymbol{x} = (33, 14, 27, 90, 12, 17)$, given in thousands. The second county has higher rates of toxic chemicals (PCBs) present in soil samples, and it is of interest to know if this town has a high disease rate as well. We will use the following hierarchical model to analyze these data:

- $Y_i|\theta_i, x_i \sim \text{Poisson}(\theta_i x_i)$;
- $\theta_1, \ldots, \theta_6|a, b \sim \text{gamma}(a, b)$;
- $a \sim \text{gamma(1,1)}$; $b \sim \text{gamma(10,1)}$.

a) Describe in words what the various components of the hierarchical model represent in terms of observed and expected disease rates.
b) Identify the form of the conditional distribution of $p(\theta_1, \ldots, \theta_6|a, b, \boldsymbol{x}, \boldsymbol{y})$, and from this identify the full conditional distribution of the rate for each county $p(\theta_i|\boldsymbol{\theta}_{-i}, a, b, \boldsymbol{x}, \boldsymbol{y})$.
c) Write out the ratio of the posterior densities comparing a set of proposal values $(a^*, b^*, \boldsymbol{\theta})$ to values $(a, b, \boldsymbol{\theta})$. Note the value of $\boldsymbol{\theta}$, the vector of county-specific rates, is unchanged.
d) Construct a Metropolis-Hastings algorithm which generates samples of $(a, b, \boldsymbol{\theta})$ from the posterior. Do this by iterating the following steps:
   1. Given a current value $(a, b, \boldsymbol{\theta})$, generate a proposal $(a^*, b^*, \boldsymbol{\theta})$ by sampling a^* and b^* from a symmetric proposal distribution centered around a and b, but making sure all proposals are positive (see Exercise 10.1). Accept the proposal with the appropriate probability.
   2. Sample new values of the θ_j's from their full conditional distributions.

   Perform diagnostic tests on your chain and modify if necessary.
e) Make posterior inference on the infection rates using the samples from the Markov chain. In particular,
   i. Compute marginal posterior distributions of $\theta_1, \ldots, \theta_6$ and compare them to $y_1/x_1, \ldots y_6/x_6$.
   ii. Examine the posterior distribution of a/b, and compare it to the corresponding prior distribution as well as to the average of y_i/x_i across the six counties.
   iii. Plot samples of θ_2 versus θ_j for each $j \neq 2$, and draw a 45 degree line on the plot as well. Also estimate $\Pr(\theta_2 > \theta_j|\boldsymbol{x}, \boldsymbol{y})$ for each j and $\Pr(\theta_2 = \max\{\theta_1, \ldots, \theta_6\}|\boldsymbol{x}, \boldsymbol{y})$. Interpret the results of these calculations, and compare them to the conclusions one might obtain if they just examined y_j/x_j for each county j.

Chapter 12

12.1 Rank regression: The 1996 General Social Survey gathered a wide variety of information on the adult U.S. population, including each survey respondent's sex, their self-reported frequency of religious prayer (on a six-level ordinal scale), and the number of items correct out of 10 on a short vocabulary test. These data appear in the file `prayer.dat`. Using the rank regression procedure described in Section 12.1.2, estimate the parameters in a regression model for Y_i= prayer as a function of $x_{i,1}$ = sex of respondent (0-1 indicator of being female) and $x_{i,2}$ = vocabulary score, as well as their interaction $x_{i,3} = x_{i,1} \times x_{i,2}$. Compare marginal prior distributions of the three regression parameters to their posterior distributions, and comment on the evidence that the relationship between prayer and score differs across the sexes.

12.2 Copula modeling: The file **azdiabetes_alldata.dat** contains data on eight variables for 632 women in a study on diabetes (see Exercise 7.6 for a description of the variables). Data on subjects labeled 201-300 have missing values for some variables, mostly for the skin fold thickness measurement.
   a) Using only the data from subjects 1-200, implement the Gaussian copula model for the eight variables in this dataset. Obtain posterior means and 95% posterior confidence intervals for all $\binom{8}{2} = 28$ parameters.
   b) Now use the data from subjects 1-300, thus including data from subjects who are missing some variables. Implement the Gaussian copula model and obtain posterior means and 95% posterior confidence intervals for all parameters. How do the results differ from those in a)?

12.3 Constrained normal: Let $p(z) \propto \text{dnorm}(z, \theta, \sigma) \times \delta_{(a,b)}(z)$, the normal density constrained to the interval (a, b). Prove that the inverse-cdf method outlined in Section 12.1.1 generates a sample from this distribution.

12.4 Categorical data and the Dirichlet distribution: Consider again the data on the number of children of men in their 30s from Exercise 4.8. These data could be considered as categorical data, as each sample Y lies in the discrete set $\{1, \ldots, 8\}$ (8 here actually denotes "8 or more" children). Let $\boldsymbol{\theta}_A = (\theta_{A,1}, \ldots, \theta_{A,8})$ be the proportion in each of the eight categories from the population of men with bachelor's degrees, and let the vector $\boldsymbol{\theta}_B$ be defined similarly for the population of men without bachelor's degrees.
   a) Write in a compact form the conditional probability given $\boldsymbol{\theta}_A$ of observing a particular sequence $\{y_{A,1}, \ldots, y_{A,n_1}\}$ for a random sample from the A population.
   b) Identify the sufficient statistic. Show that the Dirichlet family of distributions, with densities of the form $p(\boldsymbol{\theta}|\boldsymbol{a}) \propto \theta_1^{a_1-1} \times \cdots \theta_K^{a_K-1}$, are a conjugate class of prior distributions for this sampling model.
   c) The function `rdir()` below samples from the Dirichlet distribution:

```
rdir<-function(nsamp=1,a)  # a is a vector
{
  Z<-matrix( rgamma(length(a)*nsamp,a,1),
             nsamp,length(a),byrow=T)
  Z/apply(Z,1,sum)
}
```

   Using this function, generate 5,000 or more samples of $\boldsymbol{\theta}_A$ and $\boldsymbol{\theta}_B$ from their posterior distributions. Using a Monte Carlo approximation, obtain and plot the posterior distributions of $\text{E}[Y_A|\boldsymbol{\theta}_A]$ and $\text{E}[Y_B|\boldsymbol{\theta}_B]$, as well as of $\tilde{Y}_A$ and $\tilde{Y}_B$.

d) Compare the results above to those in Exercise 4.8. Perform the goodness of fit test from that exercise on this model, and compare to the fit of the Poisson model.

Common distributions

The binomial distribution

A random variable $X \in \{0, 1, \ldots, n\}$ has a binomial(n, θ) distribution if $\theta \in [0, 1]$ and

$$\Pr(X = x|\theta, n) = \binom{n}{x} \theta^x (1 - \theta)^{n-x} \quad \text{for } x \in \{0, 1, \ldots, n\}.$$

For this distribution,

$$\begin{aligned} \mathrm{E}[X|\theta] &= n\theta, \\ \mathrm{Var}[X|\theta] &= n\theta(1-\theta), \\ \mathrm{mode}[X|\theta] &= \lfloor (n+1)\theta \rfloor, \\ p(x|\theta, n) &= \text{dbinom(x,n,theta)} . \end{aligned}$$

If $X_1 \sim$ binomial(n_1, θ) and $X_2 \sim$ binomial(n_2, θ) are independent, then $X = X_1 + X_2 \sim$ binomial$(n_1 + n_2, \theta)$. When $n = 1$ this distribution is called the *binary* or *Bernoulli* distribution. The binomial(n, θ) model assumes that X is (equal in distribution to) a sum of independent binary(θ) random variables.

The beta distribution

A random variable $X \in [0, 1]$ has a beta(a, b) distribution if $a > 0$, $b > 0$ and

$$p(x|a, b) = \frac{\Gamma(a+b)}{\Gamma(a)\Gamma(b)} x^{a-1}(1-x)^{b-1} \quad \text{for } 0 \le x \le 1.$$

For this distribution,

$$\begin{aligned}
\mathrm{E}[X|a,b] &= \frac{a}{a+b}, \\
\mathrm{Var}[X|a,b] &= \frac{ab}{(a+b+1)(a+b)^2} = \mathrm{E}[X] \times \mathrm{E}[1-X] \times \frac{1}{a+b+1}, \\
\mathrm{mode}[X|a,b] &= \frac{a-1}{(a-1)+(b-1)} \text{ if } a > 1 \text{ and } b > 1, \\
p(x|a,b) &= \texttt{dbeta(x,a,b)} .
\end{aligned}$$

The beta distribution is closely related to the gamma distribution. See the paragraph on the gamma distribution below for details. A multivariate version of the beta distribution is the Dirichlet distribution, described in Exercise 12.4.

The Poisson distribution

A random variable $X \in \{0, 1, 2, \ldots\}$ has a Poisson(θ) distribution if $\theta > 0$ and

$$\Pr(X = x|\theta) = \theta^x e^{-\theta}/x! \quad \text{for } x \in \{0, 1, 2, \ldots\}.$$

For this distribution,

$$\begin{aligned}
\mathrm{E}[X|\theta] &= \theta, \\
\mathrm{Var}[X|\theta] &= \theta, \\
\mathrm{mode}[X|\theta] &= \lfloor \theta \rfloor, \\
p(x|\theta) &= \texttt{dpois(x,theta)} .
\end{aligned}$$

If $X_1 \sim$ Poisson(θ_1) and $X_2 \sim$ Poisson(θ_2) are independent, then $X_1 + X_2 \sim$ Poisson($\theta_1 + \theta_2$). The Poisson family has a "mean-variance relationship," which describes the fact that $\mathrm{E}[X|\theta] = \mathrm{Var}[X|\theta] = \theta$. If it is observed that a sample mean is very different than the sample variance, then the Poisson model may not be appropriate. If the variance is larger than the sample mean, then a negative binomial model (Section 3.2.1) might be a better fit.

The gamma and inverse-gamma distributions

A random variable $X \in (0, \infty)$ has a gamma(a, b) distribution if $a > 0$, $b > 0$ and

$$p(x|a,b) = \frac{b^a}{\Gamma(a)} x^{a-1} e^{-bx} \quad \text{for } x > 0.$$

For this distribution,

$$\begin{aligned}
\mathrm{E}[X|a,b] &= a/b, \\
\mathrm{Var}[X|a,b] &= a/b^2, \\
\mathrm{mode}[X|a,b] &= (a-1)/b \text{ if } a \geq 1,\ 0 \text{ if } 0 < a < 1\ , \\
p(x|a,b) &= \text{dgamma(x,a,b)}\ .
\end{aligned}$$

If $X_1 \sim \text{gamma}(a_1, b)$ and $X_2 \sim \text{gamma}(a_2, b)$ are independent, then $X_1 + X_2 \sim \text{gamma}(a_1{+}a_2, b)$ and $X_1/(X_1{+}X_2) \sim \text{beta}(a_1, a_2)$. If $X \sim \text{normal}(0, \sigma^2)$ then $X^2 \sim \text{gamma}(1/2, 1/[2\sigma^2])$. The chi-square distribution with ν degrees of freedom is the same as a gamma$(\nu/2, 1/2)$ distribution.

A random variable $X \in (0, \infty)$ has an inverse-gamma(a, b) distribution if $1/X$ has a gamma(a, b) distribution. In other words, if $Y \sim \text{gamma}(a, b)$ and $X = 1/Y$, then $X \sim \text{inverse-gamma}(a, b)$. The density of X is

$$p(x|a,b) = \frac{b^a}{\Gamma(a)} x^{-a-1} e^{-b/x} \quad \text{for } x > 0.$$

For this distribution,

$$\begin{aligned}
\mathrm{E}[X|a,b] &= b/(a-1) \text{ if } a \geq 1,\ \infty \text{ if } 0 < a < 1, \\
\mathrm{Var}[X|a,b] &= b^2/[(a-1)^2(a-2)] \text{ if } a \geq 2,\ \infty \text{ if } 0 < a < 2, \\
\mathrm{mode}[X|a,b] &= b/(a+1).
\end{aligned}$$

Note that the inverse-gamma density is not simply the gamma density with x replaced by $1/x$: There is an additional factor of x^{-2} due to the Jacobian in the change-of-variables formula (see Exercise 10.3).

The univariate normal distribution

A random variable $X \in \mathbb{R}$ has a normal(θ, σ^2) distribution if $\sigma^2 > 0$ and

$$p(x|\theta,\sigma^2) = \frac{1}{\sqrt{2\pi\sigma^2}} e^{-\frac{1}{2}(x-\theta)^2/\sigma^2} \quad \text{for } -\infty < x < \infty.$$

For this distribution,

$$\begin{aligned}
\mathrm{E}[X|\theta,\sigma^2] &= \theta, \\
\mathrm{Var}[X|\theta,\sigma^2] &= \sigma^2, \\
\mathrm{mode}[X|\theta,\sigma^2] &= \theta, \\
p(x|\theta,\sigma^2) &= \text{dnorm(x,theta,sigma)}\ .
\end{aligned}$$

Remember that R parameterizes things in terms of the standard deviation σ, and not the variance σ^2. If $X_1 \sim \text{normal}(\theta_1, \sigma_1^2)$ and $X_2 \sim \text{normal}(\theta_2, \sigma_2^2)$ are independent, then $aX_1 + bX_2 + c \sim \text{normal}(a\theta_1 + b\theta_2 + c, a^2\sigma_1^2 + b^2\sigma_2^2)$.

A normal sampling model is often useful even if the underlying population does not have a normal distribution. This is because statistical procedures that assume a normal model will generally provide good estimates of the population mean and variance, regardless of whether or not the population is normal (see Section 5.5 for a discussion).

The multivariate normal distribution

A random vector $\boldsymbol{X} \in \mathbb{R}^p$ has a multivariate normal$(\boldsymbol{\theta}, \Sigma)$ distribution if Σ is a positive definite $p \times p$ matrix and

$$p(\boldsymbol{x}|\boldsymbol{\theta}, \Sigma) = (2\pi)^{-p/2}|\Sigma|^{-1/2} \exp\left\{-\frac{1}{2}(\boldsymbol{x}-\boldsymbol{\theta})^T\Sigma^{-1}(\boldsymbol{x}-\boldsymbol{\theta})\right\} \quad \text{for } \boldsymbol{x} \in \mathbb{R}^p.$$

For this distribution,

$$\begin{aligned} \mathrm{E}[\boldsymbol{X}|\boldsymbol{\theta}, \Sigma] &= \boldsymbol{\theta}, \\ \mathrm{Var}[\boldsymbol{X}|\boldsymbol{\theta}, \Sigma] &= \Sigma, \\ \mathrm{mode}[\boldsymbol{X}|\boldsymbol{\theta}, \Sigma] &= \boldsymbol{\theta}. \end{aligned}$$

Just like the univariate normal distribution, if $\boldsymbol{X}_1 \sim$ normal$(\boldsymbol{\theta}_1, \Sigma_1)$ and $\boldsymbol{X}_2 \sim$ normal$(\boldsymbol{\theta}_2, \Sigma_2)$ are independent, then $a\boldsymbol{X}_1 + b\boldsymbol{X}_2 + \mathbf{c} \sim$ normal$(a\boldsymbol{\theta}_1 + b\boldsymbol{\theta}_2 + \mathbf{c}, a^2\Sigma_1 + b^2\Sigma_2)$. Marginal and conditional distributions of subvectors of $\boldsymbol{X}$ also have multivariate normal distributions: Let $\boldsymbol{a} \subset \{1, \ldots, p\}$ be a subset of variable indices, and let $\boldsymbol{b} = \boldsymbol{a}^c$ be the remaining indices. Then $\boldsymbol{X}_{[a]} \sim$ multivariate normal$(\boldsymbol{\theta}_{[a]}, \Sigma_{[a,a]})$ and $\{\boldsymbol{X}_{[b]}|\boldsymbol{X}_{[a]}\} \sim$ multivariate normal$(\boldsymbol{\theta}_{b|a}, \Sigma_{b|a})$, where

$$\begin{aligned} \boldsymbol{\theta}_{b|a} &= \boldsymbol{\theta}_{[b]} + \Sigma_{[b,a]}(\Sigma_{[a,a]})^{-1}(\boldsymbol{X}_{[a]} - \boldsymbol{\theta}_{[a]}) \\ \Sigma_{b|a} &= \Sigma_{[b,b]} - \Sigma_{[b,a]}(\Sigma_{[a,a]})^{-1}\Sigma_{[a,b]}. \end{aligned}$$

Simulating a multivariate normal random variable can be achieved by a linear transformation of a vector of i.i.d. standard normal random variables. If $\boldsymbol{Z}$ is the vector with elements $Z_1, \ldots, Z_p \sim$ i.i.d. normal$(0,1)$ and $\mathbf{A}\mathbf{A}^T = \Sigma$, then $\boldsymbol{X} = \boldsymbol{\theta} + \mathbf{A}\boldsymbol{Z} \sim$ multivariate normal$(\boldsymbol{\theta}, \Sigma)$. Usually $\mathbf{A}$ is the Choleski factorization of Σ. The following R-code will generate an $n \times p$ matrix such that the rows are i.i.d. samples from a multivariate normal distribution:

```
Z<-matrix(rnorm(n*p),nrow=n,ncol=p)
X<-t(  t(Z%*%chol(Sigma)) + c(theta) )
```

The Wishart and inverse-Wishart distributions

A random $p \times p$ symmetric positive definite matrix $\mathbf{X}$ has a Wishart$(\nu, \mathbf{M})$ distribution if the integer $\nu \geq p$, $\mathbf{M}$ is a $p \times p$ symmetric positive definite matrix and

$$p(\mathbf{X}|\nu, \mathbf{M}) = [2^{\nu p/2} \Gamma_p(\nu/2) |\mathbf{M}|^{\nu/2}]^{-1} \times |\mathbf{X}|^{(\nu-p-1)/2} \text{etr}(-\mathbf{M}^{-1}\mathbf{X}/2),$$

where

$\Gamma_p(\nu/2) = \pi^{p(p-1)/4} \prod_{j=1}^{p} \Gamma[(\nu+1-j)/2]$, and
$\text{etr}(\mathbf{A}) = \exp(\sum a_{j,j})$, the exponent of the sum of the diagonal elements.

For this distribution,

$$\begin{aligned} \mathrm{E}[\mathbf{X}|\nu, \mathbf{M}] &= \nu \mathbf{M}, \\ \mathrm{Var}[X_{i,j}|\nu, \mathbf{M}] &= \nu \times (m_{i,j}^2 + m_{i,i} m_{j,j}), \\ \mathrm{mode}[\mathbf{X}|\nu, \mathbf{M}] &= (\nu - p - 1)\mathbf{M}. \end{aligned}$$

The Wishart distribution is a multivariate version of the gamma distribution. Just as the sum of squares of i.i.d. univariate normal variables has a gamma distribution, the sums of squares of i.i.d. multivariate normal vectors has a Wishart distribution. Specifically, if $\boldsymbol{Y}_1, \ldots, \boldsymbol{Y}_\nu \sim$ i.i.d. multivariate normal$(\mathbf{0}, \mathbf{M})$, then $\sum \boldsymbol{Y}_i \boldsymbol{Y}_i^T \sim$ Wishart$(\nu, \mathbf{M})$. This relationship can be used to generate a Wishart-distributed random matrix:

```
Z<-matrix(rnorm(nu*p),nrow=nu,ncol=p)  # standard normal
Y<-Z%*%chol(M)                         # rows have cov=M
X<-t(Y)%*%Y                            # Wishart matrix
```

A random $p \times p$ symmetric positive definite matrix $\mathbf{X}$ has an inverse-Wishart$(\nu, \mathbf{M})$ distribution if $\mathbf{X}^{-1}$ has a Wishart$(\nu, \mathbf{M})$ distribution. In other words, if $\mathbf{Y} \sim$ Wishart$(\nu, \mathbf{M})$ and $\mathbf{X} = \mathbf{Y}^{-1}$, then $\mathbf{X} \sim$ inverse-Wishart$(\nu, \mathbf{M})$. The density of $\mathbf{X}$ is

$$p(\mathbf{X}|\nu, \mathbf{M}) = [2^{\nu p/2} \Gamma_p(\nu/2) |\mathbf{M}|^{\nu/2}]^{-1} \times |\mathbf{X}|^{-(\nu+p+1)/2} \text{etr}(-\mathbf{M}^{-1}\mathbf{X}^{-1}/2).$$

For this distribution,

$$\begin{aligned} \mathrm{E}[\mathbf{X}|\nu, \mathbf{M}] &= (\nu - p - 1)^{-1} \mathbf{M}^{-1}, \\ \mathrm{mode}[\mathbf{X}|\nu, \mathbf{M}] &= (\nu + p + 1)^{-1} \mathbf{M}^{-1}. \end{aligned}$$

The second moments (i.e. the variances) of the elements of $\mathbf{X}$ are given in Press (1972). Since we often use the inverse-Wishart distribution as a prior distribution for a covariance matrix Σ, it is sometimes useful to parameterize the distribution in terms of $\mathbf{S} = \mathbf{M}^{-1}$. Then if $\Sigma \sim$ inverse-Wishart$(\nu, \mathbf{S}^{-1})$, we have mode$[\mathbf{X}|\nu, \mathbf{S}] = (\nu + p + 1)^{-1}\mathbf{S}$. If Σ_0 were the most probable value of Σ *a priori*, then we would set $\mathbf{S} = (\nu_0 + p + 1)\Sigma_0$, so that $\Sigma \sim$ inverse-Wishart$(\nu, [(\nu + p - 1)\Sigma_0]^{-1})$ and mode$[\Sigma|\nu, \mathbf{S}] = \Sigma_0$.

For more on the Wishart distribution and its relationship to the multivariate normal distribution, see Press (1972) or Mardia et al (1979).

References

Agresti A, Coull BA (1998) Approximate is better than "exact" for interval estimation of binomial proportions. Amer Statist 52(2):119–126

Aldous DJ (1985) Exchangeability and related topics. In: École d'été de probabilités de Saint-Flour, XIII—1983, Lecture Notes in Math., vol 1117, Springer, Berlin, pp 1–198

Arcese P, Smith JNM, Hochachka WM, Rogers CM, Ludwig D (1992) Stability, regulation, and the determination of abundance in an insular song sparrow population. Ecology 73(3):805–822

Atchadé YF, Rosenthal JS (2005) On adaptive Markov chain Monte Carlo algorithms. Bernoulli 11(5):815–828

Bayarri MJ, Berger JO (2000) p values for composite null models. J Amer Statist Assoc 95(452):1127–1142, 1157–1170, with comments and a rejoinder by the authors

Bayes T (1763) An essay towards solving a problem in the doctrine of chances. R Soc Lond Philos Trans 5(3):370–418

Berger JO (1980) Statistical decision theory: foundations, concepts, and methods. Springer-Verlag, New York, springer Series in Statistics

Berk KN (1978) Comparing subset regression procedures. Technometrics 20:1–6

Bernardo JM, Smith AFM (1994) Bayesian theory. Wiley Series in Probability and Mathematical Statistics: Probability and Mathematical Statistics, John Wiley & Sons Ltd., Chichester

Besag J (1974) Spatial interaction and the statistical analysis of lattice systems. J Roy Statist Soc Ser B 36:192–236, with discussion by D. R. Cox, A. G. Hawkes, P. Clifford, P. Whittle, K. Ord, R. Mead, J. M. Hammersley, and M. S. Bartlett and with a reply by the author

Bickel PJ, Ritov Y (1997) Local asymptotic normality of ranks and covariates in transformation models. In: Festschrift for Lucien Le Cam, Springer, New York, pp 43–54

Box GEP, Draper NR (1987) Empirical model-building and response surfaces. Wiley Series in Probability and Mathematical Statistics: Applied Probability and Statistics, John Wiley & Sons Inc., New York

Bunke O, Milhaud X (1998) Asymptotic behavior of Bayes estimates under possibly incorrect models. Ann Statist 26(2):617–644

Carlin BP, Louis TA (1996) Bayes and empirical Bayes methods for data analysis, Monographs on Statistics and Applied Probability, vol 69. Chapman & Hall, London

Casella G (1985) An introduction to empirical Bayes data analysis. Amer Statist 39(2):83–87

Chib S, Winkelmann R (2001) Markov chain Monte Carlo analysis of correlated count data. J Bus Econom Statist 19(4):428–435

Congdon P (2003) Applied Bayesian modelling. Wiley Series in Probability and Statistics, John Wiley & Sons Ltd., Chichester

Cox RT (1946) Probability, frequency and reasonable expectation. Amer J Phys 14:1–13

Cox RT (1961) The algebra of probable inference. The Johns Hopkins Press, Baltimore, Md

Dawid AP, Lauritzen SL (1993) Hyper-Markov laws in the statistical analysis of decomposable graphical models. Ann Statist 21(3):1272–1317

Diaconis P (1988) Recent progress on de Finetti's notions of exchangeability. In: Bayesian statistics, 3 (Valencia, 1987), Oxford Sci. Publ., Oxford Univ. Press, New York, pp 111–125

Diaconis P, Freedman D (1980) Finite exchangeable sequences. Ann Probab 8(4):745–764

Diaconis P, Ylvisaker D (1979) Conjugate priors for exponential families. Ann Statist 7(2):269–281

Diaconis P, Ylvisaker D (1985) Quantifying prior opinion. In: Bayesian statistics, 2 (Valencia, 1983), North-Holland, Amsterdam, pp 133–156, with discussion and a reply by Diaconis

Dunson DB (2000) Bayesian latent variable models for clustered mixed outcomes. J R Stat Soc Ser B Stat Methodol 62(2):355–366

Efron B (2005) Bayesians, frequentists, and scientists. J Amer Statist Assoc 100(469):1–5

Efron B, Morris C (1973) Stein's estimation rule and its competitors—an empirical Bayes approach. J Amer Statist Assoc 68:117–130

de Finetti B (1931) Funzione caratteristica di un fenomeno aleatorio. Atti della R Academia Nazionale dei Lincei, Serie 6 Memorie, Classe di Scienze Fisiche, Mathematice e Naturale 4:251–299

de Finetti B (1937) La prévision : ses lois logiques, ses sources subjectives. Ann Inst H Poincaré 7(1):1–68

Gelfand AE, Smith AFM (1990) Sampling-based approaches to calculating marginal densities. J Amer Statist Assoc 85(410):398–409

Gelfand AE, Sahu SK, Carlin BP (1995) Efficient parameterisations for normal linear mixed models. Biometrika 82(3):479–488

Gelman A, Hill J (2007) Data Analysis using Regression and Multilevel/Hierarchical Models. Cambridge University Press, New York

Gelman A, Rubin DB (1992) Inference from iterative simulation using multiple sequences (Disc: P483-501, 503-511). Statistical Science 7:457–472

Gelman A, Meng XL, Stern H (1996) Posterior predictive assessment of model fitness via realized discrepancies. Statist Sinica 6(4):733–807, with comments and a rejoinder by the authors

Gelman A, Carlin JB, Stern HS, Rubin DB (2004) Bayesian data analysis, 2nd edn. Texts in Statistical Science Series, Chapman & Hall/CRC, Boca Raton, FL

Geman S, Geman D (1984) Stochastic relaxation, Gibbs distributions, and the Bayesian restoration of images. IEEE Transactions on Pattern Analysis and Machine Intelligence 6:721–741

George EI, McCulloch RE (1993) Variable selection via Gibbs sampling. Journal of the American Statistical Association 88:881–889

Geweke J (1992) Evaluating the accuracy of sampling-based approaches to the calculation of posterior moments (Disc: P189-193). In: Bernardo JM, Berger JO, Dawid AP, Smith AFM (eds) Bayesian Statistics 4. Proceedings of the Fourth Valencia International Meeting, Clarendon Press [Oxford University Press], pp 169–188

Geyer CJ (1992) Practical Markov chain Monte Carlo (Disc: P483-503). Statistical Science 7:473–483

Gilks WR, Roberts GO, Sahu SK (1998) Adaptive Markov chain Monte Carlo through regeneration. J Amer Statist Assoc 93(443):1045–1054

Grogan W, Wirth W (1981) A new American genus of predaceous midges related to Palpomyia and Bezzia (Diptera: Ceratopogonidae). Proceedings of the Biological Society of Washington 94:1279–1305

Guttman I (1967) The use of the concept of a future observation in goodness-of-fit problems. J Roy Statist Soc Ser B 29:83–100

Haario H, Saksman E, Tamminen J (2001) An adaptive metropolis algorithm. Bernoulli 7(2):223–242

Haigis KM, Hoff PD, White A, Shoemaker AR, Halberg RB, Dove WF (2004) Tumor regionality in the mouse intestine reflects the mechanism of loss of apc function. PNAS 101(26):9769–9773

Hartigan JA (1966) Note on the confidence-prior of Welch and Peers. J Roy Statist Soc Ser B 28:55–56

Hastings WK (1970) Monte Carlo sampling methods using Markov chains and their applications. Biometrika 57:97–109

Hewitt E, Savage LJ (1955) Symmetric measures on Cartesian products. Trans Amer Math Soc 80:470–501

Hoff PD (2007) Extending the rank likelihood for semiparametric copula estimation. Ann Appl Statist 1(1):265–283

Jaynes ET (2003) Probability theory. Cambridge University Press, Cambridge, the logic of science, Edited and with a foreword by G. Larry Bretthorst

Jeffreys H (1961) Theory of probability. Third edition, Clarendon Press, Oxford

Johnson VE (2007) Bayesian model assessment using pivotal quantities. Bayesian Anal 2(4):719–733

Jordan MIe (1998) Learning in Graphical Models. Kluwer Academic Publishers Group

Kass RE, Raftery AE (1995) Bayes factors. Journal of the American Statistical Association 90:773–795

Kass RE, Wasserman L (1995) A reference Bayesian test for nested hypotheses and its relationship to the Schwarz criterion. J Amer Statist Assoc 90(431):928–934

Kass RE, Wasserman L (1996) The selection of prior distributions by formal rules (Corr: 1998V93 p412). Journal of the American Statistical Association 91:1343–1370

Kelley TL (1927) The Interpretation of Educational Measurement. World Book Company, New York

Key JT, Pericchi LR, Smith AFM (1999) Bayesian model choice: what and why? In: Bayesian statistics, 6 (Alcoceber, 1998), Oxford Univ. Press, New York, pp 343–370

Kleijn BJK, van der Vaart AW (2006) Misspecification in infinite-dimensional Bayesian statistics. Ann Statist 34(2):837–877

Kuehl RO (2000) Design of Experiments: Statistical Principles of Research Design and Analysis. Duxbury Press

Laplace PS (1995) A philosophical essay on probabilities, english edn. Dover Publications Inc., New York, translated from the sixth French edition by Frederick William Truscott and Frederick Lincoln Emory, With an introductory note by E. T. Bell

Lauritzen SL (1996) Graphical models, Oxford Statistical Science Series, vol 17. The Clarendon Press Oxford University Press, New York, oxford Science Publications

Lawrence E, Bingham D, Liu C, Nair V (2008) Bayesian inference for multivariate ordinal data using parameter expansion. Technometrics 50(2):182–191

Leamer EE (1978) Specification searches. Wiley-Interscience [John Wiley & Sons], New York, ad hoc inference with nonexperimental data, Wiley Series in Probability and Mathematical Statistics

Letac G, Massam H (2007) Wishart distributions for decomposable graphs. Ann Statist 35(3):1278–1323

Liang F, Paulo R, Molina G, Clyde MA, Berger JO (2008) Mixtures of g Priors for Bayesian Variable Selection. Journal of the American Statistical Association 103(481):410–423

Lindley DV, Smith AFM (1972) Bayes estimates for the linear model. J Roy Statist Soc Ser B 34:1–41, with discussions by J. A. Nelder, V. D. Barnett, C. A. B. Smith, T. Leonard, M. R. Novick, D. R. Cox, R. L. Plackett, P. Sprent, J. B. Copas, D. V. Hinkley, E. F. Harding, A. P. Dawid, C.

Chatfield, S. E. Fienberg, B. M. Hill, R. Thompson, B. de Finetti, and O. Kempthorne

Little RJ (2006) Calibrated Bayes: a Bayes/frequentist roadmap. Amer Statist 60(3):213–223

Liu JS, Wu YN (1999) Parameter expansion for data augmentation. J Amer Statist Assoc 94(448):1264–1274

Logan JA (1983) A multivariate model for mobility tables. The American Journal of Sociology 89(2):324–349

Lukacs E (1942) A characterization of the normal distribution. Ann Math Statistics 13:91–93

Madigan D, Raftery AE (1994) Model selection and accounting for model uncertainty in graphical models using Occam's window. Journal of the American Statistical Association 89:1535–1546

Mardia KV, Kent JT, Bibby JM (1979) Multivariate analysis. Academic Press [Harcourt Brace Jovanovich Publishers], London, probability and Mathematical Statistics: A Series of Monographs and Textbooks

Metropolis N, Rosenbluth AW, Rosenbluth MN, Teller AH, Teller E (1953) Equation of state calculations by fast computing machines. Journal of Chemical Physics 21(6):1087–1092

Monahan JF, Boos DD (1992) Proper likelihoods for Bayesian analysis. Biometrika 79(2):271–278

Moore KA, Waite LJ (1977) Early childbearing and educational attainment. Family Planning Perspectives 9(5):220–225

Papaspiliopoulos O, Roberts GO, Sköld M (2007) A general framework for the parametrization of hierarchical models. Statist Sci 22(1):59–73

Petit J, , Jouzel J, Raynaud D, Barkov N, Barnola JM, Basile I, Bender M, Chappellaz J, Davis M, Delayque G, Delmotte M, Kotlyakov V, Legrand M, Lipenkov V, Lorius C, Pepin L, Ritz C, Saltzman E, Stievenard M (1999) Climate and atmospheric history of the past 420,000 years from the vostok ice core, antarctica. Nature 399:429–436

Pettitt AN (1982) Inference for the linear model using a likelihood based on ranks. J Roy Statist Soc Ser B 44(2):234–243

Pitt M, Chan D, Kohn R (2006) Efficient Bayesian inference for Gaussian copula regression models. Biometrika 93(3):537–554

Press SJ (1972) Applied multivariate analysis. Holt, Rinehart and Winston, Inc., New York, series in Quantitative Methods for Decision-Making, International Series in Decision Processes

Press SJ (1982) Applied Multivariate Analysis: Using Bayesian and Frequentist Methods of Inference. Krieger Publishing Company, Inc.

Quinn K (2004) Bayesian Factor Analysis for Mixed Ordinal and Continuous Responses. Political Analysis 12(4):338–353

Raftery AE, Lewis SM (1992) How many iterations in the Gibbs sampler? In: Bernardo JM, Berger JO, Dawid AP, Smith AFM (eds) Bayesian Statistics 4. Proceedings of the Fourth Valencia International Meeting, Clarendon Press [Oxford University Press], pp 763–773

Raftery AE, Madigan D, Hoeting JA (1997) Bayesian model averaging for linear regression models. J Amer Statist Assoc 92(437):179–191

Raiffa H, Schlaifer R (1961) Applied statistical decision theory. Studies in Managerial Economics, Division of Research, Graduate School of Business Administration, Harvard University, Boston, Mass.

Rao JNK (1958) A characterization of the normal distribution. Ann Math Statist 29:914–919

Ripley BD (1979) [Algorithm AS 137] Simulating spatial patterns: Dependent samples from a multivariate density. Applied Statistics 28:109–112

Robert C, Casella G (2008) A history of markov chain monte carlo–subjective recollections from incomplete data `arXiv:0808.2902 [stat.CO]`, arxiv:0808.2902

Robert CP, Casella G (2004) Monte Carlo statistical methods, 2nd edn. Springer Texts in Statistics, Springer-Verlag, New York

Roberts GO, Rosenthal JS (2007) Coupling and ergodicity of adaptive Markov chain Monte Carlo algorithms. J Appl Probab 44(2):458–475

Rubin DB (1984) Bayesianly justifiable and relevant frequency calculations for the applied statistician. Ann Statist 12(4):1151–1172

Rubinstein RY, Kroese DP (2008) Simulation and the Monte Carlo method, 2nd edn. Wiley Series in Probability and Statistics, Wiley-Interscience [John Wiley & Sons], Hoboken, NJ

Savage LJ (1954) The foundations of statistics. John Wiley & Sons Inc., New York

Savage LJ (1962) The foundations of statistical inference. Methuen & Co. Ltd., London

Savage LJ (1972) The foundations of statistics, revised edn. Dover Publications Inc., New York

Severini TA (1991) On the relationship between Bayesian and non-Bayesian interval estimates. J Roy Statist Soc Ser B 53(3):611–618

Smith JW, Everhart JE, Dickson WC, Knowler WC, Johannes RS (1988) Using the adap learning algorithm to forecast the onset of diabetes mellitus. In: Greenes RA (ed) Proceedings of the Symposium on Computer Applications in Medical Care (Washington, 1988), IEEE Computer Society Press, pp 261–265

Stein C (1955) A necessary and sufficient condition for admissibility. Ann Math Statist 26:518–522

Stein C (1956) Inadmissibility of the usual estimator for the mean of a multivariate normal distribution. In: Proceedings of the Third Berkeley Symposium on Mathematical Statistics and Probability, 1954–1955, vol. I, University of California Press, Berkeley and Los Angeles, pp 197–206

Stein CM (1981) Estimation of the mean of a multivariate normal distribution. Ann Statist 9(6):1135–1151

Sweeting TJ (1999) On the construction of Bayes-confidence regions. J R Stat Soc Ser B Stat Methodol 61(4):849–861

Sweeting TJ (2001) Coverage probability bias, objective Bayes and the likelihood principle. Biometrika 88(3):657–675

Tibshirani R (1989) Noninformative priors for one parameter of many. Biometrika 76(3):604–608

Tibshirani R (1996) Regression shrinkage and selection via the lasso. J Roy Statist Soc Ser B 58(1):267–288

Welch BL, Peers HW (1963) On formulae for confidence points based on integrals of weighted likelihoods. J Roy Statist Soc Ser B 25:318–329

White H (1982) Maximum likelihood estimation of misspecified models. Econometrica 50(1):1–25

Zellner A (1986) On assessing prior distributions and Bayesian regression analysis with g-prior distributions. In: Bayesian inference and decision techniques, Stud. Bayesian Econometrics Statist., vol 6, North-Holland, Amsterdam, pp 233–243

Index